German Science in the Age of Empire

This seminal study explores the national, imperial and indigenous interests at stake in a major survey expedition undertaken by the German Schlagintweit brothers, while in the employ of the East India Company, through south and central Asia in the 1850s. It argues that German scientists, lacking in this period a formal empire of their own, seized the opportunity presented by other imperial systems to observe, record, collect and loot manuscripts, maps and museological artefacts that shaped European understandings of the East. Drawing on archival research in three continents, von Brescius vividly explores the dynamics and conflicts of transcultural exploration beyond colonial frontiers in Asia. Analysing the contested careers of these imperial outsiders, he reveals significant changes in the culture of gentlemanly science, the violent negotiation of scientific authority in a transnational arena and the transition from Humboldtian enquiry to a new disciplinary order. This book offers a new understanding of German science and its role in shaping foreign empires and provides a revisionist account of the questions of authority and of authenticity in reportage from distant sites.

MORITZ VON BRESCIUS, PhD, is Lecturer in Modern History at the University of Bern and Fellow of the Munich Centre for Global History, Ludwig Maximilian University of Munich.

SCIENCE IN HISTORY

Series Editors
Simon J. Schaffer, University of Cambridge
James A. Secord, University of Cambridge

Science in History is a major series of ambitious books on the history of the sciences from the mid eighteenth century through the mid twentieth century, highlighting work that interprets the sciences from perspectives drawn from across the discipline of history. The focus on the major epoch of global economic, industrial and social transformations is intended to encourage the use of sophisticated historical models to make sense of the ways in which the sciences have developed and changed. The series encourages the exploration of a wide range of scientific traditions and the interrelations between them. It particularly welcomes work that takes seriously the material practices of the sciences and is broad in geographical scope.

German Science in the Age of Empire

Enterprise, Opportunity and the Schlagintweit Brothers

Moritz von Brescius

University of Bern

CAMBRIDGE
UNIVERSITY PRESS

University Printing House, Cambridge CB2 8BS, United Kingdom

One Liberty Plaza, 20th Floor, New York, NY 10006, USA

477 Williamstown Road, Port Melbourne, VIC 3207, Australia

314-321, 3rd Floor, Plot 3, Splendor Forum, Jasola District Centre, New Delhi - 110025, India

79 Anson Road, #06-04/06, Singapore 079906

Cambridge University Press is part of the University of Cambridge.

It furthers the University's mission by disseminating knowledge in the pursuit of
education, learning and research at the highest international levels of excellence.

www.cambridge.org
Information on this title: www.cambridge.org/9781108446068
DOI: 10.1017/9781108579568

First published 2018
First paperback edition 2020

A catalogue record for this publication is available from the British Library

Library of Congress Cataloging in Publication data
Names: Brescius, Moritz von, author.
Title: German science in the age of empire : enterprise, opportunity, and the
Schlagintweit brothers / Moritz von Brescius, University of Bern
Description: Cambridge, United Kingdom; New York, NY:
Cambridge University Press, 2019. |
Series: Science in history | Includes bibliographical references and index.
Identifiers: LCCN 2018049106| ISBN 9781108427326
(hardback) | ISBN 9781108446068 (pbk.)
Subjects: LCSH: Scientific expeditions – India, North – History –
19th century. | Scientific expeditions – Asia, Central – History –
19th century. | Schlagintweit, Adolph, 1829–1857. | Schlagintweit-
Sakünlünski, Hermann von, 1826–1882. | Schlagintweit,
Robert von, 1833–1885. | East India Company – History –
19th century. | India–History–Sepoy Rebellion, 1857–1858.
Classification: LCC Q115.B835 2019 | DDC 910.92/55 [B]–dc23
LC record available at https://lccn.loc.gov/2018049106

ISBN 978-1-108-42732-6 Hardback
ISBN 978-1-108-44606-8 Paperback

To my parents

Contents

Figures and Maps

Figures

Maps

Acknowledgements

Like the exploratory ventures discussed in the present study, my own research and writing also became a 'voyage of discovery' that depended on the personal generosity, expertise and critical support of many hands and minds. I am glad to acknowledge and thank many colleagues and friends, who have accompanied the journey towards publication with their unfailing support and enthusiasm. This book is the revised outcome of my PhD thesis, written between 2010 and 2014 and submitted to the European University Institute (EUI), Italy. There, I had the great pleasure to work with Antonella Romano, my first supervisor, who has generously and expertly accompanied the work from rocky start to completion. Our long discussions over coffee in Florence opened up new horizons for me and I remain deeply grateful for her formative influence and personal kindness. I am also grateful to Jorge Flores, my second reader, who provided me with the right mixture of support, criticism and academic freedom to realise this work. My deepest thanks also go to Jürgen Osterhammel, who generously acted as my external supervisor during the time in Florence. He later kindly agreed that I could use part of my postdoctoral position at his chair in Konstanz to revise the monograph for publication. The close and humbling exchange with him over the years has shaped my historical thinking in many ways, and I feel honoured to have had the chance to further develop the manuscript by profiting from his perceptive insights and our team's discussions on 'Global Processes'.

From the beginning, the lively international community of the EUI offered valuable occasions to present and refine my findings. The Institute also generously funded numerous research missions to various countries, and allowed me to co-organise academic conferences – especially that on 'Colonial Careers: Transnational Scholarship Overseas in the Nineteenth and Twentieth Centuries' (2012). I am thankful to my co-organisers and the senior scholars for their important contributions to the event, above all Ulrike Lindner and David Arnold. Our discussions have lastingly shaped this book. So did many encounters at the other academic institutions where this book ultimately took shape.

The intellectual universe I encountered at Cambridge University as a visiting PhD student between 2012 and 2014 was exceptional, and I enjoyed lively debates with leading south Asianists, world historians and the members of the German Academic Exchange Service (DAAD) research programme 'Germany and the World in the Age of Globalization'. Henning Grunwald generously integrated me into this group. Besides Richard Evans and his German History Workshop, I would like to thank in particular Christopher Clark for our stimulating and cheerful meetings. The feedback I received at the World History and the Modern European History workshops was also invaluable.

This book also greatly benefited from the comments, interest and support of colleagues too numerous to name them all. I am, however, particularly indebted to Stephanie Kleidt, Felix Driver, Daniel Midena, David Motadel, Devyani Gupta, Dietmar Rothermund, Etienne Stockland, Friederike Kaiser, Bianca Gaudenzi, Stefan Esselborn, Daniel Robinson, Robrecht Declercq, Daniel Hedinger, Nimrod Tal, Adam Storring, Ulrich Päßler, Tilmann Kulke, Erica Wild, Sebastian Conrad, Emma Martin, John Darwin, Niklas Thode Jensen, Christof Dejung, Irmtraut Stellrecht and Anthony La Vopa. Special thanks go to Harry Liebersohn and the other brilliant SIAS group members for cherished discussions on 'Cultural Encounters: Global Perspectives and Local Exchanges, 1750–1940' at both the Wissenschaftskolleg, Berlin, and the National Humanities Center, North Carolina. I am also most thankful to Ali Khan, James White, Hermann Kreutzmann, Shekhar Pathak and Christoph Cüppers, who kindly helped with the translation of difficult sources from Persian, Russian, Hindi and Tibetan. Ella Müller assisted with the footnotes and bibliography. I warmly thank my series editor and outstanding historian Simon Schaffer, who also acted, during the viva, as a most stimulating president of the examining board. I am furthermore grateful to Lucy Rhymer at Cambridge University Press for encouraging me and guiding the work towards publication and to others involved in seeing this book through the press, especially Christopher Feeney for copy-editing. Of course, all remaining errors are my own. Finally, no other person took a livelier interest in, or was kind enough to engage in arguably too many discussions about, those German scientific travellers than Meike von Brescius, a beautiful mind who has been reliably wonderful in reading the chapters at various stages of imperfection.

The staff at the history department and the patient librarians in Florence also receive my warmest gratitude, as they efficiently dealt with my requests for numerous and often rare books, journals and maps. The same appreciation goes to the helpful archivists in other countries: in Britain especially those at the British Library, The National Archives, the Royal Botanic Gardens Archives and the University Library in Cambridge; at various archives in the United States, India, France, the Netherlands, Switzerland and Germany, in particular those at the Bayerische Staatsbibliothek, the Bayerisches Hauptsaatsarchiv and the Alpine Museum, Munich (and Nora Janitzki for her generous hospitality during my prolonged stays in Munich), the Bundesarchiv and the Geheime Staatsarchiv, Berlin, without whom the archival research could not have been so efficient and pleasant.

It is a pleasure to gratefully acknowledge several grants. Undertaking research in fifty-seven archives and museums on three continents and seven countries was crucial, as it allowed me to use 'the veto power of the sources' (Koselleck) that lured me time and again into new directions and beyond perceived wisdom about the *dramatis personae* of this work. Among them were a generous Gerald D. Feldman Travel Grant of the Max Weber Stiftung at the German Historical Institutes in London and Paris, a Herzog-Ernst-Scholarship from the Fritz Thyssen Foundation in Gotha and a one-year PhD scholarship from the Schoeneberg-Stiftung. The thesis was furthermore supported by a PhD grant from the German National Academic Foundation (*ideelle Förderung*), a three-year PhD scholarship from the DAAD and a Completion Grant from the EUI. Their warm support was much more than financial. I also thank the Zukunftskolleg at Konstanz for allowing me to pursue further work on the brothers' visual materials and for the monetary contribution to this work's colour illustrations. I am finally obliged to the Indo-German Society for the Gisela Bonn Award, which allowed me to explore further archival materials on the subcontinent.

Lastly, I am deeply indebted to my family. As regards Meike, without her loving support, her untiring cheer and encouragement my years at Florence, Cambridge and Konstanz would not have been half as fun, warm and bright. The same thanks go to my parents and my brother, who provided assistance from afar and turned several relocations between England, Italy and Germany into enjoyable road trips. This book is dedicated to them.

Brisbane, Australia, December 2017

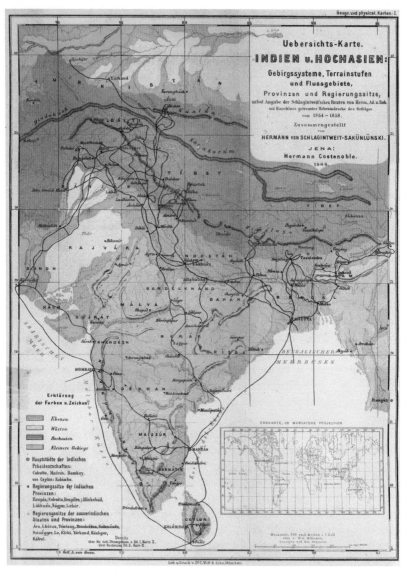

Map 0.1 Expedition routes of the Schlagintweits and their establishments map.

Introduction: Empires of Opportunity

When the German explorer Adolph Schlagintweit embarked with a group of indigenous guides and assistants on his last and ill-fated journey from northern India to central Asia in the spring of 1857, his excursions deep into the frontier regions of the British Empire coincided with the eruption of the Indian rebellion in this most important British overseas colony. The Sepoy rebellion brought British rule on the subcontinent close to collapse. It also instigated the dissolution of the East India Company, in whose service the three Munich-born brothers Adolph, Hermann and Robert Schlagintweit had formally travelled. Chinese Turkestan, the destination of Adolph's final excursion, was then regarded with interest by adjacent powers, which sought to expand their knowledge, trade and influence into the region.[1] A more thorough scientific scrutiny of central Asia by European travellers and officers had only started in the first half of the nineteenth century, when pioneering expeditions began to identify and chart the main routes and patterns of trade in the vast, highly complex and often dangerous environments within and beyond the trans-Himalaya and Turkestan.[2]

Adolph Schlagintweit and a number of his indigenous companions ultimately met their deaths in August 1857 at the hands of a Muslim warlord in Kashgar, who had rebelled against Chinese rule over this town, an important crossroad that connected the northern and southern arteries of the ancient Silk Roads. At the site of Adolph's beheading, a monument was later erected in 1889, tellingly to the sole commemoration of this 'heroic' German explorer and 'martyr of science' (Figure 0.1).[3] On

---

[1] Waller, *Pundits*, 14; Alder, *Bokhara*; Fletcher, 'Sino-Russian Relations', 329–32; Bayly, *Information*, 134–5, Raj, *Relocating Modern Science*, 184–5.

[2] Withers, 'Enlightenment's Margins'; Waller, *Pundits*; a critical and highly suggestive treatment of the 'Great Game' perspective, which often tends to ignore central Asian interests and agency by focusing almost exclusively on assumed Anglo-Russian antagonism and rivalry over *their* lands, is Hopkins, *Modern Afghanistan*, 34–60; also Morrison, 'Beyond the "Great Game"'.

[3] E. Schlagintweit, 'Bericht über das Denkmal'.

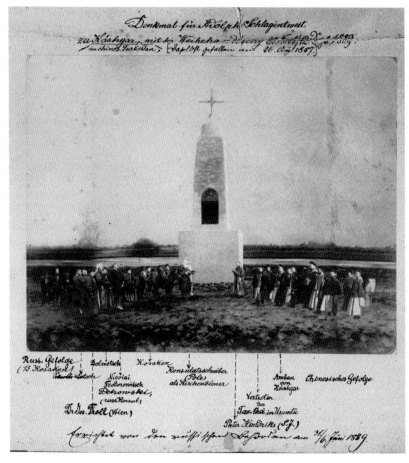

Figure 0.1 Adolph Schlagintweit's monument in Kashgar, annotated photograph of its opening ceremony on 30 November/12 December 1889, with the attendance of Russian and Chinese representatives and the Austrian traveller Dr Troll, who was coincidentally in the area.
© DAV.

the face of it, it might seem odd that the memorial for a German scientific traveller on the payroll of the British East India Company was initially conceived and erected by members of the Imperial Geographical Society of St Petersburg, and that at first only their French counterparts, of the Geographical Society of Paris, were willing to contribute a commemorative plaque and a more splendid cross on top. With it, as the president of the society made clear, however, he also wanted to celebrate

France's own role as 'the first and least self-interested protector of civilisation and Christianity in the Orient' by graciously acknowledging the
contributions of the traveller 'who first brought into those regions the
light of modern science'.[4]

Only belatedly, it seemed, did Ferdinand von Richthofen, the eminent traveller to China and then president of the Berlin Geographical
Society, try to join the lead of other powers in also suggesting the preparation of an additional plaque to formally honour German contributions
to the exploration of extra-European lands.[5] Administrative hurdles
meant that the German plate would only be finished a few years later, by
which time the stand-alone monument in Asia had been washed away by
severe floods. It was never re-erected – fittingly symbolising the fleeting
glory of the Schlagintweit brothers, who remain largely forgotten today
in all these national communities of remembrance.[6] British authorities,
by contrast, which had largely financed the Schlagintweits' exploration
of India, the Himalayas and parts of central Asia at the time, remained
indifferent to the erection of the monument. In fact, no sign of British
patronage or tutelage of this venture appeared on the memorial, leaving
the site to be appropriated by others.

The monument episode captures some of the key issues that provide
the rationale for this study, which together lend much wider historiographic significance to the Schlagintweits' enterprise. It puts into focus
the high status of German expeditionary science in the mid nineteenth
century and an increased German presence in the extra-European world.
Through the brothers' official employment by British authorities, it raises,
however, also important questions over the logics and contradictions of
transnational science, and the mechanisms of collaboration and competition at play. The commemorative site further points to the role of
the state and learned institutions in providing patronage for scientific
enquiry and for shaping (that is, advancing or undermining) 'heroic'
memories of exploration. It also alludes to the different ways colonial
and metropolitan societies could draw meaning from such expeditionary

---

[4] *St. Petersburger Herold*, 8 August/27 July 1887, Bayerisches Hauptstaatsarchiv, Munich
[BayHStA], Abt. II Geheimes Staatsarchiv, MA 53157 'Denkmalerrichtung für den
Asienforscher Adolph Schlagintweit in Kaschgar, 1890', appendix to doc. 3. All
translations from German and French into English are my own.

[5] See the correspondence of the German Consul in Peking, Max von Brandt, Politisches
Archiv des Auswärtigen Amtes, Berlin [PA AA], 'Berichte der Gesandtschaft Peking',
Peking II 891, fols. 77ff., series 'Wissenschaftliche Bestrebungen', vol. 5, 'Juli 1887 bis
September 1892'.

[6] Richthofen to the 'Reichskanzler, Baron von Caprivi', 24 February 1892, Bundesarchiv,
Berlin [BArch], R901, 37418.

work, and how the scientific reputation of its 'leaders' was negotiated, and how at times it sharply diverged, within a transnational arena.

Finally, the monument entirely ignored the role of the indigenous partners that made much of the Schlagintweits' excursions beyond Company terrain possible in the first place. This recurrent denial of indigenous agency and ambition at work in schemes of European exploration strongly suggests the need to overcome the myth of the western solitary traveller by taking a new and multi-perspective look at the inner life of expeditions. Such an approach requires probing the role of transcultural encounters in processes of knowledge production in the field, but also looking critically at contemporary conflicts over trust and scientific authority in reportage over distant sites and cultures at a time of great social and political crisis in Britain and its empire.[7]

This is the first monograph on the contested careers of German scientific travellers in a foreign empire in the nineteenth century. It offers a detailed analysis of significant facets of the programme launched by the Schlagintweit brothers in and beyond the Company realm in south Asia.[8] Their enterprise is significant not least for the vast quantity of materials and documents it accumulated and the ambiguous relationship it maintained throughout with its main sponsor, the East India Company, and other agents and patrons of imperial and European sciences. The multiple contexts in which the German naturalists realised their mission offer rich and unusual opportunities for the historical examination of major themes in the study of imperial knowledge and of transnational and cross-cultural engagement. Using a rich archival story, this work also examines the diverse roles of European and indigenous agents in the forming of accounts of nature, territory and society at a critical period in the development of the field sciences and of museology.[9] The brothers' more peculiar story was thus connected with much broader processes readable through their personal experiences across the scientific, social

---

[7] See the suggestive works by Raj, *Relocating Modern Science*; Withers, 'Voyages et crédibilité'; Safier, *Measuring*; Driver and Jones, *Hidden Histories*.

[8] This work builds on but also significantly advances previous scholarship on their expedition. A first useful overview of their recruitment and activities was given by Polter, 'Nadelschau'; S. Schlagintweit, 'Abriß'; Körner, 'Brüder Schlagintweit'. Another important introductory article by Finkelstein focused mainly on generalities and the aesthetics of German science, yet with less consideration of the imperial context of the enterprise, Finkelstein, 'Mission'; also Felsch, 'Söhne'. New research on the Schlagintweits was undertaken in preparation for a museum exhibition in 2015, for which the author's PhD thesis provided a more developed contextualisation of the brothers' researches within an imperial setting; Brescius, 'Empires of Opportunity'. The exhibition was accompanied by a richly illustrated catalogue, Brescius et al. (eds.), *Über den Himalaya*.

[9] An important series of case studies on the precarious status of fieldwork is given in Kuklick and Kohler (eds.), *Science in the Field*; for German museology, Sheehan, *Museums*.

and political contexts that connected the German states with British, European and south Asian patrons, institutions and interests.[10] While fully conscious of the great ideological and administrative power of modern and fully fledged 'nation states', this book uses a transnational perspective not in the narrow sense of referring only to such relatively recent polities in the western world (which would give transnational history also a problematic Eurocentric bias). A transnational perspective is here rather deployed as a methodical move intended to transcend various *internalist* frameworks applied to single states or empires, and to look for transversal connections, exchanges and mobilities also in the non-European world from the late eighteenth century onwards.[11]

The Schlagintweits' case of imperial sojourning, however controversial and fraught with tension, suggests that the pursuit of expeditionary and colonial science in south Asia was, to a degree, a shared European project.[12] Such an approach calls into question and 'complicates our understanding of "Europe" and "empire"', as it presents a significant departure from orthodoxy.[13] Historians have traditionally worked within a framework of mutually conflictive imperial powers. That logic of internally self-contained and internationally antagonistic states engaging in a kind of zero-sum-game over imperial expansion and profit was often exemplified by the longstanding Anglo-French rivalry.[14] Outsider careers such as those of the Schlagintweits under the aegis of a foreign empire, with the attendant transfer of scientific and technical skills, expertise and ideologies, certainly never challenged the political or military supremacy of the sitting power. They did not work towards a form of 'informal rule' by trying to establish alternative power structures to undermine

---

[10] Significant works that deploy singular lives as a narrative strategy to analyse macro-processes include, Davis, *Trickster Travels*; Lambert and Lester (eds.), *Colonial Lives*; Colley, *Ordeal*; Ogborn, *Global Lives*.

[11] Cf. the intervention by Isabel Hofmeyr in Bayly et al., 'On Transnational History', 1444; see also the methodological discussions in Conrad, *Globalgeschichte*, 17–18.

[12] This was also the subject of a conference on 'Colonial Careers: Transnational Scholarship Overseas in the 19th and 20th Centuries', EUI, Florence, May 2012, organised by the author with Stefan Esselborn. The analytical focus of most studies of 'imperial biographies' is still confined to the personal and professional networks within one distinct imperial formation, which this work advances; see Rolf, 'Einführung'.

[13] Rüger, 'Writing Europe', 47.

[14] Exemplary are the Anglo-French imperial competition on several continents during the Seven Years War, their scientific rivalry over Pacific expeditions later in the eighteenth century and the Fashoda Crisis in the late nineteenth century. The same assumption of essential conflict applies to the large historiography on various East and West India Companies, including the Dutch East India Company (Vereenigde Oostindische Compagnie: VOC) as the key rival of the British East India Company. Much in line with my approach is Arnold, 'Contingent Colonialism'. Lindner, *Koloniale Begegnungen*, focuses on inter-imperial exchanges and cooperation in Africa.

British sovereignty.[15] However, as this work will demonstrate, the hosting empire's internal workings, as concerns scientific practices and outlooks, and at times also their institutional arrangement, were clearly shaped by external participation. That in turn questions the historiographic tendency to narrate the history of science compartmentalised in distinct imperial polities.[16]

There were altogether six Schlagintweit brothers, who pursued publicly engaged careers in a world dominated by European imperial powers. They were born between 1826 and 1849 and fathered by Joseph Schlagintweit, a famous eye surgeon from Munich, with two different wives.[17] While they all led remarkable lives in their own right, at the heart of this book are the training, early Alpine excursions and especially the Indian and Himalayan expeditions in the mid 1850s of three of the brothers: the physical geographers Hermann (1826–82) and Robert Schlagintweit (1833–85), and the geologist Adolph Schlagintweit (1829–57). The three brothers' eastern journeys also deeply marked, however, the life and work of the younger Emil Schlagintweit (1835–1904), who used the observations, artefacts and Buddhist manuscripts of his siblings to turn himself into an internationally renowned and prolific Tibetan scholar.[18] Emil first played a crucial role as a mediator between his itinerant brothers and the German sponsors and mentors of their scheme. He was later also deeply involved in the analysis, publication and dissemination of the scientific and material legacies of the Schlagintweits' Asian travels. The armchair orientalist with excellent contacts to imperial authorities and local informants in India will therefore feature in this work. His career sheds light on the significant yet little studied aspect of kinship and the negotiation of scientific authority.[19]

A fifth brother, the Bavarian officer Eduard Schlagintweit (1831–66), also experienced European expansion firsthand by accompanying the Spanish military invasion of Morocco in 1859–60. Eduard contributed

---

[15] Arnold, 'Contingent Colonialism'. This can be compared with the role of hired German government advisers in Meiji Japan, who were often instrumentalised by their national government to advance commercial and political influence; Meissner, 'Oyatoi'.

[16] Hodge, 'Science and Empire'; Sujit Sivasundaram early warned about the possibility that the global history of science might become dominated by such 'structural frameworks'; Sivasundaram, 'Introduction', 95.

[17] These were Rosalie Seidl and Johanna Prentner. One of their sisters died after birth, and Mathilde Schlagintweit remained largely detached from her brothers throughout her life.

[18] E. Schlagintweit, *Buddhism in Tibet*; E. Schlagintweit, *Indien*. He also published on contemporary political, military and social concerns in the British Empire in India and its border regions.

[19] For an introduction to Emil's little-studied work, see Neuhaus, 'Tibet-Forschung'.

to the ethnographic collections of his brothers from India by taking plaster casts of the local inhabitants he encountered in northwestern Africa, and published intricate cartographic work of the lands he passed through.[20] To complete the family portrait, by far the youngest was Max Schlagintweit (1849–1935), who inherited a passion for travel from his half-brothers. In his early adulthood, he experienced the emergence of a unified Germany with global ambitions of its own, and decided, as a Bavarian officer, to engage himself in the colonial circles and societies of the Kaiserreich, not least by completing an expedition to Asia Minor in 1897, a region he afterwards forcefully suggested for future German economic and political penetration.[21]

An unusually large number of members of the Schlagintweit family thus encountered and became actively involved in different national-imperial projects through their own mobility, as they searched for and found diverse opportunities in an age of empires. They were intermediaries and brokers of knowledge: Their firsthand experiences filtered back into Europe through their erudite writings, museological displays, pedagogical materials and public performances.[22] The Schlagintweit trio who ventured into south and central Asia produced several monographs on their expeditionary travels and the history, cultural varieties and current state of British India.[23] Through the vital support of a small army of indigenous assistants and porters, the brothers also acquired a vast collection of around 40,000 natural historical and ethnographic artefacts. Originals and replicas found their ways into various museum collections and archival repositories across Europe, India and North America, after the brothers' plan to permanently establish an 'India museum' in Berlin – filled with their Asiatic booty and modelled on the famous East India Company museum in London – was frustrated by the Prussian ministry of culture.[24] Besides publishing a long list of popular works and producing a wealth of elaborate visual representations of eastern sceneries,

---

[20] He later went to England and France; before being killed in the Austro-Prussian War in 1866, he published *Der spanisch-marokkanische Krieg* and 'Militärische Skizzen'.

[21] A writer on almost exclusively colonial and military themes and a vocal supporter of the Berlin–Bagdad Railway, Max published *Deutsche Kolonisationsbestrebungen in Kleinasien*; *Militärische und topographische Mitteilungen*; *Verwaltung des Kongostaates*; *Afrikanische Kolonialbahnen*; *Routen-Aufnahme*; *Verteidigungsfähigkeit Konstantinopels*, among other works.

[22] Highly suggestive are the contributions in Schaffer et al. (eds.), *Brokered World*.

[23] Hermann, Adolph and Robert initially set out to publish a nine-volume work, *Results of a Scientific Mission to India and High Asia*, but finished only the first four. It was accompanied by a striking *Atlas*. A more popular German travelogue and political-social account of British India was later compiled by H. Schlagintweit, *Reisen*.

[24] Brescius, 'Empires of Opportunity', ch. 7: 'Conflicts of Collecting'; Kleidt, 'Sammlungen'.

the two surviving brothers, Hermann and Robert, also recruited various popular audiences to consume the results of their travels through public events such as lectures and exhibitions that effortlessly combined academic research with entertainment spectacles.[25] Noteworthy was their participation in several international congresses and colonial fairs in the 1860s and 70s, which provided a privileged stage for showcasing both their acquired artefacts and personal exploratory feats.[26]

Through their imperial sojourning, the Schlagintweits formed, of course, part of a much larger, indeed centuries-long movement of German naturalists, travellers, doctors, soldiers, scribes and other occupational groups who had accompanied the expansionist designs by various European powers and numerous African, Levantine and East and West India Companies since the late fifteenth century.[27] The decades from 1760 to 1820 in particular had seen a significant wave of German scientific travellers undertaking exploratory journeys with foreign state support. Among the most notable cases were the Forsters, father and son, who had accompanied Captain James Cook on his second voyage to the Pacific in the 1780s with the official sponsorship of the British Admiralty – a journey that inspired Alexander von Humboldt's American voyage (1799–1804) after he had to give up his plans to explore Egypt as a result of the French Egyptian campaign under Bonaparte in 1799.[28] Carsten Niebuhr realised his extensive travels under the protection of the Danish Crown as part of the Royal Danish Arabia Expedition in the 1760s.[29] And besides those well-known writers there were several prominently employed (Baltic) Germans in Russian service who explored parts of Siberia and other corners of the Tsarist Empire and beyond in the eighteenth and early nineteenth centuries.[30]

---

[25] Daum, *Wissenschaftspopularisierung*. Robert alone made over 1,300 public appearances.

[26] Among them the Colonial Exhibition in London, 1862; the World Fair (Exposition Universelle) in Paris, 1867; the Second International Geographical Congress in Paris, 1875; and various meetings of the British Association for the Advancement of Science.

[27] A useful overview is Denzel (ed.), *Deutsche Eliten*, esp. the editor's introduction, and the contributions by Michael Mann and Jürgen G. Nagel on Germans in India in the early modern period and their leading positions in the Dutch VOC. On German soldiers in 'French' and 'British' India, and their large presence in the Dutch East and West India Companies, as in the War of American Independence, see Tzoref-Ashkenazi, *German Soldiers*, 4–5. For a helpful overview in the longue durée, see also Blackbourn, 'Germans Abroad'.

[28] Liebersohn, *Travelers' World*; Gascoigne, *Encountering*; Berman, *Enlightenment*; Beck, 'Forster'; Goldstein, *Forster*.

[29] Baack, *Curiosity*.

[30] Duchhardt (ed.), *Russland*. On the travels of Peter Simon Pallas in Russian Asia as part of the Academic Expeditions (1768–74), see Vermeulen, *Before Boas*, 306–10. After the turn of the century, the German naturalists Johann Tilesius (von) Tilenau and the intrepid Georg Heinrich von Langsdorff both accompanied the Baltic German admiral

The decades around the mid nineteenth century offered increased chances of exploration for a new cohort of elite university-trained Germans because of the ongoing imperial advances by various European powers into continental heartlands. The interior of Australia, Africa, the vast extents of south and inner Asia, China and parts of South America all attracted western scientific scrutiny.[31] Different metropolitan authorities and colonial establishments often turned their attention to recruiting skilled German naturalists and explorers on a contractual basis for numerous survey works, technical and advisory missions or administrative positions.[32] One reason for this was that Germans, coming from territorially fragmented homelands and lacking formal overseas territories until the 1880s, were not perceived as a threat – as agents with their own national-imperial designs. It is no coincidence, therefore, that while Germans were appointed to lead large scientific missions or even found and head entire government departments in British India, French scientists were in contrast treated with greater scepticism and hardly ever acquired such elevated administrative positions – given France's longstanding imperial interests in this world region.[33] The Company also regarded French travellers and those from other colonial powers with suspicion because it feared they could either join the service of various Indian princely states, which at times offered more lucrative terms of employment than the British, or undermine British sovereignty in other ways.[34] From the late eighteenth century British officials therefore signed a number of discriminatory treaties with Indian states that forbade the employment of European personnel in the military, civil and scientific services of the Company's Indian 'allies'.[35]

Adam Johann von Krusenstern on the first Russian circumnavigation of the world (1803–6); Tucker Jones, *Empire of Extinction*, 198.

[31] See, e.g., Kennedy, *Spaces*.

[32] On the survey work of the brothers Schomburgk in British Guyana Burnett, *Masters*; on Ludwig Leichardt and his famous explorations into the interior of Australia, Thomas, 'Expedition'; several cases of German travellers in Africa are mentioned in the splendid work by Naranch, 'Beyond the Fatherland'.

[33] A prime example of such a high office holder is the German forestry expert Dietrich Brandis, who helped establish the Indian Forestry Department in 1864 and which he subsequently led for the next two decades – only to be succeeded by two other German scientists he had personally trained who led this branch of government until 1900; see the partly autobiographical account by Berthold Ribbentrop, who replaced Wilhelm Schlich as Inspector-General of Forests in 1885–6, Ribbentrop, *Forestry*, 76; and Rajan, *Modernizing Nature*; on French influence on EIC science, Arnold, 'Contingent Colonialism'.

[34] See on the Company's fear in relation to the presence, in particular, of French engineers, military experts and soldiers in Indian Princely states, Jasanoff, *Edge*.

[35] Alexandrowicz, 'Asian Treaty Practice', 189. On Anglo-French rivalries at Hyderabad and the detailed genesis of the discriminatory treaty of 1798, see Bührer, 'Intercultural Diplomacy'.

What further added to the appeal of appointing German specialists to scientific positions was the fact that German universities had undergone important changes in the way their graduates were trained. It was in the German states that the first modern research laboratories in the natural sciences had been established, first by Justus von Liebig in Giessen in 1826, and subsequently at different universities in the competing landscapes of the German principalities, kingdoms and free cities.[36] Young men from Britain and elsewhere in continental Europe, the United States and India, came to the German states for a laboratory-based training in chemistry from the 1830s onwards, but also to receive humanistic education and training in the field sciences.[37] There existed indeed a specific notion of 'German science' [*deutsche Wissenschaft*] in the nineteenth century, which nurtured national feelings of scientific and natural philosophical supremacy.[38]

The established overseas infrastructures of other powers can thus be understood as *empires of opportunity* for a large spectrum of scientific travellers and specialists from a country without colonies keen to satisfy their curiosity, adventurous impulses and learned interests. Crucially, this was a case of *mutual instrumentalisation*: while imperial institutions profited from the influx of foreign manpower and expertise, Germans could expand their opportunities to travel and research. Given the surplus of skilled personnel in the German states, and the attraction of paid employment in an imperial labour market, such migration patterns may indeed be seen 'as a balancing process, a process of redistribution of human resources according to the interests of those involved ... and according to the interests of the sending and receiving societies'.[39]

The resulting presence of German philologists, missionaries, travellers, administrators, scientists, engineers and other occupational groups in Britain and its Empire has attracted significant interest of late.[40] This forms part of a wider reorientation of attention towards the remarkable permeability and openness of imperial systems for indigenous and other groups of actors, who originated from beyond the boundaries of the respective imperial motherland.[41] Most recent studies claim a more

---

[36] Holmes, 'Complementarity'; Allen, 'Parallel Lines', 363.
[37] Blackbourn, 'Germany'; exploring a whole set of Indo-German intellectual connections in the period *c.* 1880–1950 is Manjapra, *Entanglement*.
[38] See Chapter 8 for a thorough treatment of this charged notion as concerns exploration and the expeditionary sciences, dimensions less considered in Schiera, *Laboratorium*.
[39] Hoerder, 'Macrosystems', 81.
[40] See for a literary approach, Zantop, *Colonial Fantasies*; Home, 'Science'.
[41] A pioneering work on the appropriation of imperial networks, even for anti-colonial objectives, is Alavi, *Muslim Cosmopolitanism*; Schär, *Tropenliebe*; on the 'permeability of imperial domination', Oppen and Strickrodt, 'Biographies'.

or less successful integration of Germans in foreign overseas projects and exchanges through cross-border networks held together by shared cultural and political ties and Protestant leanings.[42] For instance, Kris Manjapra unambiguously declared that '[t]he role of German-speakers as managers of British imperial science raised few eyebrows in the metropole in the period of Victorian self-confidence'.[43]

The present work, by contrast, takes much more seriously the manifold and at times vicious conflicts of scientific authority and loyalty that surrounded Germans' imperial sojourning. It explores in detail the tensions between different scientific epistemes at work in transnational science, and studies the significant political ramifications of international controversies over such imperial outsiders.[44] That is, while the cosmopolitan tradition of enlightened voyagers moving across national and imperial boundaries may have continued into the nineteenth century, often the terms of employment, the public reception and political appropriation of German travellers and their overseas achievements had tangibly changed in an age of increased nationalist sentiment and discourse. Overseas exploration became, in a sense, a significantly more politicised phenomenon, as debates over national achievements and entitlements to well-funded scientific appointments in the British Empire were waged in a transnational arena. The reason was that, by mid-century, recruited Germans also often ceased to be mere detached guests of foreign empires. Rather, in order to realise their formal and salaried commissions, they generally received full administrative and institutional support and were often, though not always, identified with imperial service by other European authorities and the indigenous peoples they encountered alike.[45]

Since the late eighteenth century, 'British' India in particular had become a key site for hundreds of university-trained Germans to help explore – and sometimes administer – distant lands, and also to investigate areas that still lay outside the East India Company's (EIC) formal

---

[42] A wealth of biographical details on Germans of various occupational backgrounds in Britain is provided in the excellent work by Kirchberger, *Aspekte*; Kirchberger and Ellis (eds.), *Networks*. Useful on different occupational groups of Germans overseas is also Naranch, 'Beyond the Fatherland', Oesterheld, 'Germans in India' and, most recently, Panayi, *Germans*, who makes a thorough examination of, in particular, the German missionary experience in south Asia and of the fate of Germans in India after the outbreak of the Great War.

[43] Manjapra, *Entanglement*, 32.

[44] The controversial repercussions of transnational science are alluded to, but not fully explored, in Finkelstein, 'Mission' and Kirchberger, 'Naturwissenschaftler', 647.

[45] On the case of German African travellers negotiating commercial treaties with African statesmen in the name of the British government, as in the case of Heinrich Barth, Naranch, 'Beyond the Fatherland', 242.

dominion, including the Himalayan regions to the north of the expanding empire. The Company's Court of Directors, seated in London and jealously guarding access to their territories, was then one of the greatest patrons for scientific personnel from Britain and northern Europe – including many Scandinavian naturalists working in the tradition of systematic botany as advanced by Carl Linnaeus and his pupils.[46]

The scientific pursuits of the Schlagintweits in Asia were mainly sponsored by and formally stood under the 'orders' of the EIC's Court of Directors, but they also received significant support from the Prussian king Frederick William IV and the Bavarian monarch Maximilian II. It was in reality a series of different expeditions through India, the Himalayas and central Asia, undertaken individually and conjointly by the brothers in constantly changing constellations with indispensable indigenous guides, translators and assistants. Some of the latter, however, significantly enlarged the scope of the entire enterprise by realising their own independent survey and collecting excursions that could last several months – and were later inscribed into the official cartography of the mission. While it has been the convention to date the mission to 1854–7, the years the brothers were present in Asia, some indigenous partners continued their contractual work into 1858, when the surviving Schlagintweits were long back in Europe preparing their publications. (Unlike later historians, the brothers themselves officially dated the mission to 1854–8.) It is thus a convenient misnomer to speak of the 'Schlagintweit expedition', as this highlights only the role of its supposed European leaders, and problematically silences what was fundamentally a collaborative project of expeditionary science at a great scale of operation. It is more apposite to describe the scheme as a whole cluster of expeditions with no clearly defined centre. Indeed, the itineraries of those involved constantly pointed in different directions – only to briefly converge at pre-arranged meeting points to facilitate the comparison of the gathered material collections, scientific instrumentation and corresponding observations. Such moments of concentration, exchange and planning of future routes and objectives were immediately followed by another round of dispersion of its main, transcultural workforce.[47]

The brothers employed several dozens of porters, collectors, runners, informants and other agents along the way, who constantly injected new local know-how, language skills, social contacts and geographical orientation into the travelling party – understood here as a 'moving colony'

---

[46] Arnold, *Science*, 19–20; Hodacs, 'Circulating Knowledge'.
[47] Brescius, 'Innenleben', 98.

with diverse internal social and religious hierarchies.[48] The miscellaneous expeditionary workforce ultimately came to include also some well-known figures in the history of exploratory science in high Asia – some of whom had assisted British travellers before. Among those who accompanied the Schlagintweits on significant sections in the Himalayas and beyond were Mani Singh and his cousin Nain Singh – later the famous 'Pundit No. 1' of the larger group of indigenous surveyors who secretly explored Tibet and much of central Asia for British authorities in the wake of the Indian Rebellion.[49] The brothers' close but conflictive relationship with Nain Singh is reconstructed from the latter's own writings and fittingly illuminates the intricacy of mutual dependence and exploitation at play in the mission – especially when it moved through non-British territories where the leadership would often shift to the indigenous personnel. The case of Nain Singh also suggests that several assistants operated well beyond the confines of 'local knowledge' and territory, but rather became significant explorers in their own right, 'thereby destabilising the very categories that sustained exploration as a European endeavour.'[50] Crucially, while the brothers – and several Indian partners – were free to identify suitable guides for their excursions, colonial officials also repeatedly ordered skilled south Asian surveyors and draughtsman to join the staffing of the expedition. This reflected official interests, such as that of the Office of the Quarter Master General at Madras, and the Indian Surveyor-General's Bureaus in Dehra Dun and Calcutta (Kolkata), to use the scheme for topographical and military reconnaissance of little-studied territories, especially along and past the volatile northern frontier of British India.

Indeed, the travelling parties constantly collaborated with and drew on the expertise of different external agents, scientific institutions and imperial authorities. These were incessantly mobilised to provide political intelligence and to collect further instrumental data and observations on numerous natural phenomena – from climatic anomalies and geological analyses to magnetic forces, with all of which the Schlagintweits were concerned. The existing literature has predominantly focused on the large if not utopian ambitions behind the brothers' scientific programme

---

[48] Useful are the reflections by Thomas on an expedition as being significantly more than a mere grouping of individuals, but rather a 'distinct socio-cultural formation', Thomas, 'What is an Expedition?', 7; see also Gräbel, *Expeditionen*.

[49] Waller, *Pundits*; Alcock, 'Pioneers', was the first to draw attention to their indigenous establishments; Armitage, 'Schlagintweit Collections'; see also Brescius, 'Forscherdrang'; Driver, 'Intermediaries'. The names of south and central Asian individuals in this work generally follow the Schlagintweits' spellings.

[50] Kennedy, 'Reinterpreting', 12.

and often ignores entirely the multiple collaborations of the expeditions on the ground. In contrast, this work demonstrates that the colonial context and the deep institutional embeddedness of this scheme fundamentally shaped its scientific concerns, execution and published outcomes.[51]

Recent work on exploration has pointed to the complexities that were at play when multiple sponsors and interests tried to initiate and influence expeditionary work. Torn between, but also actively shaping, the different expectations of multiple patrons and audiences across Britain, Europe and India, the Schlagintweit enterprise generated an unusual amount of public scrutiny at the time. The arresting constellation and transnational sponsorship behind their scheme requires a nuanced analysis of the interpretative flexibility over the relation between utility, science and empire in the work of the itinerant naturalists. The Schlagintweits could have been a mere footnote in the longer history of contractual survey work at the far corners of the British Empire. Yet, their own aspirations to become great explorers in the mould of their mentor Alexander von Humboldt, and their eagerness to publish with imperial support a large interdisciplinary oeuvre, put them on a different track, and ultimately exposed them to significant hostile and hagiographic reactions in Britain and the German states.[52]

The brothers' personal and intellectual connection with Humboldt, whom they befriended and collaborated with after moving from Munich to Berlin in 1849, was significant in several regards. Humboldt's enthusiastic support of their researches not only opened the door for their later employment in India, but it also shaped the brothers' research agenda in significant, but by no means exclusive, ways. Humboldt's American voyage and the influential scientific oeuvre built on it provided both a model and a taxing challenge to the Schlagintweits' own travel plans.[53] This pressure only increased since Humboldt's own ambitions to complement his New World excursions with a Himalayan expedition had been left unfulfilled, as British authorities never granted access to their territories to this outspoken critic of colonialism.[54] The closest Humboldt got was a trip to Siberia up to the Chinese border in central Asia courtesy of the Tsarist government in 1829.[55] This meant there was a special

---

[51] For example, Finkelstein, 'Mission', did not consider the colonial setting in which the travels took place. The same applies to analyses by other European scholars interested in the collecting mania of the Schlagintweits, both of instrumentation and artefacts; Felsch, '14.777 Dinge'; Felsch, 'Söhne'; Finkelstein, 'Berge'; Finkelstein, 'Headless'.

[52] Brescius, 'Rezeption'.

[53] Körner, 'Photographien'; Finkelstein, 'Mission'; Felsch, 'Söhne'.

[54] Théoderides, 'Humboldt'.

[55] Osterhammel, *Entzauberung*, 43.

attraction for younger Humboldtians to go and explore what the master had been unable to see himself.

The brothers' eastern scheme offered their critics a kind of moving target, as they changed and enlarged its concerns and objectives several times. Having proven themselves as capable naturalists in the Alps in the late 1840s and early 50s, the Schlagintweits nevertheless at first failed to secure support in 1852 for an exclusively Prussian-funded expedition to India and the Himalayas that would have made them independent observers in British domains – detached from EIC interests and guidance. Chiefly on Humboldt's recommendation and through the tireless efforts of the savant and Prussian consul in London, Christian Karl von Bunsen, the three Schlagintweits were then formally employed by the EIC two years later, but at first merely to complete the geomagnetic survey of the Indian archipelago left vacant after the death of its British leader.[56] This commission reflected current interests of the metropolitan avant-garde in global physics.[57] Yet, even prior to departure, this narrow objective was already transformed by the Schlagintweits in negotiation with the Court of Directors into an interdisciplinary Humboldtian enterprise of physical geography. The substantially revised programme was now to combine large-scale data gathering through a vast array of modern precision instruments with numerous local observations on geology, meteorology, botany and zoology and also included an intensely cartographic and aesthetic orientation.[58]

Yet, in view of the brothers' imperial sponsorship and their hope for a renewed contract from the Company upon return – to publish their observations and prepare their collections as museological artefacts – the Schlagintweits were equally keen to meet, not always successfully, the imagined material interests of their British patrons. In the field, the travellers thus sought to undertake, besides more disinterested enquiries, macro-geographical surveys of natural resource deposits and trading potentials across India, the Himalayas and central Asia. Later, they also suggested ambitious, and at times unfeasible, schemes of large-scale economic and social colonisation of south Indian hills and parts of the Himalayas to the Company and its successor, the India Office. Not all of their proposals were original; they only added their voice to existing

---

[56] While no mention is made of the initial quest for exclusively Prussian assistance, otherwise useful recapitulations of their way into foreign employment are given in Polter, 'Nadelschau'; Finkelstein, 'Mission'.

[57] Cawood, 'Magnetic Crusade'; Ratcliff, 'Geomagnetism'.

[58] By also considering the aesthetic dimension of Humboldt's work and philosophy of science, I diverge from scholars who have tended to neglect this dimension in their definition of Humboldtian science. See, however, Werner, *Naturwahrheit*.

and ongoing discussions about exploitable opportunities in Indian and metropolitan circles. These included uses for large amounts of valuable Himalayan timber going to waste every year; the chance of even larger cultivation schemes of tea, indigo and other staples than those already in operation for decades; and the desirability of establishing health sanitaria and future white settlement in the massive mountain chain that marked British India's northern boundary.[59]

Scholars have, however, tended to study the Schlagintweit scheme in splendid isolation – as purportedly detached from previous colonial travels or current British concerns.[60] Yet, to understand what scientific fields the brothers chose to engage with and what they were ultimately able to publish in terms of available records and data, it is essential to place their excursions and accounts in a much larger sequence of former and ongoing British and European observations, explorations and technical surveys on which they explicitly build and that they hoped to advance.[61] Is it also in this way that we fully see the truly European nature of their enterprise. There was indeed a serial logic to such exploratory ventures: 'Every expedition exists in relationship to a greater set of other expeditions: past and future, real and imagined.'[62]

Yet the context of the Schlagintweit enterprise in colonial India mattered in yet another fundamental regard. This book argues that in consequence of the extended presence of British observatories, military stations and medical and technical services stretching across the subcontinent from the coastlines into the Himalayas by the mid nineteenth century, empire came to shape the very concerns and practicalities of European science in south Asia.[63] As such, empire was decisively more than a mere socio-political backdrop for the activities of the Schlagintweits and other German scientists. In reality, the existing imperial infrastructure

---

[59] On the potential of timber resources from the 'Lower Himalaya', Joseph Hooker had already informed the Governor-General in 1848; see his letter to William Hooker, Archive of the Royal Botanic Gardens, Kew [RBG], JDH/1/10, July 1848, 61; Kennedy, *Stations*; for British tea and other natural commodity production in India since the early decades of the nineteenth century, Drayton, *Government*. On the brothers' colonial proposals, House of Commons, *Select Committee on Colonization and Settlement*; Schlagintweit, 'Practical Objects connected with the Researches of the MM. Schlagintweit, under the Orders of the Hon. Court of Directors', 21 September 1857, British Library, London [BL], MSS EUR, F 195/5.

[60] In line with the contextualisation of the Schlagintweit enterprise provided in Brescius et al., *Über den Himalaya*, Sarkar, 'Science', has confirmed links between the brothers and colonial institutions such as the Great Trigonometrical Survey.

[61] The brothers opened their first book with this positioning in a much larger continuum of previous exploration: H. Schlagintweit et al., *Results* I, 8.

[62] Thomas, 'Expedition', 81.

[63] Vetter, 'Introduction', 9; Reidy, 'Spatial Science'.

decisively determined what natural phenomena could be simultaneously observed over the vast British possessions, and hence what insights into larger natural systems and the variation of physical forces could be gained and published by many (foreign) observers of Indian geography and climatic physiognomy.[64] Even such progressive scientific projects now conveniently subsumed under the label of 'Humboldtian science' in fact heavily depended on 'the mobilization of symbolic and political resources which extended well beyond the realms of disinterested knowledge'.[65] Exemplary is the study of the Earth's magnetic field – tellingly called the 'Magnetic Crusade' by its British proponents in the 1830s – a project to which the Schlagintweits contributed and that inevitably drew, besides the involvement of esteemed scientific bodies, on the vital support of the Royal Navy, the Admiralty and the EIC's overseas possessions.[66]

The existing imperial infrastructure could also determine what foreign visitors and employees in British India could measure, collect and bring back to Europe as anthropological data or material artefacts. The power asymmetries of the colonial context deeply shaped at times those sites of cultural encounters from which many such measurements and objects were produced or acquired. This applied, for instance, to facial plaster casts and physical anthropological studies taken by German scientists like the Schlagintweits from the bodies of convicts in Indian prisons.[67] It also enabled their accumulation of human remains from colonial hospitals, or the sanctioned plundering of tombs for entire skeletons (see Chapter 4). Such activities, where the boundary between collecting and looting was blurred, exemplified how British institutional arrangements in India were not the passive background of independent enquiries but, recurrently, the facilitating condition.[68] Of course, the reliance of many German scientific travellers on established imperial structures does not imply their consent to or even active support of the ruling power's designs and ambitions. It rather requires a case-specific analysis to establish to what extent, and why, the imperial identification of German scientific experts in British service might have shifted over time.

---

[64] For significant reflections on how scientific results bear the marks of their place of production, Livingstone, *Geographies*.

[65] Driver, *Geography*, 26; Patricia Fara has shown how interests in geomagnetism had a longer imperial connection, *Attractions*, ch. 4.

[66] Morrell and Thackray, *Early Years*, 353–70, 523–31.

[67] Bayerische Staatsbibliothek, Munich [BSB], Schlagintweitiana [henceforth: SLGA], vol. II.1.37–8, 'Messungen von Menschenraçen'.

[68] Cohn, *Colonialism*, 76–105; on prisons, barracks and hospitals as important sites of physical anthropology, Sysling, *Racial Science*, 49–54.

What underpinned transnational science and expert mobility in the long nineteenth century was the existence of cross-border networks of patronage.[69] Their efficacy lay in British trust in the recommendation of scientific peers from the European continent. This, in turn, owed much to the fact that German exploratory science had enjoyed wide international esteem since the late eighteenth century, not least due to two towering figures of physical geography: Humboldt and the Berlin professor and armchair geographical polymath, Carl Ritter. These eminent men of science from Berlin's geographical circles and societies were not only prolific authors of great synthesising works up to the 1850s in their own right; they were also adept managers, pulling strings with gentle insistence for numerous protégés who then acted as their informants on hitherto unvisited sites.[70] This led to an effective 'channelled mobility' of young German talent into foreign service through private skeins of patronage, which in Britain had, at times, been more oblivious to nationality in terms of entitlement to scientific appointments but had long shown a predilection for natural philosophical travellers and continental naturalists since the days of Joseph Banks.[71]

Despite the benefits that arose from the co-optation of foreign expertise to the hosting empire, on an individual level the great influx of outsiders to privileged scientific and administrative positions in British India (and elsewhere) was fraught with tension. This peaked with the appointment and generous sponsorship of the Schlagintweits in the 1850s. The contested careers of these German travellers in British service thus do not suggest a frictionless pan-European project of overseas exploration, colonisation and domination. On the contrary, in terms of methodology, this work narrows in on specific moments of inner-European tensions, which occurred throughout the contingent process of European expansion and concurrent disciplinary specialisation in the nineteenth century.[72] Such a perspective forces us to view exploration, especially when undertaken by non-nationals, 'through a different lens, as a site of conflict and controversy, rather than a synonym for the triumphal progress of European science'.[73]

---

[69] From a wealth of studies on social networks, Derix, 'Netzen'; see also the seminal work by Schulte Beerbühl, *Kaufleute*; Kirchberger and Ellis (eds.), *Networks*. Therein, as in the present work, 'the network functions as a key conceptual tool with which to assess the coherence of Anglo-German scholarly relations and to reflect on the dynamics which characterized them'. Kirchberger, 'Introduction', 3.

[70] Päßler, *Mittler*.

[71] Grove, *Imperialism*, 331; Huber, *Mobilities*.

[72] For an investigation of how western scientific disciplines and practices were shaped by colonial expansion and encounters, see, e.g., Fabian, *Out of Our Minds*; Endersby, *Nature*.

[73] Driver, 'Critics', 142.

At the heart of the Schlagintweit controversy, multi-layered and long running as it was, stood debates over the respectable pursuit of science and the slowly changing nature of scientific patronage from a private affair towards public, and putatively more accountable, forms of national support. It thus throws into sharp relief significant aspects of the culture of gentlemanly science and its conflictive transformation. In Britain, the brothers were mostly seen as greedy and self-interested poachers, breaking with the social conventions of pursuing science not for the sake of knowledge or public welfare, but for personal financial gain. Yet, the older ideal still prevailed that personal wealth and hence independence would ensure the respectability and trustworthiness of natural historical enquiry – as typified by many well-heeled members of the Royal Society since the seventeenth century.[74] The Schlagintweits' clever negotiations with the Company, and their private enrichment through foreign employment and multiple sponsorships, thus illuminate deeper trends and changes in the social fabric of science. What emerges is a detailed study of the general mechanisms of how scientific experts with no independent means had to engage in a competitive (imperial) market for resources, both material and symbolic.

Intriguingly, the polemic over what were seen as mercenary and corrupt strategies of the brothers involved some of the most illustrious names, journals and institutions in the history of science at the time. It brought together prominent representatives of the Royal Society, the Royal Geographical Society of London (RGS), its counterparts in Paris and Berlin, and also the Royal Botanic Gardens at Kew, together with other central sites of knowledge production and disciplinary debate. Crucially, the lines of friction did not neatly run along national borders, but also along epistemic communities and, above all, patronage circles. The 'scientist of empire' and powerful RGS president, Sir Roderick Murchison, for instance, belonged to a British clique with close contacts to Berlin who cherished the tradition of German expeditionary science.[75] He and others adopted a pragmatic stance on external recruitment – a means to secure available talent from abroad 'provided we have no better & fitter men ready'.[76] He thus opposed the exclusionary rhetoric of some British peers, including the highly respected imperial botanist, Joseph Hooker, who like Murchison cherished close ties to Humboldt in Prussia

---

[74] This points also to the shortcomings of the 'professionalisation model' for explaining mid-century science, a time of great insecurity about the propriety to engage in scientific activity as paid work; for a powerful critique, Endersby, *Nature*.

[75] Stafford, *Scientist*.

[76] Murchison to Joseph Hooker, 19 January 1854, RBG, DC, 96.

but nonetheless acted as one of the Schlagintweits' fiercest critics. The level of competition ran especially deep since Hooker, later long-serving director of Kew, initially struggled to secure an adequate scientific position of his own, and had undertaken a scientific expedition to India and the Himalayas only a few years prior to the Schlagintweits, collecting scientific specimens in overlapping regions but failing to secure Company support for its projected publication.[77]

One of Britain's leading artistic and learned journals, *The Athenaeum*, likewise responded to the Schlagintweits' employment in more alarming tones, since 'This desire to promote science through the instrumentality of foreigners, so insulting to our Indian service, is not now for the first time indulged in.'[78] Yet the paper did more than give prominent voice to a general resentment to such transnational career-making. It also alluded to the important role contemporaries attached to British scientific activity in India, presented here as a form of imperial legitimation in the eyes of the colonised: 'The employment of these young Prussians will scarcely tend towards the maintenance of that belief in our superiority on which the Government of India by Great Britain is said to hang.'[79]

Once the controversy over the Schlagintweits had gathered momentum and public significance, it ultimately prompted discussion of further fundamental issues including the funding and justification of such projects, and the systems of favouritism behind them. Those British officials and men of science who had supported their scheme, such as the influential Royal Society man and military engineer Edward Sabine, or the naturalist William Henry Sykes from the EIC's Court of Directors, all had to defend their actions.[80] What had been smooth-running channels of recruitment, based on trust, became suddenly politicised, pulled into the spotlight and exposed to external criticism. The Schlagintweit controversy was thus inextricably linked with the rhetorical nationalisation of empire in reaction to a perceived wave of well-funded foreigners, whose 'scandalous' appointment called for public accountability. Scientific networks of patronage from Berlin to London and Calcutta suddenly appeared corrupt, became fragile and ultimately defunct. Yet more was at stake than a nationalist sense of entitlement among British men of science and the wider public.

---

[77] On Hooker's struggles and his critique of the EIC's unsystematic support of the sciences, even of projects of evident use for its utilitarian interests, Endersby, 'Philosophical Botanist'.

[78] *The Athenaeum*, 1379, 1 April 1854, 408.

[79] Ibid. All the Schlagintweits were Bavarians, but given the powerful support they received from Humboldt and the Hohenzollern king, were often identified as Prussians.

[80] Finkelstein 'Mission'.

Much transnational and global history writing has shown a tendency to see transversal networks as appealing harbingers and facilitators of cross-border integration and concord – less so as sites of conflict and debate over unwarranted privilege and the omission of meritorious others who felt ignored and left behind. What this work thus adds to the rich field on connectivity through social and professional ties is an in-depth study of their external perception and social logics, especially in moments of crisis. In reconstructing personal and epistolary links between members of Berlin's scientific community and their British peers and Indian officials, it considers what happens to such links once they become targets of private and public pressure and critique – indeed when they break down in the context of violent controversy over their very efficiency.[81] In paying close attention to the mechanisms of inclusion and exclusion at play in transnational scientific ventures, and Britain's systematic reliance on foreign expertise to staff lucrative positions at home and abroad, the brothers' case neatly demonstrates how, in fact, 'more interdependence can yield more conflict'.[82]

In this study, the unfolding of the controversy is carefully reconstructed from the standpoint of different actors and not only because it illuminates national antagonisms in the field and practice of science. Rather, a number of transnational developments such as an emerging crisis of Humboldtian models of interdisciplinary enquiry, and the changing ways in which individual travellers sought to establish their right to speak about distant lands and societies, are pulled into focus. The Schlagintweits' struggle to establish their own credentials and authority over the vast and diverse Indian and Himalayan regions across which they had moved can only be understood within the much longer history of European encounters and the resulting humanistic and scientific engagement with that world region since the sixteenth century.[83] Despite what the brothers hoped to project of their pioneering achievements in front of popular audiences, they thus operated within a crowded field of rich previous scientific exploration and a competitive market in the European metropoles for cultural representations. Theirs was, in other words, mostly not a voyage into the unknown; there rather existed a wide

---

[81] I have benefited from a symposium in Konstanz on 'Flows and Orders: A Tension in Global History', organised by Jürgen Osterhammel with the author and Cornelia Escher (October 2016), as part of a wider Leverhulme network on Global Nodes, Global Orders: Macro- and Micro-histories of Globalization. I am most thankful to James Belich and John Darwin for the generous chance to contribute to its events.

[82] Adelman, 'Global History'.

[83] For a post-Saidian analysis of the different visions and representations of India across the entire early modern period, Subrahmanyam, *Europe's India*.

variety of people of European and Indian origin – among them migrants, former travellers, returned expatriates, oriental scholars, lecturers, printmakers and other voices – who themselves claimed particular knowledge in Europe of India's past, geography and religious and cultural variation.[84]

The brothers' scientific approach and practices sat somewhat uncomfortably within larger developments such as the slow and by no means orderly differentiation of various fields within observational-instrumental science and interpretative scholarship – until the formation of more rigid modern scientific disciplines later in the nineteenth century. Both the disciplinary boundaries and the status and appropriate methods of fieldwork were in flux and subject to intense debate at the time. Felix Driver has argued that contemporary manuals and instructional 'hints to travellers' should therefore not be read as authoritative descriptions of fixed and widely accepted practices; they rather reflected the levels of uncertainty that prevailed among advocates and practitioners of the field sciences.[85]

In this conjuncture, the Schlagintweit mission sought to bring extremely diverse areas of activity together under the umbrella of a single collecting and documenting enterprise. Their choice of operation, partly symptomatic of wider contemporary practices in the field sciences, was to amass incredibly large amounts of documentation. This took the form of over forty volumes filled with instrumental recordings, natural historical and ethnographic notes, and various other including philological observations.[86] The brothers and their large set of Indian and Himalayan assistants complemented this textual accumulation with vast material collections, several hundred photographs and around 750 charcoal and aquarelle sketches of architectural and holy sites, natural scenes and ethnographic types.[87] These rich materials were to serve as the basis for their anticipated multi-volume work that was to cover at least eleven distinct fields of study.[88] Perhaps unsurprisingly, this sweeping enterprise attracted strong reactions from different communities of specialists. Yet, disciplinary politics cannot easily account for the simultaneously

---

[84] Fisher, *Counterflows*; Leask, 'Orientalism'; White, *Little Bengal*.
[85] Driver, *Geography*, ch. 4; Cooper, 'Voyaging'.
[86] H. Schlagintweit et al., *Results* IV, 6.
[87] On their visual strategies, a wealth of new insights is provided by Kleidt, 'Kunstwerk'.
[88] These were: 'Astronomy and Magnetism; Hypsometrical and Trigonometrical Observations; Topical Geography', a 'Route Book of the Himalaya, Tibet, and Turkistan', with additional volumes on 'Meteorology; Geology; Natural History: Botany and Zoology; Ethnography', and finally wider 'Geographical Aspects' of parts of south and central Asia; H., A. and R. Schlagintweit, *Prospectus: Results of a Scientific Mission*.

critical and triumphant reactions to the Schlagintweits' programme and publications across scientific communities in Britain, continental Europe and south Asia. Rather, this book argues that disciplinary orders and expertises were defined and consolidated during the process of managing the responses to their ambitious expeditionary scheme and its written and material legacies.

Only a fraction of the brothers' large-scale documentation was ever published; the Schlagintweit enterprise nonetheless allowed numerous other European scholars, natural scientists, museum directors and commercial promoters to use their accumulated materials as quarries for their own pursuits and interests.[89] As it stands, the existing literature underplays the role of their expeditions in generating, for instance, a wider framework for the study of eastern religions and especially Buddhism through their texts and museological specimens. To be sure, prior to their arrival in India, none of the brothers was a trained orientalist, in so far as they had no more than a scant knowledge of the languages, ethnicities and religions of the subcontinent. Yet, a common interest in the history, character and cultural diversity of the populations and religions they met grew steadily – leading the brothers on to acquire data, impressions and artefacts that they hoped would capture and represent different creeds, rituals and epistemologies. While there was a strong strategic and commercial dimension to such collecting behaviour, it was also a tangible expansion of interest brought about by the experience of overseas travel and cultural encounter – a significant instance of geographical displacement spurring disciplinary transgressions.

The two surviving brothers never identified themselves as orientalists in a more narrow sense.[90] Hence, while Hermann and Robert Schlagintweit remained largely detached from the specialist debates of German orientalism at the time[91], they did provide oriental artefacts, stunning

---

[89] For the use of the Schlagintweit collections for commercial promotion, see the analysis in Driver and Ashmore, 'Mobile Museum', which the present work expands to the German states.

[90] The recent wealth of valuable studies on German orientalism includes Mangold, *Orientalistik*; also on the development of university structures, Sengupta, *Indology*; McGetchin, *Indology*; Berman, *Enlightenment*; a more literary historical approach is taken in Kontje, *Orientalisms*.

[91] Marchand, *German Orientalism*, xxiv. Edward Said problematically left out German orientalists and their pioneering oriental studies from his analysis in *Orientalism*. This shortcoming has provoked a series of studies that demonstrate that the German relationships with the Orient was, well before the unified Kaiserreich and its formal colonies in the East, protracted – with a variety of religious, commercial, political encounters over the centuries; see, e.g., Berman, *German Literature*. While, as Kirchberger and others have shown, some German orientalists were indeed involved in imperial affairs, as for instance by training future colonial servants of the Raj in Britain, it is clear after Marchand's seminal study that 'German orientalism' was never

visualisations and knowledge about the East to various audiences, including a 'geographical glossary'[92] of Indian and Tibetan geographical designations, or eclectic insights about Indian and Himalayan religious and social practices both ancient and modern, often gleaned from personal encounters with spiritual leaders and monks. For some time, they also retained the services of an Indian munshi who accompanied them back to Europe and stayed on for eighteen months as an important philological adviser to their work.[93]

There is some irony in the fact that while much of their contracted researches on natural history and magnetism were later attacked, especially by British peers for a perceived lack of significance and originality, the brothers achieved, by contrast, considerable praise for their more spontaneously developed pursuits of artistic representation of nature and culture, and their detailed studies in physical anthropology. Their masterful sketches and images of partly unknown regions in the Himalayas and the Karakoram and the Kunlun chains in central Asia, and their production of visual depictions of racial and confessional types, all situate the Schlagintweits' work in a significant pattern of contemporary field sciences and comparative ethnography (Chapter 5). While some sections of their observations may appear unsatisfactorily transient and superficial, dominated, as it were, by the logic of a rapid transit through Asian territories and cultures, their work can nonetheless be seen as an early and significant example of survey ethnography and the individual collection of detailed information about local populations.

Here, it also becomes evident how the diverse set of employed indigenous companions ultimately served a double function. While they acted as indispensable interpreters and assistants that independently managed entire legs of the expedition's routes, and through their intermediation enabled the Schlagintweits to purchase many valuable objects even from sites inaccessible to Europeans, the same Indian and Himalayan personnel acted also as models for physical-anthropological measurements and sources for ethnographic enquiries. The brothers produced, among other things, a widely acclaimed set of 275 ethnographic plaster casts of

a single, monolithic discourse intrinsically and perpetually tied to imperialist concerns; Kirchberger, 'Contribution'; also Wokoeck, *German Orientalism*; cf. with Arnold, 'Contingent Colonialism', and Manjapra, who problematically portrays German-speaking scientists, philosophers and oriental scholars, sedentary and itinerant, almost exclusively as collaborators, and falls short of proving the purportedly 'overwhelming importance of German philosophies for the consolidation of British rule in India'; Manjapra, *Entanglement*, ch. 1, esp. 22.

[92] *Glossary*, in H. Schlagintweit et al., *Results* III, part 2, 133–293.

[93] H. Schlagintweit et al., *Results* I, 68–9.

indigenous faces, among them several taken from their hired companions, including Nain Singh. The individual recognition and appreciation of their assistants as real persons in field encounters, warmly described in their published travelogues and, at the same time, the representative function of their plastered faces and measured bodies in larger series on ethnic variation, point to a significant tension in the enterprise's scientific programme and the logics of transcultural exploration more generally, in which cooperation and exploitation were not mutually exclusive.[94]

The Schlagintweits' production and dissemination of texts, objects and images about indigenous cultures and Asian topography further point to the important entrepreneurial aspect of exploration. While they frequently downplayed this aspect in their published accounts, the Company commission was not only a well-paid job, especially considering the scarcity of salaried scientific positions at the time, but also one that opened up further financial opportunities that the brothers exploited to the full. They used their collections, for instance, as a versatile asset to enhance their scientific standing and prestige. Parts of their collections were gifted to aristocratic and royal supporters or scientific bodies in return for further patronage. Others were directly sold or mechanically reproduced to multiply their potential as a source of income. Due to the Schlagintweits' position 'in between' multiple royal supporters and national and scientific publics, this work traces how the brothers skilfully managed their material collections and legacy and sought to manipulate, not always successfully, numerous patrons, informants and colleagues across political boundaries and within different epistemic communities.

The book ultimately explores how the international dispute over the brothers' foreign employment, the ambiguous value of their scientific results and their controversial personal conduct, eventually led to highly divergent reputations in India, Britain and in continental Europe. By analysing a significant corpus of diverse sources, including newspaper and journal articles, private correspondence and books, diaries, contracts and lecture manuscripts, material artefacts and iconographic works connected with the enterprise, it is demonstrated that the diverging culture of commemoration that emerged in pre-1871 Germany about the Schlagintweits' overseas travels can reveal crucial aspects of the reinforcement of an imperial ideology in a non-colonial country.[95] That

---

[94] Bührer et al. (eds.), *Cooperation*; Brescius, 'Innenleben'; Driver, 'Face to Face'.
[95] The work is based on archival research in over fifty holdings, in seven countries, on three continents. It works with sources in French, German and English and, on a selective basis, in Dutch, Russian, Hindi, Tibetan and Persian. I gratefully acknowledge the help of several colleagues with the latter materials.

is, precisely because the German states were then neither unified as a nation nor possessed any overseas territories, the intrepid travellers could be glorified in the 1860s as trailblazers and 'heroes' of a projected *future* German Empire that ought to be built on their overseas achievements and personal martyrdom.[96] To be sure, German-born sailors, settlers, soldiers and scientific voyagers had died in foreign imperial service from the outset of European commercial and colonial expansion in the Americas, Asia and Africa in the wake of Columbus and Vasco da Gama. Yet, the growing German national aspirations of the mid and late nineteenth century domesticated, in a sense, the German mind. In this new imperial vision of the historical place of 'Germany' in the world, it was claimed that the earlier 'sacrifices' made by German travellers for the exploration and opening up of extra-European lands to western science, commerce and colonisation had earned their homelands the status of a formal overseas power as well.[97]

This book combines convergent modes of contextual analysis around the case of the Schlagintweits and loosely follows a chronological structure with each chapter addressing distinct topics of larger significance. Chapter 1 focuses on the training and early Alpine careers of Hermann, Adolph and Robert Schlagintweit, their entry into transnational patronage networks and EIC service, and also their place within the Humboldtian tradition of scientific enquiry. Chapter 2 shifts the attention to Britain and its century-long history of exploration towards and beyond the north Indian frontier – the background against which to assess the achievements and perceived scientific shortcomings of the Schlagintweits' expeditions. The discussion then moves to the broader discourses that accompanied the recruitment of German specialists into British service in the middle decades of the century – and the recurrent patterns of adverse reactions to this reliance on external expertise.

Chapter 3 demonstrates that although nationalist discourses and personal competition were at play in the mid-century reception of German specialists in British overseas territories, it was the daring behaviour of the Schlagintweits and their astute manipulation of transnational sponsors and publics that further fuelled the controversy over these imperial outsiders. Paying close attention to their clever exploitation of cross-border communication channels, it shows how they engaged in

---

[96] Friedel, *Gründung preußisch-deutscher Colonien*, 82–3; see also Chapter 8.
[97] The study thus makes a contribution to the growing literature on 'heroes of empire', which until now has merely focused on 'exceptional' figures within already existing colonial societies; MacKenzie, 'Heroic Myths'; Sèbe, *Heroic Imperialists*.

numerous double games to maximise the benefits from the co-financed Anglo-Prussian-Bavarian expedition.

Situated in Asia, Chapters 4 and 5 take a closer look at the Schlagintweits' scientific practices in dealing with the unfamiliar human and natural landscapes of India and central Asia. Chapter 4 explores the making of science in the field and how the Schlagintweit expeditions could be realised *in situ*, not least through official privileges and their adroit mobilisation of British colonial infrastructures and scientific-technical services. Chapter 5 shifts attention to the heterogeneous group of indigenous helpers, porters, translators and assistants, whose vital functions within the complex social configuration of the expedition party have until now been largely ignored. The existing literature has thus conveyed a misleading picture of the inner life of this exploratory scheme, which experienced a veritable 'role reversal' between the German explorers and their Indian assistants.[98] Going beyond a mere appreciation of individual 'contributions' that the non-European helpers are said to have offered to their European 'leaders', the chapter offers a radically different understanding of what this expedition meant for different agents involved in its execution, and how it was shaped by significant conflicts of authority between its members. The chapter thus further seeks to chart the slippery ground that itinerant scientists had to navigate in order to establish, maintain and fabricate personal authority and scientific reputation in and outside the colonial realm.

Chapter 6 returns to the European stage and traces how the brothers and their mentors sought to promote their careers and the legacy of their expedition amidst the imperial crisis of the Great Indian Uprising. It provides detailed discussion of scientific reputation and its negotiation through public controversy, and the role of the press in particular is examined. It also shows how the brothers sought to attract different national audiences through the publication of their scientific results, data and images from Asia, combining the newest technologies of print and photography to further develop the visual expressions of Humboldtian science. While their enterprise was never entirely controlled by imperial authorities, the Schlagintweits, out of strategic considerations, showed a remarkably flexible definition of its aims and scope, at times and in secret communications closely associating their mission with utilitarian interests while downplaying them in public accounts.

Chapter 7 demonstrates that the brothers' perhaps most ambitious project for the permanent foundation of their own India museum – envisioned

---

[98] Driver, 'Hidden Histories'.

as a site of international learning, memory and the propagation of commercial opportunities – ultimately failed, due to the loss of their strategic position within patronage networks between Prussia and London. It thus traces the history of their significant natural historical and ethnographic collections between scientific display, commodification and ultimate dispersal – all processes closely linked to the Schlagintweits' quest for social mobility and prestige.

Chapter 8 moves forward in time to explore the wider repercussions of the Schlagintweit expedition and its asymmetric commemoration. It shows how their pioneering work on Indian meteorology shaped the institutional structure of European science in the British colony, and how longstanding epistolary ties connected the brothers, including the sedentary orientalist Emil Schlagintweit, to western and indigenous informants and specialists in south Asia. The existing knowledge gap between German and British popular audiences about India and the long history of prior discovery is key in explaining their rise to popular adoration. While France, Russia and the United States also accepted the authority of the brothers in their claims to have opened up the trans-Himalayan regions to western understanding and commerce, by the 1860s we can detect a sharp change in the general assessment of the Schlagintweits' contributions to science and exploration – with important political ramifications for their 'heroic' legacy in a nascent German Empire on the rise.

1 Entering the Company Service: Anglo-German Networks and the Schlagintweit Mission to Asia

Building (a) Reputation, Building Networks: The Early Careers of the Schlagintweit Brothers

On 12 August 1850, the eminent German naturalist and overseas explorer Alexander von Humboldt penned a letter of recommendation for two of his most treasured scientific protégés, the brothers Hermann and Adolph Schlagintweit. Hoping to open the doors and opportunities of the world of Victorian science to his pupils, whom he described as 'very amiable and modest young people', Humboldt addressed one of the leading British naturalists of the time:

Dare I ask for your benevolence in favour of two of my compatriots, Physicists and Naturalists, the two ... Messrs. Schlagintweit, who have long since lived among us and who are currently preparing an excellent work (similar to the one by Saussure) on the Eastern Alps. They have accomplished very interesting research on the geography of Alpine plants, on magnetism and the meteorology of the high strata of the atmosphere.[1]

The recipient of this letter praising the brothers' mountainous accomplishments was the director of the Royal Botanic Gardens in Kew, Sir William Hooker, who then presided over one of the most prestigious botanical institutions in the world and maintained an empire of patronage over aspiring British and continental naturalists. Humboldt knew that Hooker was a highly respected man of science within London's scientific community, whose support – or rejection – could make or break a career.[2] The 'benevolence' towards these foreign naturalists, for which Humboldt politely begged, could translate into many things: from

---

[1] Humboldt to William Hooker [WH], 12 August 1850, RBG, DC 51, letter 254, 330. Humboldt addressed a similarly flattering letter to Michael Faraday, 13 August 1850, in James (ed.), *Correspondence* IV, letter 2313, 173.

[2] On WH's prestige and influence, Drayton, *Government*, 146; on his vital role in training a small legion of German naturalists in Kew at the recommendation of Humboldt and the Prussian Envoy to London, Christian Carl (von) Bunsen, RBG, DC 51, letters 52, 53, 56, 57.

29

further introductions in London's many scientific societies to Hooker's support for a potential future employment in Britain.

Nathaniel Wallich, another leading botanist in England at the time, was also aware of the Kew director's far-reaching influence, when, in early 1854, he discussed with Hooker the Schlagintweit brothers' plans – this time to travel to India. Yet, Wallich approached the director with very different intentions:

Two German arch-puffers, yclept Schlagintweit brothers were recommended in 1852, by Baron Humboldt through the Pruss[ia]n Gov[ernmen]t and Consul Bunsen to [accompany] a surveying party vacant by the sad death of [Captain Elliot]. The case went through the Council of the Royal Soc[iet]y. I put a stop to the Soc[iet]y's direct recommendation. ... As I expected it to happen: the request was granted and it was stated, that in case an efficient officer in the Comp[an]ys Service not being found, or not being to be spared for that peculiar work, the brothers Schlagenze would be employed.[3]

Clearly enraged by this pending appointment of the German naturalists and seeming impostors to a plum position in British India, and determined to sabotage the scheme through a backdoor intervention, Wallich upped the ante by proposing a purportedly more able substitute for the brothers. Seeking support from the director for his plans, he openly mused: 'Why does not [Thomas] Thomson ask for an interview with the Chairman and offer himself as a candidate for the survey vacated by the death of Captain Elliot?'[4] Adding fuel to the fire, Wallich explained that the British naturalist Thomas Thomson was certainly 'better qualified in all respects than ten Schlagintweyts, or 10 similar German puffers, carrying large sails with little ballast'.[5]

In 1854 these 'German arch-puffers' were at the beginning of their scientific careers, having only recently reached the age of majority; nevertheless, opinions were already deeply divided over their talents, future prospects and personal character. The Bavarian brothers grew up in a respectable social milieu. Their father, Joseph Schlagintweit (1791–1854), had in some sense already anticipated many of his sons' later traits and passions. He had himself been a keen traveller and self-made 'improver'.[6] Joseph had studied medicine, gaining a doctoral degree from the Ludwig Maximilian University in Munich. His qualification as a surgeon was followed by extensive travels throughout the German states,

---

[3] Wallich to WH, 28 January 1854, RBG, DC 55.
[4] The British officer formerly in charge of the geomagnetic survey of the Indian Empire.
[5] Wallich to WH, 31 January 1854, RBG, DC 55.
[6] The family account is mostly based on E. Schlagintweit, 'Schlagintweit'; S. Schlagintweit, 'Abriß'.

where he operated in numerous hospitals. His experiences culminated in a well-received treatise on eye surgery, complemented by a description of a new medical instrument he himself had invented for operating purposes.[7]

Following his central European travels, which brought Joseph Schlagintweit from Vienna to Prague, and from Berlin to Frankfurt am Main, he put down roots in Munich, where he founded a private hospital for eye surgery in 1822. Over time he greatly improved this field, while also writing numerous studies on childbirth, medical treatments for the poor and epidemic diseases. He assumed the directorship of Munich's Institute for the Blind (1837), and received not only the title of Royal Councillor in 1839, but also the Order of St Michael in 1842. Perhaps nothing better reflects the confidence that the Bavarian monarch Maximilian II placed in his skills than the fact that he was entrusted with operating on the king's mistress, Lola Montez.

Besides Joseph Schlagintweit's social and specialist advancement, he managed to improve his finances to such an extent that he could afford an excellent education for his growing family. His marriage to Rosalie Seidl, the daughter of a well-heeled brewer, had brought in an attractive dowry, thus cementing the family's bourgeois status. (Rosalie Seidl, born 1805, died after a prolonged illness in 1839.) The young Schlagintweit brothers attended the Königliche Alte Gymnasium (since 1849, the Königliche Wilhelmsgymnasium) in Munich. In a short time, they emerged as outstanding pupils, with top marks, especially in the field of geographical science.[8] German teachers at the time sought to provide a deeper understanding of the field of geography, which meant putting an emphasis on the relationship between the earth and its human inhabitants – an anthropocentric approach clearly influenced by the works of the eminent German armchair geographer Carl Ritter.[9] The latter had provided a classical account of this approach in his monumental work on 'Comparative Geography', whose original volumes can still be found in the old library of the brothers' former school, suggesting that they had encountered Ritter's oeuvre at a young age.[10]

In addition to their schooling, the ambitious father further improved his sons' *Bildung* by hiring private tutors, so that the young Schlagintweits acquired a privileged training in modern languages and the natural

---

[7] J. Schlagintweit, *Pupillenbildung*.
[8] See Hermann's and Adolph's school certificates at the Archiv des Deutschen Alpenvereins, Munich [DAV], and the final school examination of Robert, BSB, SLGA, VI. 8.3.1–11.
[9] Lüdecke, 'Einfluß'.
[10] Ibid., 144; Ritter, *Erdkunde*.

Figure 1.1 Hermann, 'Brunnthal', 13.2 × 19 cm.
© StBB, HVG. 47/1–200, no. 47/5, photograph: Gerald Raab.

sciences.[11] From 1844, Franz Joseph Lauth, later the first professor of Egyptology at the University of Munich, also tutored them as official 'Hofmeister' (instructor) of the family.[12] This thorough education was complemented by an early engagement with the art of painting. Hermann once noted that the celebrated Munich artist Anton Zwengauer had instructed him in his first studies of nature.[13] Two surviving pencil drawings of the environs of Munich by Hermann and Adolph suggest that the two had indeed received early training to nurture their talents (Figure 1.1).

Young Humboldtians in the Alps

Their visual studies *en plein air* point to another crucial aspect of their education: their early impetus to examine nature *in situ* – further spurred by their reading of Humboldt's *Cosmos*, whose first parts were published

[11] E. Schlagintweit, 'Schlagintweit'.
[12] Körner, 'Brüder Schlagintweit', 62.
[13] H. Schlagintweit, *Reisen* II, 164f.; A. Schlagintweit and H. Schlagintweit, *Untersuchungen*, 444; Baud et al., *Haute-Asie*, 52.

in 1845. From 1846 to 1847, the closely attached brothers Hermann and Adolph embarked on their first two major Alpine excursions, which resulted in the publication of their first treatises.[14] Crucially, their extensive research in the Alps allowed them to acquire a substantial stock of practical knowledge and experience in the 'field'. Yet, their trips also formed part of a thorough physical training (Figure 1.2). Consequently, the two were soon able to achieve some remarkable feats of mountaineering, nearly accomplishing the first ascent of Monte Rosa (4,634m) in August 1851.

While the scientific travellers continued their explorations of the German, Swiss and Italian Alps from the mid 1840s for almost a decade, this period of study in nature coincided with the start of their university education in Munich. Hermann, first encouraged by his father to follow in his footsteps, started to study medicine yet soon abandoned the subject to follow his passion for the natural sciences, and completed his geographical studies in July 1848 with a doctoral dissertation on angular measurements.[15] Adolph, by contrast, received his PhD in 1849 in the field of geognosy, a branch of geology that investigates rocks and minerals in the study of the layers of mineral matter.[16] The third brother, Robert, who joined the Alpine travels only in 1852 with a trip to the Zugspitze, undertook independent excursions in the autumn of 1853; he explored the mountain mass of the Kaisergebirge, a work that earned him a doctoral degree in geography in 1854.[17]

The Schlagintweits were part of a new wave of scientific specialists in the mid nineteenth century who started to take natural historical studies up to the highest regions of the central European mountain chain. Since the Middle Ages, there had prevailed a strong belief among European peoples in the existence of supernatural phenomena in the massive mountain system. These beliefs, which included myths about dragons and ghosts, were so forceful that a more thorough exploration of the Alps had been impeded until the late seventeenth century.[18] Only then did naturalists gradually start to take measurements and to collect natural specimens and species at ever-new altitudes. One of them was the Swiss naturalist Johann Jacob Scheuchzer (1672–1733), who travelled extensively through the Swiss Alps at the turn of the eighteenth century. His works proved highly influential for future geological, meteorological,

[14] Hermann, 'Gletscher'; for other early works, E. Schlagintweit, 'Schlagintweit'.
[15] Hermann, 'Messinstrumente'.
[16] Adolph, *Ernährung der Pflanzen*.
[17] Robert, *Bemerkungen*; his diploma is held in BSB, SLGA, VI. 8.3.1–11.
[18] Ireton and Schaumann, 'Introduction'.

Figure 1.2 Photographic collage of Adolph and Hermann Schlagintweit in the Alps, *c.* 1850.
© DAV.

historical and cartographical studies of the mountain system, and were also a reference for the Schlagintweits.[19] Scheuchzer had no difficulty in reconciling his empirical approach with a conviction in God's creation of the mountain chain as part of his physico-theological programme, indeed, he also maintained a 'lingering belief in the existence of dragons'.[20]

[19] Scheuchzer, *Natur-Historie.*
[20] Ireton and Schaumann, 'Introduction', 10.

Nineteenth-century itinerant geographers and geologists like the Schlagintweits, by contrast, sought to portray themselves as rational, scientific investigators of these elevated regions. The images and treatises they produced on their travels found a ready market, not just in the German-speaking world. Indeed, the mid-century witnessed a European-wide craze for Alpinism, reflected in a nascent tourism industry and the foundation of several Alpine societies throughout the continent and in the British Isles.[21] The brothers profited from this contemporary interest among mountaineers and natural historians alike, who then rightly regarded the Alps as one of the last understudied regions *within* Europe.

The Schlagintweits' early scientific approach was heavily influenced by the work of Alexander von Humboldt (1769–1859), whose writings they had thoroughly studied and to whom they dedicated their first monograph, published in 1850.[22] Humboldt acted as a role model for a whole generation of naturalists during the early decades of the nineteenth century. Charles Darwin and the son of the Kew director, Joseph Hooker, both acknowledged the influence of Humboldt's overseas expedition on their careers as itinerant naturalists. Humboldt's *Personal Narrative* – his most famous American travelogue – remained for them both a constant source of inspiration and crucial point of reference.[23]

'Humboldtian science', a term introduced by Susan Faye Cannon to define the scope and main features of Humboldt's intellectual programme, was among other things based on personal observations in the 'field' and the extensive and increasingly exact measuring of the natural world through an array of the most up-to-date instruments.[24] There was, to be sure, nothing particularly 'German' about conducting empirical science out in the open.[25] Rather, Humboldtian science combined a set of practices and scientific interests with often global reach (as in the fields of plant geography, terrestrial magnetism and meteorology) that were shared by a community of scientists that cut across national-political boundaries. Humboldt himself wrote the bulk of his American opus whilst residing in Paris from 1804 to 1827, in close exchange and discussion with Parisian scientific communities while he manifested his scientific-aesthetic paradigm in over twenty volumes. One aim of

[21] Hansen, 'Founders'.
[22] A. Schlagintweit and H. Schlagintweit, *Untersuchungen*. Finkelstein, 'Mission', 182.
[23] Werner, 'Verhältnis'; for further examples of Humboldtianists, also in Britain, Kirchberger, 'Contribution'.
[24] The best contextualisation and critique of the concept, initially advanced by Cannon in *Science in Culture*, are Dettelbach, 'Humboldtian Science'; and Ratcliff, 'Geomagnetism'.
[25] Porter, 'Gentlemen', 820–1; Dettelbach, 'Humboldtian Science'; Olesko, 'Humboldtian Science'.

Humboldt's rigorously trans-disciplinary approach to physical geography was to capture the specific character of a given landscape by collecting as much detailed data as possible, which could in turn be compared trans-regionally, indeed trans-continentally. In a sense, the scale of Humboldtian science was always local and global at the same time.[26] The overarching concern was to formulate and then to depict, in entirely novel ways, general physical laws out of a wealth of observational data, and thus to detect the 'interaction of forces' in nature that in Humboldt's view formed a 'general equilibrium'.[27] As he famously stated before his American travels, '[m]y single true purpose is to investigate the confluence and interweaving of all physical forces'. He thus sought to combine the data collecting and classifying practices of the naturalist in order to achieve a holistic approach to 'terrestrial physics [as] a master-science'.[28]

Humboldt's conviction that a good naturalist also had to be an inspired physicist was accepted by some, but certainly not all, practitioners of natural history. The disorderly process of disciplinary specialisation in the sciences was, by the mid nineteenth century, underway, and there were many contemporaries of the Schlagintweits who did not appreciate their all-encompassing approach towards the study of nature. German universities, especially when compared with their British counterparts, underwent important reforms in the first half of the nineteenth century, and tended to place a stronger emphasis on rather specialised fields of research. This resulted in the foundation of chairs in newly circumscribed fields such as forestry, chemistry, mineralogy and other branches of science.[29] Their holders tended to be critical of Humboldtian approaches in field sciences. Hence, despite the long shadow that Humboldt cast upon European science in the first half of the nineteenth century, ideals and scientific practices were gradually changing throughout the German states and other parts of continental Europe and Britain.[30] In that sense, the brothers were transitory figures between competing scientific paradigms.

What further connected the Schlagintweits' studies with the works of their role model and later mentor Humboldt was their eagerness to visualise nature and its inherent forces in an aesthetically pleasing way

[26] On the different scales at work in Humboldtian science, and the importance of trans-continental analogies of forms and types, Daston, 'Humboldtian Gaze'.
[27] Dettelbach, 'Humboldtian Science', 289f.
[28] Ibid., 290.
[29] Still useful is Allen, 'Professionals'; Allen, Naturalists.
[30] For important changes in the German university system, see Nipperdey, Deutsche Geschichte, 470–82; Wehler, Deutsche Gesellschaftsgeschichte, 3, 417–29; Cittadino, Nature, 22–5; Dorn and McClellan, Science, 309.

(Figure 1.3).[31] Already in their first book, the brothers included a variety of diagrams and lithographed watercolours of beautiful panoramas, yet always with a specific object of study in focus, most often Alpine glaciers. These views were accompanied by a wealth of observations and data, and an explanatory sheet – a visual technique later repeated for their images from Asia.[32] Their views from the Alpine glaciers had such a quality in the use of colours and contrasts that many depictions even managed to convey a sense of the depth and direction of the slowly moving masses of ice.[33]

Yet, to produce even greater *Anschauungsmaterial* (illustrative material) of the topographical forms they encountered, the Schlagintweits collaborated with a Berlin zinc plaster company to produce three-dimensional mountain reliefs.[34] These objects provided a tangible sense of the shapes of ranges and valleys to the viewers (Figures 1.4, 1.5). The reliefs or 'galvanised models', which the brothers not only presented as gifts to royal benefactors but also sold to scientific institutions and private collectors, give us a sense of them as science popularisers.[35] Not only could these models be ordered and used for pedagogic purposes, the brothers also provided a cheaper series of stereoscopic photographs of these reliefs for the wider public.[36] The use of new techniques and visual aids indeed became a pillar of their research and scientific reputation. In their later careers, too, the Schlagintweits never tired of experimenting with the most recent instruments and new photography and print technologies in order to enhance the appeal of their work, which was otherwise heavily based on columns of data and dry prose.

Unlike their later publications on the Indian mission, parts of their Alpine treatises were immediately translated into other European languages.[37] The apparent appreciation of their early works was also reflected by the invitations to deliver papers at scientific societies and royal courts in Berlin, Paris, London and elsewhere, including two lectures at London's Royal Society in January 1851, during their first visit to England.[38] Their initial explorations within Europe, together with

---

[31] Daston, 'Humboldtian Gaze'; Werner, *Naturwahrheit*; Kraft, *Figuren*.

[32] See, e.g., A. Schlagintweit and H. Schlagintweit, *Untersuchungen*, 52–3; also useful as an introduction are Fritscher, 'Panoramas', 606; Fritscher, 'Praxis', 77; Finkelstein, 'Mission', 182–3; indispensable is Kleidt, 'Kunstwerk'.

[33] The Schlagintweits' studies of the structures and properties of Alpine glaciers, their velocity and motion heavily drew on the work of the Swiss-born naturalist Louis Agassiz; Agassiz and Bettannier, *Études*; Lüdecke, 'Hochgebirge'.

[34] Brogiato, Fritscher and Wardenga, 'Visualisierungen', 239.

[35] Felsch, '14.777 Dinge'.

[36] On the pedagogic function of their visual materials, Fritscher, 'Praxis'.

[37] For example, H. Schlagintweit, *Observations*.

[38] A. Schlagintweit to WH, 10 January 1851, RBG, DC 51, 549.

Figure 1.3 Adolph and Hermann Schlagintweit, 'Vergleichende Darstellung der physicalischen Verhältnisse der Alpen', a visualisation of plant geography and Alpine meteorology clearly influenced by Humboldt's work on the varying distribution of the Chimborazo's flora in the Andes. Adolph and Hermann Schlagintweit, *Atlas* accompanying their second book, *Neue Untersuchungen* (1854). © ETH, Rare Books.

Figure 1.4 Relief of the Zugspitze and the Wetterstein in the Bavarian Alps, 'galvanised zinc cast by M. Geiss in Berlin'.

© Reliefsbestand der Erdwissenschaftlichen Sammlungen der ETH Zürich.

Figure 1.5 Relief of the Zugspitze and the Wetterstein in the Bavarian Alps, 'landscape view'.
© Reliefsbestand der Erdwissenschaftlichen Sammlungen der ETH Zürich.

the skills and professional acquaintances they had thus acquired, opened up the potential for overseas employment.[39]

One important stepping-stone for the Schlagintweits' future was to have attracted the attention of a group of eminent geographers in Berlin, then one of the leading German scientific centres. Many of them had close ties to its Geographical Society (Gesellschaft für Erdkunde zu Berlin), founded in 1828 as the second oldest in Europe. Notably the Society's president, Carl Ritter (1779–1859), who also held the chair in geography at Berlin University, and one of its honorary members, A. v. Humboldt, acted as significant patrons of German geographical talents and overseas explorers.[40] The less illustrious Carl Ritter was perhaps just as important as Humboldt in promoting transnational scientific collaborations. Ritter had played a crucial role, for instance, in arranging Heinrich Barth's employment in a British-backed African exploration of 1849 by mobilising his various diplomatic and scientific acquaintances in London and Berlin.[41] Crucially, both Humboldt and Ritter had developed a strong interest in Asia's geographies and natural histories, and it is certain that the Schlagintweits' life-long engagement with that continent was strongly influenced by the works of these Berlin-based

---

[39] Roquette, 'Note'.
[40] Päßler, *Mittler*; Suckow, 'Rußland'.
[41] Naranch, 'Beyond the Fatherland', 237.

mentors.[42] To be sure, there also existed a large community of German oriental scholars in the Prussian capital, who engaged in important religious and philological studies of the East, sometimes through similarly close contact with British peers and East India Company men in south Asia. Yet, the Schlagintweits mingled less with Berlin's humanistic scholars but regularly attended and contributed to its geographical and natural scientific societies, including the Physikalische Gesellschaft zu Berlin (founded in 1845), full members of which they became in 1849.[43]

Berlin as a Hub of Indian and Central Asian Geography

In the first decades of the nineteenth century, a number of systematic accounts of Indian and central Asian geography, natural history and mineralogical resources were compiled and published in Berlin, and in the nearby Saxon town of Gotha near Erfurt. Gotha was the centre of the publishing house of the Justhus Perthes Anstalt, where August Petermann (1822–78) produced his widely read journal, *Petermanns Geographische Mitteilungen*. Crucially, geographic and cartographic works by famous geographers such as Petermann, Heinrich Kiepert and Heinrich Berghaus drew heavily on the accumulated data, observations and collections made by Russian and French travellers and missionaries, and also by EIC servants in south Asia. These, in line with the by no means static hierarchies of science at the time, would often provide the materials and observations for further analysis and compilation in Europe – not least by German *savants*, who in most cases had no direct duty to the British East India Company.[44] Their works were then often retransmitted to the scientific and imperial establishments of other European states. Humboldt's treatises on central Asia's geography were widely consulted among scientific and administrative circles in Britain, the Russian Empire and India. The same applied to the armchair geographer and master synthesiser Carl Ritter. Years after the publication of the 'Comparative Geography', Ritter's volumes on Asia were still considered important enough that Peter Semenov, Secretary of the Imperial Russian Geographical Society, was sent to Berlin, following

---

[42] Ritter's monumental work, *Erdkunde*, almost exclusively treats the continent of Asia; see also Humboldt, *Fragments*; Humboldt, *Asie Centrale*.

[43] See also Finkelstein, 'Mission', 183.

[44] Endersby, 'Herbarium'. On these hierarchies and their critical reconsideration, see also Arnold, 'Plant Capitalism'. Other works of history and geography have further challenged and refined any simple conceptualisation of 'centres' and 'peripheries' in the history of science; Schaffer et al. (eds.), *Brokered World*; Raj, 'Beyond Postcolonialism'; Burke, *History of Knowledge*; works often in direct opposition to older diffusionist models as classically advanced by Basalla, 'Spread'.

the Society's decision to have his oeuvre updated and translated into Russian. Semenov remained in Berlin for three semesters, closely collaborating with Ritter on the translation while also preparing his own journey into central Asia.[45]

Perhaps the most striking case of a scientific interlocutor between empires was Humboldt. In view of his unfulfilled desire to travel in the British territories in India and into the Himalayas, he had found a way to complement his American travels with a mission into parts of central Asia on behalf of the Russian Empire in 1829. This was the second expedition that the Prussian explorer had undertaken within the colonial framework of a foreign power. Similar to his former journey through the Spanish Empire in the Americas (1799–1804), Russian officials expected that Humboldt would provide useful and commercially applicable knowledge on the regions he traversed. The terms of his employment set out by Tsar Nicholas I and his minister of finance, Georg Cancrin, made clear that the Prussian naturalist was expected to deliver information on 'exploitable resources'; Humboldt ultimately agreed 'to report more on products and institutions than on people'.[46]

Even though Humboldt might be best known for his American opus, he was, however, deeply involved in and respected for his work on the trans-Himalayan and central Asian natural histories, in particular those regions' massive and complex mountain chains: the Himalayas, the Karakoram and the Kunlun Shan.[47] In fact, he personally regarded his book *Asie Centrale*, published in French in 1843, as 'a work, which has never been translated into English, but which is that in which, I think, I have brought forward more novel information than in any of my other publications'.[48]

To test his own assumptions and interpretations of the physical character of south and central Asia against the accounts of itinerant naturalists, Humboldt was indefatigably concerned with securing firsthand observations from Company servants and other European travellers in those regions.[49] He was acquainted with a number of Anglo-Indian officials and naturalists in the 1840s and 50s. Among them were the

---

[45] Freitag, 'Atlas', 121.

[46] Sachs, *Current*, 83. Humboldt was accompanied by other German naturalists, including Gustav Rose (1798–1873), who published an important study of Russia's mineralogical treasures, *Reise*.

[47] For Humboldt's crucial role in 'rediscovering' South and Central America, Kennedy, *Spaces*, 6ff.; and Pratt, *Imperial Eyes*.

[48] Humboldt to WH, 11 December 1850, RBG, DC 51, 217.

[49] The Prussian-born soldier-naturalist Leopold von Orlich reported to Humboldt during his time in a British regiment fighting against the Sikhs (1842–3); von Orlich, *Reise*.

British Resident in Darjeeling, Brian Houghton Hodgson ('an impartial and well-informed eye-witness', according to Humboldt), and the eminent Himalayan traveller Joseph Hooker, son of the Kew director William Hooker.[50] The detailed correspondence between the younger Hooker and his aged Prussian confrère not only testifies to how Humboldt's *Asie Centrale* was indeed a widely read and authoritative source on the region's geography for British naturalists and Company servants.[51] It also provides insights into how the Berlin-based geographer subtly influenced ongoing explorations in the East. Humboldt and the British botanist managed to exchange long and detailed letters during the latter's travels. Their correspondence dealt with a number of scientific conundrums in fields as diverse as plant geography, Indian topography, meteorology, mineralogy and glaciology. Humboldt regularly supplied Hooker with long lists of unresolved questions that he urged the traveller to address whilst still in Asia, thus subtly guiding the occupations and studies of his 'close friend'.[52] Partly flattered, partly stimulated, Hooker was eager to meet Humboldt's demands, and spared no time or effort to send long elaborations, sometimes illustrated with topographical sketches.[53]

Indeed, after the publication of his travelogue, Hooker wrote to Humboldt about the immense influence the latter had exercised on his scientific pursuits.[54] 'I have felt so much the influence of your career, from my childhood, & owe so much to all you have done for science generally & for myself in particular that I do feel it a great privilege to have been permitted to write a book that has especially interested you.'[55] Humboldt, in turn, regarded some of Hooker's letters from the field as so important that he secured their publication in British journals, relying on his close relations with a number of metropolitan men of science. In doing so, he self-consciously acted as a scientific intermediary between India and Britain.[56]

Another close collaborator of Humboldt's was the geographer and editor August Petermann, who acted as a crucial intermediary between Britain and the German states. Petermann had been a member of the Royal Geographical Society since 1847 and had lived in Britain for many

---

[50] Humboldt to WH, 11 December 1850.
[51] Joseph Hooker [JH] to Humboldt, Khassya, 23 September 1850, RBG, JDH/1/9, 482–4.
[52] Ibid.
[53] The archives in Berlin and London are filled with their field correspondence, besides JDH/1/9, see especially the letters (of up to twenty pages each) in Staatsbibliothek, Berlin [SBB], Nachlass Humboldt.
[54] Hooker, *Himalayan Journals*.
[55] JH to Humboldt, SBB, Nachlass Humboldt, gr. Kasten 11, no. 10, 21 September 1854.
[56] Humboldt to WH, 11 December 1850.

years before returning to Germany following a dispute over his loyalty with other members of the RGS in 1854.[57] His career is an intriguing example of the role of personalised knowledge, as his mobility entailed a transfer of skills from the British imperial centre to the European imperial periphery. As Bradley Naranch put it: 'Petermann's relocation to Germany, following years of extensive experiences in Britain with leading scientific societies and research facilities, provided an important impetus for the development of cartography, overseas exploration, and scientific imperialism in German society during the later 1850s.'[58] Humboldt considered Petermann a vital source of information from the centre of the British Empire: after Petermann's departure, Humboldt thought that 'it is a great loss for German geography that he did not stay close to the source on the happy island'.[59]

The willingness of German experts in Indian and central Asian geography to collaborate with British colleagues – both at home and in the colonies – found a ready expression in a number of projects that helped to further integrate the German states into the knowledge networks of British imperialism.[60] One was a joint publishing scheme in 1849–51. Preliminarily termed *Traité de géographie, destiné à l'instruction des écoles de l'Indoustan*, this work – intended as schoolbook for Indian pupils – was compiled by the Berlin-based Heinrich Berghaus (1797–1884), Joseph Hooker, Hodgson and Humboldt.[61] While ultimately never published, it is evidence that Germans were respected authorities on Indian geography, and that joint publications created strong bonds between them and British men of science that could be mobilised for shifting purposes. One of them was to secure employment for talented German naturalists and explorers.

When the Schlagintweit brothers entered the stage and made a name for themselves in the late 1840s and early 50s, they could readily tap into these long-established networks. In particular, in London Humboldt, Ritter and Bunsen maintained close ties with a number of leading British men of science active in the Royal Society and other institutions, including the militarily, politically and commercially inclined Royal Geographical Society. The RGS had become the largest and most powerful of all

---

[57] Naranch, 'Beyond the Fatherland', 243.

[58] Ibid., 244.

[59] Humboldt to Bunsen, 30 December 1854, in Schwarz (ed.), *Briefe*, 184. On Bunsen's life, his standing and connections in England: Höcker, *Vermittler*; Foerster, *Bunsen*; Kirchberger, *Aspekte*.

[60] Kirchberger, 'Naturwissenschaftler'.

[61] The episode is comprehensively reconstructed in Brescius, 'Empires of Opportunity', ch. 1.

London societies by mid-century, and vigorously promoted the cause of British overseas exploration and empire under its president, Sir Roderick Murchison.[62]

It is unclear when precisely the brothers formed the idea to embark on an Indian and Himalayan expedition. Robert Schlagintweit claimed in the 1860s that his wish to explore the Himalayas reached back to his 'earliest youth'.[63] Such a statement has to be taken with a pinch of salt, however, since Robert was always keen to portray himself as a true *conquistador* in front of popular audiences. It is more reasonable to assume that the plan emerged between 1849 and 1850. In May 1849, Adolph and Hermann left Munich. To pursue their *Habilitation*, they settled down in Berlin, the Mecca for geographical science in the German-speaking world. Humboldt, in his meetings with the brothers since their acquaintance in June 1849, made no secret of the vast opportunities awaiting naturalists in the Himalayas, especially those who were experienced mountaineers. While Humboldt enthusiastically endorsed the brothers' pursuits, others were far more critical about their abilities, especially when set against their extensive ambitions, as their *first* and forgotten attempt to embark on an Indian scientific voyage in 1852 demonstrates. However, the episode sheds light on the ambiguous perceptions that German scientists also had of the brothers, long before they would become the focus of an international polemic over their Asiatic travels and results. Many of the tropes of the later Schlagintweit controversy, in fact, already appeared in previous years, albeit on a smaller scale.

On 12 May 1852, the Prussian monarch Frederick William IV received a 'direct submission' (*Immediateingabe*) by Adolph Schlagintweit.[64] The purpose of the submission was twofold. First, Adolph again sought to obtain his *Habilitation* from the Philosophical Faculty of Berlin University, which had been declined the year before. The second objective was to petition for the monarch's support for a scientific expedition to the Himalayas, to be carried out by Adolph and Hermann 'at the public expense'. Both dimensions of the petition were inextricably linked, since the king's granting of his financial patronage for the voyage essentially depended on a positive evaluation of Adolph's qualifications. To enquire about these matters, Karl Otto von Raumer, the conservative minister

---

[62] Bell et al., 'Introduction', 8; Driver, *Geography*, ch. 2; on its strong 'military emphasis', Stoddard, 'RGS and the "New Geography"', 191.

[63] Robert Schlagintweit [RS], Lecture notes for 'English Lectures on High Asia Delivered during the Years 1868 and 1869 in Various Towns of the United States of America', BSB, SLGA, V.2.2.1, 21.

[64] Adolph Schlagintweit [AS] and Hermann Schlagintweit [HS] to the king, 12 May 1852, Geheimes Staatsarchiv Preußischer Kulturbesitz, Berlin [GStAPK], I. HA Rep. 76 Ve, Akt Kultusministerium, Sekt. 1 Abt. XV Nr. 189, 'Wissenschaftliche Reisen ...' [WR].

of education in Prussia, requested a formal report to be issued both on Adolph's renewed application for the *Habilitation*, the licence to teach at university level, and on the naturalist's general competence, not least with a view to completing such a strenuous overseas exploration.

The advisory scientist charged with compiling the official report was Christian Samuel Weiss (1780–1856), a notable German mineralogist born in Leipzig, who had become professor in mineralogy at Berlin University and also director of the Cabinet of Mineralogy. Unfortunately for Adolph, Weiss had also sat on the committee that had previously declined his *Habilitation*.[65] In his formal reply to the government, Weiss provided a lengthy assessment of the petition, and of the perceived scientific qualifications of the brothers – or, in some regards, a lack thereof:

Concerning the individual aptitude of the two brothers Schlagintweit for such a travel scheme, it has to be fully acknowledged that both are able, persevering and experienced mountaineers, who do not shy away from pains and hardships. [Both] are precise observers in the field of physical geography, whose painstakingly compiled observations ... are useful and worthwhile contributions to physical geography, without being able to claim a rank amongst important discoveries.[66]

In restating the reasons for Adolph's previous failure to obtain his *Habilitation*, Professor Weiss continued by acknowledging the unusual 'physical endurance' of the brothers; yet, he also pointed to the perceived gaps in their scientific competence. 'When ... their joint work on the physical geography of the Alps was carefully examined last year, it seemed that, despite the fact that their talent and their achievements as observers were duly praised, their professional qualifications seemed not to be without fault, and not everywhere thorough enough, [especially] for lecturers at a university.' This was a direct critique of a colleague, since Hermann Schlagintweit, who had successfully received his *Habilitation* under Carl Ritter in 1850, had started to lecture as a *Privatdozent* for 'physical geography' at Berlin University two years later, giving classes especially on meteorology.[67]

In focusing on Adolph's works in the field of geology, Weiss's report further stated that 'they, too, provide evidence of painstaking and meticulous observing; the general description, however, was ... merely a repetition of the already known conditions'.[68] Hence, while the professor stressed above all the Schlagintweits' skills as travelling and observing naturalists,

---

[65] Raumer to Weiss, 19 June 1852, ibid.
[66] The report by Weiss to Raumer, 16 August 1852, ibid.
[67] HS, *Vertheilung*.
[68] Report by Weiss to Raumer, 16 August 1852, GStAPK, WR.

he subtly criticised their failure to adequately use these empirical results to alter general understandings of, in this instance, Alpine geology. In other words, no higher scientific theories were gained from the mosaic of local observations the brothers had gathered in the field. In hindsight, this was the Achilles' heel of the Schlagintweits, as they were to be confronted with the same criticism regarding the results of their future travels to the East.

In a next step, Professor Weiss turned to a lecture 'On the Geological Structure of the Alps' that Adolph had attached to his application. Weiss saw it as merely 'an attempt to synthesise foreign accounts of the most recent times', thus indicating that Adolph possessed only 'an ephemeral personal acquaintance with Switzerland'. Even worse, Adolph seemed perfectly unacquainted with, or had failed to acknowledge, the results of leading naturalists such as the Zurich-born geologist Johannes Konrad Escher (1767–1823).[69] Here, too, Weiss raised a point of criticism that would later play a significant part in the international controversy over the Schlagintweits' Indian expedition, as it relates to the lack of acknowledgement that many British travellers felt the Schlagintweits had given to their predecessors.

To complete his judgement on the brothers' ineptitude, Weiss considered the work Adolph had submitted for his 'renewed application for the *Habilitation*' in the field of geognosy. Adding insult to injury, Adolph had apparently failed 'to provide a clear and commanding understanding of the incredibly fragmented mountain range of the Monte Rosa', because he had not become sufficiently familiar with the area itself. 'It would … require a considerably longer and more often repeated stay' on the spot to gain a thorough knowledge – a knowledge 'that was not to be hypothetic, but grounded in actual observation'.[70] In short, it seemed that Adolph had a tendency to literally cover too much ground; to spread his scientific investigations over too large an area, leading him to make judgements on regions he was less familiar with, and to attempt more scientific disciplines than his academic education had prepared him to succeed in.

Not surprisingly, Weiss's report to the Prussian government ended with a negative evaluation of the petition. He even claimed that Adolph required further scientific training 'before he embarks on such an important geognostic voyage'.[71] Building on this dissection of the brothers' scientific qualities and Alpine research, the Berlin professor

[69] Ibid.
[70] Ibid.
[71] Ibid.

suggested that their 'analysis in situ, and the scientific opinions [that would be later] grounded in them, would not sufficiently guarantee scientific results worthy of the modern progress in the sciences'.[72]

Weiss was an expert in the field of mineralogy, and his more narrow expertise sat uncomfortably with the much broader aspirations of the brothers and their Humboldtian tendency to engage in holistic studies of a given region. This tension is reflected in the report that the ministry of education compiled for the Prussian king, which copied entire passages of Weiss's statements.[73] In the eyes of administrators, the brothers' proposal for a loosely defined trans-disciplinary investigation of the enormous mountain chain was not sufficient; rather, a clear-cut geographical 'problem' – an *explicandum* – was needed to justify the large sums of money necessary for such an undertaking in an age of advancing scientific specialisation, and growing bodies of literature to be mastered in each field of study.[74]

Yet, before Frederick William IV decided to decline the petition, he informed von Raumer that 'I wish first and foremost that you also obtain the opinion of … von Humboldt, who is acquainted with the brothers Schlagintweit'.[75] This request reflects the status Humboldt enjoyed as scientific adviser to the monarch. The fact that the brothers had earlier strategically nurtured their relation with the king also played a role. To establish their names with the king, they had used Humboldt as their go-between to present him with scientific gifts, which they hoped would reflect their scientific achievements and potential.[76]

When Humboldt was consulted on the matter in mid October 1852, his reply was more than a recommendation for the Schlagintweits' projected scheme; it was also a defence of his own scientific paradigm – in view of the specialising ambitions and attacks of his contemporaries. Humboldt first stated that '[t]he opinions, which I hereby [express?] about these so scientifically excellent and multi-talented young men, are not based on personal contact and impressions acquired through individual conversations' – a rather blatant lie. Adolph Schlagintweit would soon even assist Humboldt in preliminary works for the later parts of his *Cosmos*.[77] Humboldt nonetheless claimed to have based his judgement

[72] Ibid.
[73] Draft of the Ministry's report to the king, 24 September 1852, GStAPK, I. HA Rep. 76 Ve, Kultusministerium, WR.
[74] Weiss's report, 16 August 1852, ibid.
[75] King to Raumer, 6 October 1852, ibid.
[76] HS to Humboldt, 19 March 1852, SBB, NAvH, gr. K. 11.
[77] SBB, NAvH, gr. K 11, 53, Berlin, 18 December 1851.

of the brothers only on his 'close acquaintance' with their published Alpine works.

Since Humboldt regarded the Schlagintweits as his talented pupils, his report was overly imbued with praise. He purported that the brothers' 'important treatise on "The Physical Geography of the Eastern Alps" encompasses more [findings] than any other recent [work] on a specific mountain range'.[78] Against the charge that the brothers were intellectually overreaching, Humboldt argued that their 'great range of miscellaneous observations gives a satisfying impression of the current state of the sciences' and further praised their 'talent for graphic depiction, [and] a long and proven experience in mountaineering'.[79]

The aged polymath further backed the idea of a Schlagintweitian Himalayan expedition, one that would necessarily take them through British territories, by alluding to the international reputation the brothers had purportedly secured. The brothers had 'acquired not through recommendation, but through published work – the only source of impartial evaluation – an outstanding esteem in a country [Britain], where one is overly parsimonious with praise, especially to foreigners'. In view of their supposed 'industriousness ... and fondness for thorough research', Humboldt concluded that the brothers had chosen a suitable and promising object of study: 'The Himalayan mountains will present a stimulating area for scientific investigations for another century', and 'one can expect that their stay in India would prove of great value to the sciences.'[80] The expedition in this format, framed as a Prussian-only initiative and funded by Frederick William IV, who held no political stakes in south Asia, was never to take place.[81] However, it introduced the general idea of such a scheme to the monarch and, given Humboldt's words of praise, also established the brothers' names with the king.

This initial failure did little to quell the Schlagintweits' eagerness to undertake an ambitious Himalayan expedition, for which they had already nurtured ties with the British scientific and political establishment. The brothers were acutely aware that no scientific expedition into British India was feasible without the concession of free passage by the Court of Directors in London. The East India Company carefully channelled and restricted access to their colonial possessions, fearing the intrusions of disruptive outsiders who might undermine their commercial privileges

---

[78] Humboldt's report, 27 November 1852, GStAPK, WR.
[79] Ibid.
[80] Ibid.
[81] For Humboldt's frustration with Raumer's influence over the king, see Humboldt, *Letters to Varnhagen von Ense*, diary entry Ense, 9 September 1853, 214–16.

or hegemony in India – the most important British overseas colony at the time. To realise a Himalayan expedition under whatever flag in the future, it had thus been crucial that the brothers established a reputation in the British Isles as capable naturalists. For this purpose, too, the long-established networks of their mentors could be mobilised.

Given the high esteem in which Carl Ritter was held in British circles, the Schlagintweits turned to him prior to their first trip to England in 1850 to 'kindly ask you to provide us with some recommendations … especially to the directorates of the great ethnographic and other collections'.[82] 'The most perfect admiration' Ritter enjoyed there meant, they assured him, that 'only a few words' would secure them 'an excellent reception'.[83] Equipped also with letters from Humboldt and Heinrich Wilhem Dove (1803–79), professor of physics and, since 1849, director of the Prussian Meteorological Institute, the brothers received a warm welcome from many eminent British men of science during their stay in the winter of 1850/1. They frequented scientific institutions in London, but also visited Oxford and Cambridge, meeting there such distinguished polymaths as William Whewell, who had coined the term 'scientist' in 1833 at a meeting of the recently founded (1831) British Association for the Advancement of Science.[84]

Adolph and Hermann Schlagintweit presented themselves in 1850/1 to many of the same British naturalists and science administrators who would later support their appointment to British India.[85] A meeting with the royal family also proved successful. Adolph Schlagintweit noted that 'Lord Palmerston', the future prime minister, 'had the great kindness of offering us to be presented to Prince Albert' in 1850. Albert, Queen Victoria's consort, acted as an important patron for German musicians and artists in England but also supported scholars and naturalists.[86] However, the Schlagintweits also encountered those who sought to obstruct their scheme. For instance, they were introduced to the renowned Danish-born surgeon-naturalist Nathaniel Wallich (1786–1854). Wallich had made a long and successful career in colonial India himself, above all by clinging to the position of superintendent of the

---

[82] Hermann to Ritter, 20 July 1850, SBB, Nachlass Ritter [NR], Hermann Schlagintweit, 85–85a.

[83] Ibid.

[84] Hermann to John Couch Adams, St John's College, Cambridge [JC], Adams Papers, 13/32/1, 5 January 1851.

[85] Neuhaus, *Tibet*, 70.

[86] Adolph to John E. Gray, 19 December 1850, American Philosophical Society [APS]; on the Hanoverian connection and its role for German engagement with south Asia, Tzoref-Ashkenazi, 'Voices'.

Calcutta Botanic Garden from 1817 to 1846, then one of the most prestigious scientific offices outside Europe.[87] Following his prolonged period in the Company's service, Wallich had settled down in Britain in 1847 as a respected naturalist. His elevated status was cemented in his appointment as vice-president of the Linnean Society, and of the Royal Society in 1852 – the same institution that would later back the recruitment of the Schlagintweits.[88]

Adolph lectured in December 1850 at the Linnean Society and the brothers at first profited from the acquaintance with Wallich.[89] This turned into more than a superficial contact, and he still offered later in 1851 to 'charge himself with making the abstract' for a Schlagintweit publication, destined for *Hooker's Journal of Botany and Kew Garden Miscellany*.[90] Yet, the brothers' acquaintance with the Danish-born naturalist turned sour, as Wallich's harsh judgements on the brothers' lack of ability and social comportment clearly show. His rejection was arguably the result of the brothers' presumptuous and often rude behaviour, which would estrange them from numerous colleagues over their careers.

Given his influential position and botanical knowledge, Kew director William Hooker was another crucial figure the Schlagintweits were keen to meet. In letters of introduction to him, Michael Faraday and others, Humboldt lent scientific authority to the brothers by flatteringly comparing their work with Horace-Bénédict de Saussure's oeuvre (1740–99), the alleged founder of Alpine studies.[91] It proved even more significant to meet those scientific administrators directly involved with British surveying projects overseas. First and foremost among these was the physicist and army officer Colonel Edward Sabine (1788–1883), whose acquaintance the Schlagintweits made during a guided visit to the New Observatory at Kew in January 1851.[92] Sabine had gradually climbed the ladder of office at the prosperous Royal Society, whose elected treasurer and vice-president he had become in 1850. His influential role at the Royal Society was complemented by his position as general secretary of the British Association for the Advancement of Science.[93] Sabine

---

[87] Arnold, 'Plant Capitalism'.
[88] Boulger, 'Wallich'.
[89] A. Schlagintweit, 'Summary'.
[90] AS to WH, RBG, DC 51, 10 January 1851, 549.
[91] Humboldt to WH, 12 August 1850, RBG, DC 51, 254.
[92] On Sabine's influence, Good, 'Sabine'.
[93] On the rivalry between the Royal Society as an elite, traditional scientific club, and the much younger and reform-oriented British Association for the Advancement of Science, which organised public lectures and events across Britain and represented a more dynamic approach with a pronounced interest in applied science, Cawood, 'Magnetic Crusade'.

also acted as scientific adviser to the Admiralty, and maintained close relations with the War Office.[94]

Besides his excellent contacts with political and scientific authorities in Britain, Sabine was also an active member of the networks that linked London to British overseas colonies, and to Berlin. He was a personal friend of Alexander von Humboldt, whom he had known since 1818. Humboldt was highly appreciative of Sabine's research, and had twenty-five of his works in his private library in Potsdam; no other scientist is more often cited in the index of Humboldt's *Cosmos*.[95] The fact that both shared a strong interest in the field of geomagnetism had already led to collaborative projects. His wife, Elizabeth Juliana Sabine, had translated Humboldt's *Cosmos* into English. Edward had annotated her translation with personal notes and instructive explanations that had received high praise from the Prussian *Naturforscher*.[96] Humboldt, in turn, secured a number of foreign medals for the British colonel, such as the Cosmos medal in 1848.[97] Humboldt furthermore secured Sabine's honorary membership of the prestigious Berlin Academy in 1855, and paved the way for his friend's admission into the esteemed order Pour le Mérite for Sciences and Arts in 1857. These personal collaborations and mutual favours generated a feeling of obligation between the two men of science. Unsurprisingly, Humboldt could later count on Sabine's outspoken support for the Schlagintweits to realise their Himalayan expedition after all – this time through a different and unexpected window of opportunity.[98]

The 'Magnetic Crusade' in Britain

Edward Sabine owed his Prussian and other foreign decorations to his achievements in the field of geomagnetism, and he spent almost a lifetime forcefully promoting its study. It was during the first decades of the nineteenth century that an increasing interest in the earth's magnetic sphere had emerged among European physicists. Whereas European imperial powers undertook topographical surveys of their national and imperial

[94] Ibid., 518.
[95] Biermann, *Miscellanea*, 103–5.
[96] Humboldt to Bunsen, 28 September 1846, in Schwarz (ed.), *Briefe*, 88.
[97] Humboldt to Bunsen, 29 July 1848, ibid., 107–13, 109; Biermann, *Miscellanea*, 103–5.
[98] Humboldt to Bunsen, in the midst of the Crimean War, 20 February 1854, 'The king ... had instructed me to thank you wholeheartedly for the useful vividness [*Lebendigkeit*], with which you in a time of tense political conflicts keep on supporting the travel of these young men. It needs your powerful protection to initiate and carry out the scheme', in Schwarz (ed.), *Briefe*, 170–8.

territories as separate state-backed projects, measuring the magnetic field and its variation at specific moments, by contrast, had necessarily to be carried out simultaneously over a wide area, at best on a global scale. It was therefore only feasible through the collaborative effort of several European states and empires.

Partly initiated by Humboldt, magnetic observatories were set up in a number of European countries and overseas colonies in the first decades of the nineteenth century: in Germany, Russia, Italy, Sweden, England, the United States and Australia.[99] Since Britain lagged considerably behind in this field, Humboldt had specifically addressed the president of the Royal Society in 1836, stating that being 'in possession of the most extensive commerce and the largest navy in the world', it would be crucial for the advancement of the discipline for Britain to establish magnetic stations in its overseas possessions.[100] At that time, British territories already spanned the globe, ranging from Canada, over St Helena, the Cape of Good Hope to Asia and Australia. Humboldt's proposal received a favourable response from the British government. Consequently, fixed magnetic observatories were established in a number of British colonies. The accumulated data from those stations were then often compiled in Europe, with the view of formulating scientific theories in terrestrial physics, thereby exposing the 'geography of scientific production in the British Empire'.[101]

There was only a narrow dividing line between the pure and applied aspects of the study of terrestrial magnetism. It was for this reason that the British War Office, the Admiralty and the East India Company soon heavily financed the systematic study of the earth's magnetic field. Shortly after Britain had launched what was called its global 'magnetic crusade' in the 1830s, it had become 'a scientific enterprise ... of a magnitude never obtained before' in that country.[102] Above all, a more comprehensive understanding of the factors that caused the variation of the magnetic north had wider implications for the art of navigation. First, it was hoped that a more thorough knowledge of geomagnetic forces and their troublesome variations could help to improve navigational skills in case of bad weather conditions at sea. Second, there was a growing need in the nineteenth century to handle the problems that the construction of

---

[99] Chapman, 'Geomagnetic Science'.

[100] Reich et al., *Geniestreich*.

[101] Ratcliff, 'Geomagnetism', 327. Beside Sabine and Humboldt, a leading theorist in the field was the German mathematician Carl Friedrich Gauß (1777–1855), who, in 1838, published the path-breaking treatise 'General Theory of Geomagnetism'; on his significance, Schröder and Wiederkehr, 'Geomagnetic Research', 1651.

[102] Cawood, 'Magnetic Crusade', 517.

iron-hulled ships caused for reading a compass bearing.[103] An enhanced understanding of geomagnetic forces promised to yield highly useful knowledge for any seafaring nation – or so the British supporters of the 'crusade' argued to secure renewed financial support. The British promotion of large-scale geomagnetic studies was also bound up with inter-imperial rivalries and notions of scientific prestige. John Herschel, a leading lobbyist for the crusade, even claimed that 'Great physical theories, with their chains of practical consequences, are pre-eminently national objects, whether for glory or utility.'[104]

In 1846, the Court of Directors of the EIC launched a major 'Magnetic Survey of the Eastern Archipelago' under the leadership of Captain Charles M. Elliot.[105] This survey extended the area of research from the oceans and coastlines deep into the interior of British India. Sabine had by then assumed effective control of the magnetic mission. He presided over the resources through which the Company and the Royal Society financed the magnetic survey in south Asia.[106] However, the project came to an abrupt halt in 1852 with the unexpected death of Captain Elliot, who had only commenced the survey. The EIC let the project lie, until it was reinvigorated by a Prussian initiative. Bunsen, from London, informed Humboldt about the willingness of the Company and the Royal Society to conclude the eastern magnetic survey.

What followed was a masterwork of scientific diplomacy that underlined the importance of scientific networks and transnational systems of patronage. Eager to send his close protégés to India, who could provide him with crucial observations for his treatises on Asia, Humboldt seized the moment. The right timing was crucial. He first arranged another meeting between the Schlagintweits and the Prussian king. The royal audience was successful and convinced Frederick William IV to now support a *different* Indian mission of the Bavarian naturalists. Since the consent, and considerable financial support, of the EIC were the *sine qua non* for this undertaking, a concerted effort had to be made. On 27 February 1853, Frederick William IV dispatched a letter to Bunsen. Therein, the envoy was informed that the king would commit himself to subsidise an Indian expedition of the Schlagintweits on the condition that the East India Company would grant their permission, but also share the burden of the expenses.[107]

---

[103] Headrick, *Tentacles*, 18–24.
[104] Herschel, 'Memorial', 39.
[105] Elliot, 'Magnetic Survey'.
[106] Cawood, 'Magnetic Crusade', 515.
[107] Polter, 'Nadelschau'; Finkelstein, 'Mission'.

Bunsen, in office since 1841, was a man of considerable qualities in both political and scientific terms – as well as being another close ally of Humboldt's with excellent contacts to the British scientific community.[108] Their appreciation of the liberal Bunsen went so far that Joseph Hooker wrote to Humboldt after the envoy's resignation in 1854: 'We all feel the departure of Mr Bunsen as a national loss.'[109] For numerous German travellers and naturalists seeking employment in Britain or her colonies, Bunsen's embassy at Carlton Terrace in London was the first calling point.[110] The Prussian diplomat was a respected orientalist scholar himself, who mastered Persian and Arabic and had long held the wish to travel to India, expressing his great interest in oriental scholarship in his correspondence with William Hooker.[111]

It now proved an asset that Bunsen maintained excellent relations with the directors of the EIC, with most of whom he was personally acquainted. Once informed by the Prussian monarch about the projected scheme, Bunsen first set up a meeting with Edward Sabine in April 1853. Soon, the two had worked out an initial agreement. A formal proposal was sent to William Parsons (the Earl of Rosse), an Irish astronomer and president of the Royal Society from 1848 to 1854.[112] In the communication, Bunsen proposed that the Schlagintweits might be employed 'for the purpose of exploring the Himalayan range on behalf of a more complete knowledge of telluric magnetism, and many other branches of terrestrial physics, for the purpose of which the King of Prussia proposed to grant them pecuniary allowances'.[113]

The Royal Society president approved of the scheme and had it transmitted to the Company's Court of Directors. Among them was Colonel William Henry Sykes (1790–1872), a former EIC servant in south Asia. During his colonial service, he had completed a number of statistical and natural history surveys in India.[114] Sykes eagerly supported the cause within Company circles. Backed by a phalanx of international authorities in magnetic studies, the EIC subsequently approved to continue India's magnetic survey. Sykes was to remain a crucial supporter of the brothers over the coming years. His relationship with these imperial outsiders reflects an important fact: 'While science in this period was certainly

---

[108] For a valuable treatment of him as both a political and scientific interlocutor between Britain and Germany, Kirchberger, *Aspekte*, ch. 6.

[109] JH to Humboldt, 21 September 1854, SBB, NAvH, gr. K. 11, Nr. 10, 16.

[110] Kirchberger, *Aspekte*, 394.

[111] Bunsen to WH, 4 December 1849, RBG, DC 51, 64.

[112] Sabine to Murchison, *The Athenaeum*, 1767, 7 September 1861, 320.

[113] Ibid.

[114] Woodward, 'Sykes'.

not reducible simply to politics by another means ... science and empire seldom failed to communicate with each other, and indeed, often found patronage and agency in one and the same person.'[115] But even now, the Company had still not decided to hire the Schlagintweit brothers for the mission.

In a resolution taken on 18 May 1853, it said that 'the East India Company regard all such missions with great satisfaction'. Therefore, 'the Court of Directors propose to instruct the Government of India, in the event of their having no officer available for carrying out the objects left unfinished by Capt. Elliot, to apply to the Messrs. de Schlagintweit to ascertain if one of those gentlemen would undertake the duty; and if so to place the instruments at his disposal, and to grant him a suitable allowance for the purpose'.[116] The reasons for the Company's initial wish to appoint a British officer in India are evident. First, there was the general expectation by British scientists and officers that they, not foreigners, were entitled to the position. Second, it was more expensive to employ a naturalist from the continent for the scheme because he had to be specifically trained in London and then brought over to India. The Court of Directors had already enlisted a large number of skilled British officers and surgeons on the Indian subcontinent perfectly capable of finishing the scheme, and who often longed for properly paid Company employment.[117]

High Aspirations

For the Schlagintweits to turn British India into their promised land, they depended on their own initiative, since the Company was willing to pay only one of them, and merely for completing the magnetic survey. However, their personal ambitions went much further. They succeeded in securing the appointment of more than one brother in 1854 when Lord Dalhousie, governor-general of India, informed the Directors that 'no officer competent to such an undertaking could ... be spared from military duty' so that an 'application was made by the Court to the Messrs. de Schlagintweit'.[118]

The brothers also managed to fundamentally change both the scientific objectives and the financial grounding of their commission. Whereas

---

[115] Arnold, *Travelling Gaze*, 10.
[116] Sabine to Murchison, 7 September 1861.
[117] On the many EIC servants engaging in scientific studies in India without Company support, see Arnold, *Science*, ch. 2.
[118] *The Athenaeum*, 1767, 320.

the Prussian monarch had agreed in 1853 to grant £200 per annum for three years, the Schlagintweits and their advocates succeeded in having the king more than double the amount, £350 per year for two brothers (roughly equivalent to around £37,000 today). Each of them also received another £100 for the purchase of books and instruments.[119] It was furthermore 'upon the highest order of His Majesty' that the king issued travel passes to the Bavarians.[120] Therein, he asked any foreign military or civil authorities, but formally 'ordered' any Prussian subject and servant to provide the brothers full support. Hence, they could draw on the networks of Prussian consuls in India, an important asset that would later provide them with space for manoeuvring between their British and German benefactors.

The EIC directorate had underlined in the appointment letter to Adolph the more circumscribed task of completing above all the magnetic researches.[121] In the end, however, the nature of the enterprise dramatically changed as the result of negotiations driven by the Schlagintweits and their supporters with the Court, senior metropolitan scientists, and the Imperial Government of India. There is no better evidence for understanding this transformation of terms than a document called 'Operations Proposed to the India House', which survives as a copy with annotations by Sabine.[122] Adolph submitted this list of operations on behalf of (at first) two brothers to the directors on 28 March 1854.[123] Its objective was to expand their initially minor employment into a major scientific investigation of south Asia and its northern frontier regions.

The brothers first laid out a plan of the routes they would take in India and central Asia in 1854–7. The Court was informed that the brothers intended to proceed 'from Bombay to Madras, if possible on two different routes' after their landing in the second half of 1854. The summer of 1855 would be spent exploring the region of Darjeeling and the 'eastern Himalaya, perhaps if under favourable circumstances with a journey to Nepal'.[124] Nepal was then hardly accessible to Europeans and jealously guarded by the Chinese; Hooker had earlier experienced great troubles regarding his planned journey into the country.[125] While

---

[119] Polter, 'Nadelschau', 79; see the correspondence leading up to the order of 8 July 1854, at GStAPK, I. HA Rep. 162, Verwaltung des Staatschatzes, 'Reisezuschuß'.
[120] BSB, SLGA, IV.6.2.
[121] J. D. Dickinson, East India House secretary, to AS, 10 June 1854, BL, IOR, E/1/300, 1854, 1715.
[122] The National Archives, Kew [NA], BJ 3/53.
[123] Ibid.
[124] Ibid.
[125] Letter to Wallich, 30 March 1850, RBG, JDH/1/9/1, 489–92.

the following winter would again force the Schlagintweits to descend into northern India (through Bihar to Agra and as far south as Nagpur in central India), they proposed to spend the summer of 1856 venturing into the central regions of the Himalayas, with travels through Kumaon, Almora, Tibet and Simla. After another separation during the winter months, the brothers would finish their researches in the summer of 1857 in the 'Western Himalaya' with separate journeys through the 'valley of the Indus at Ladak' and to the famous valley of Kashmir. From there, the plan was to proceed from the mountain chain to the Indian coast, and to return via steamship from Bombay (Mumbai) to Europe.[126] In short, the Schlagintweits hoped to combine their surveying project in Company-controlled territories with explorations in lands beyond British sovereignty, which promised significantly more public interest.

The brothers also managed to reinterpret the nature of their entire mission: 'With the magnetical observations in different parts of India we propose ourselves to unite [a] Regular Series of Observations on … the Physical Geography of the country'. Adolph added that 'I myself will direct my particular attention to collect as complete a series as possible of observations on the Geology of India and of the Himalaya.' In alluding to what would later become large-scale surveys of resource deposits, the German geologist also suggested that:

We shall be also very happy to give our closest attention to any question of practical interest, as for instance to the determination of the Geological age of the various Coal deposits of India to the occurrence of salt or to the practical use of the sulphur occurring in western Scinde.

Such a comprehensive geological survey of the vast natural landscapes of India and the mountain chain at its northern border would have been an undertaking of several years in itself. The fact that Adolph hoped to achieve such an objective literally *en passant* reflects their high aspirations, but equally their delusion about what was achievable. This impression of over-ambition is confirmed throughout the document, as the brothers declared their intention to complement their geological, geographical and geomagnetic studies over those vast regions with additional investigations in several other fields of enquiry such as meteorology, hydrology, potamology, botany, mineralogy, palaeontology and zoology. To carry out extensive studies in all of these departments, and over such

---

[126] A. Schlagintweit, 'Proposed Operations', NA, BJ 3/53. Sabine thought the plan unfeasible, noting on their proposed routes during the 'Winter 1855/56': 'impossible in one winter. There are no roads & much of the country is very unhealthy even in winter.' Subsequent quotations in this and the next paragraph, ibid.

vast and diverse landscapes as proposed, was the exact opposite of their initially narrow magnetic mission.

Yet, the Company granted permission to the Schlagintweits 'to undertake this [geomagnetic] duty in connection with the other objects mentioned'.[127] To maintain a degree of control over the expedition, and to ensure that the brothers' pursuits would meet the interests of imperial authorities, the Court, however, established that 'your proposed plan of operations will be communicated to the Government of India, and will be subject to such modifications as from time to time may seem to the Government desirable or requisite'.[128] While initially only Adolph was to be employed, ultimately, the Court allowed first Hermann, and then also Robert, to join the scheme.[129] Like Adolph, Hermann was also enlisted in paid employment for the duration of four years for the eastern mission. Robert, by contrast, was to accompany his brothers without a salary from British pockets, but secured a 'pension' from the Bavarian king Maximilian II.[130] The Schlagintweits thus tripled their manpower. As a later critic summoned it up, the Company's final 'contract with them was so loosely drawn up that they had practically a roving commission in science, to make researches ... in all manner of subjects', to which, during the mission, the troika would add further fields of enquiry and observation, including languages, ethnography and racial sciences.[131] This disciplinary overreach, henceforth built into the expedition's software, was indeed the birth defect of the entire enterprise under which their controversial careers and reputation would later suffer greatly.

From the outset, the Schlagintweits thus sought to become more than a mere episode in the century-long history of British surveying of south and high Asia. Many surveys in the Indian Empire during the first half of the nineteenth century had gone largely unnoticed by the general public. Indeed, very little was known about the precise activities of the legions of Company servants engaged in measuring and classifying India's natural landscapes. As the *Calcutta Review* put it in 1851: 'We believe that there

---

[127] Dickinson to AS, 10 June 1854.

[128] Ibid.

[129] The current Surveyor General of India, A. S. Waugh, noted on the 'liberal' agreement: 'The salary assigned to Mr. Schlagintweit is nearly equivalent to that of a first class Surveyor who receives no travelling allowances & exceeds by 100 Rs the salary of an Executive Engineer of the first class.' Waugh to Colonel R. J. H. Birch, Secretary to Government of India, Fort William, in National Archives of India, New Delhi [NAI], Military Department, Controlling Agency, Consultation, 15 September 1854, no. 445. After negotiations with the imperial authorities at Calcutta, the brothers ultimately received 'Rs 600 per Ms with Rs 150 per Ms for travelling expenses', NAI, Letters from the Court of Directors, 1955-09, 19 September 1855, no. 127, 919.

[130] Humboldt to Ritter, 11 March 1856, in Päßler, *Briefwechsel*, 172–3.

[131] Huxley, *Letters of Huxley* I, 329.

are very few persons, even in India, who have any notion whatever of what the Trigonometrical Survey really is, or what it does for geography or science: or who can comprehend what has been already done, and why it has not long since been brought to a conclusion.'[132] In direct contrast to such obscure surveying, the Schlagintweits were eager to turn into great explorers in the public eye and longed for scientific breakthroughs and geographic feats. Their (self-)promotion as 'heroic travellers' into the supposed unknown was soon launched and promised a very different public recognition and career opportunities in Europe.

As this example of imperial recruitment shows, securing funds for scientific expeditions in foreign empires did not necessarily follow established ways. Yet, the Schlagintweits' case also demonstrates how much depended, as a recurrent pattern for such cross-border appointments, on the interests and power of patrons, whose connections reached deep into both the intellectual centres of London and Company science in India. The brothers' ingenious careers launched in the mid nineteenth century thus expose the slow, extremely uneven and conflictive transformation of private patronage towards more public, and allegedly more answerable, systems of national support of scientific enquiry.

---

[132] Walker, 'Trigonometrical Survey', 514–15.

2   Imperial Recruitment and Transnational Science in India

In order to achieve a multifaceted analysis of the controversies surrounding the Schlagintweits' mission, and to understand how the brothers had to negotiate their reputation within the context of transnational scientific collaboration and competition, this chapter is divided into several inter-linking parts. First, the brothers' enterprise is set against a longer chronology of British and continental European explorations towards and beyond the north Indian frontier. The British already looked back on a century of surveys and exploratory missions when the brothers entered the stage with their ambition to essentially reshape scientific understandings of this world region. The discussion then shifts to the discourses that accompanied the recruitment of German naturalists and advisers in the mid nineteenth century, a period in which Germans were not yet seen as agents of their own state-driven imperial ambitions and were thus deemed suitable for employment in the British territories. The chapter demonstrates how British rule and expansion in India required a greater store of experts than the home population could provide. Numerous German scientists and technicians not only helped to fill positions within British India's establishment, but they also shaped the very practices of colonial governmentality in fields such as forestry and agricultural chemistry. Yet, this reliance on foreign expertise was harshly criticised by British officers and naturalists whose achievements may have been acknowledged but were scarcely rewarded by the East India Company with paid employment. The generous funding of non-British personnel, recruited over the heads of many proven British officers and men of science, puts the conflictive careers that naturalists without hereditary fortune had to pursue at the time into sharp relief and sheds light on contemporary debates over the unsystematic patronage of science in Britain and its empire.

British Exploration and the North Indian Frontier

The English East India Company had been founded in 1600, when a group of London merchants sought to compete with the Portuguese in the lucrative spice trade in Asia, and had thus successfully acquired a monopoly charter from the English Queen Elizabeth I. As part of the global conflict between the English and the French in the Seven Years War (1756–63), it was in the wake of the battle of Plassey, in 1757, that the British had assumed territorial control of Bengal and the Mughal office of *diwan*, the right to collect revenues. Bengal then ranked among the richest provinces of the Mughal Empire. This conquest, achieved as much by military might as by intrigue, instigated a new phase of British commercial and political engagement with the subcontinent. William Pitt's India Act of 1784 only confirmed the much earlier transition of the EIC 'from a mercantile corporation to a major territorial power', yet whose rise to dominance on the subcontinent was by no means inevitable or linear.[1]

From the late eighteenth to the mid nineteenth century, British military power and administrative control spread across south Asia, forcing the Company time and again to negotiate and establish clearly marked frontiers and well-regulated political and economic relationships with Indian rulers. In the aftermath of these territorial conquests, which soon extended British possessions to the foothills of the Himalayas, colonial officials became increasingly troubled over their lack of knowledge about the geographical, political and military situation in those regions that lay within and beyond the mountain frontier north of the Indian empire.[2] British fears of a French-Russian invasion of India during the Revolutionary Wars equally heightened the sense of urgency not only to master the sea but also the land routes into India.[3] Although Napoleon's (and some Russian Tsarists') ambitions of conquering this most precious British colony were never realised, knowledge of the plans of invasion by the arch-enemy France nonetheless triggered Company efforts to explore and map out potential routes for such an inroad.[4]

---

[1] Edney, *Mapping*, 4–8; on the longstanding resistance of Mysore to British expansion in southern India, and the role of French influences and military assistance in the kingdom, Jasanoff, *Edge*. Philip J. Stern has shown how several defining features of a 'Company state' had already been developed well before Plassey, as the EIC had become 'a polity in itself: a self-sustaining global system built upon sound civic institutions and informed by a coherent if composite political ideology', Stern, 'Politics', 1; Stern, *Company-State*.

[2] Waller, *Pundits*, 1.

[3] The ill-fated French Egyptian campaign (1798–1801) was also intended to undermine British hegemony in India; Jasanoff, *Edge*.

[4] Wood, *Silk Road*, 149; despite British fears, on the unlikelihood of a Russian invasion of India, see, however, Morrison, 'Logistics'.

Yet, increased knowledge of the trans-Himalayan regions was also of strategic and commercial importance for the Company. The prospects of trade with Tibet particularly captivated British officials, leading to various attempts to establish commercial relations with the hard-to-access state under Chinese suzerainty.[5] In 1826, Nathaniel Wallich, then superintendent of the Calcutta Botanic Garden (itself a hotspot for 'colonial science'[6]), waxed lyrical about the 'glorious' Himalaya chain, and the 'general desire to become nearer acquainted with the vegetable treasures of those hitherto so little known, and yet so immensely interesting regions'.[7]

To remedy their ignorance, the British intensified and institutionalised earlier efforts to traverse and map the subcontinent by launching one of the most ambitious scientific projects of this period. In 1802, the Great Trigonometrical Survey (GTS) was initiated in the presidency of Madras, quickly gaining momentum and extending its sphere of operation. The colonial government of India took charge of the GTS in 1818 – at a time when British territorial gains for instance in Assam were constantly opening up further areas for systematic study.[8] The GTS was organised by the Military Department of the Government in India. The surveyors were asked not only to provide accurate maps, but also to 'collect information on the peoples of the area, the nature of their livelihood, the available crops, and other details, military and commercial'.[9] This major surveying project was still running – then under the orders of Major-General Sir Andrew Scott Waugh (1810–78) as Surveyor-General of India – when the Schlagintweit brothers were appointed to lead their scientific mission from 1854 to 1858. The GTS had by then covered significant parts of British India and stretches of the Himalayan frontier region. In 1850, Waugh could thus speak in a self-congratulatory manner of '[t]his magnificent geodetic understanding, which at the present time extends from Cape Comorin to Tibet, and from the meridian of Calcutta to that of Cashmere'.[10]

In addition to the systematic northward expansion of British knowledge of the subcontinent, a number of British travellers, some of whom

---

[5] Waller, *Pundits*, 7–13.

[6] Vicziany, 'Imperialism', 641–2.

[7] Quoted from Arnold, 'Plant Capitalism', 923.

[8] Note that the dynamic of further British expansion originated often from schemes of individual Company officers or governor-generals in India, not with policy-makers in London, who at times directly opposed annexationism; Darwin, 'Dynamics of Territorial Expansion', 625.

[9] Waller, *Pundits*, 17.

[10] Waugh, 'Trigonometrical Survey', 285.

would be personally involved in the Schlagintweit controversy, also set their sights on exploring the trans-Himalayan region. The hardly accessible countries and kingdoms of the mighty mountain barrier had already attracted great interest among European travellers and missionaries since the early modern period. The Jesuits António de Andrade, Johann Grueber and Albert D'Orville were among the first who had penetrated into the Himalayas in the early and mid seventeenth century. Later travellers included the missionaries Manoel Freyre and Ippolito Desideri, who jointly travelled to Tibet in the 1710s.[11]

While acknowledging the travels of these early pioneers, by the mid nineteenth century, reflecting a sense of dominance over the subcontinent, the British had drawn up their own heroic chronology of Company officers, naturalists and 'systematic geographers' who had shed light on 'the apparently inextricable labyrinth of snowy peaks'.[12] They included William Webb of the Bengal Infantry, who in 1808 had visited the upper Ganges in search of the source of the river. Later he also undertook an extensive survey of the recently conquered region of Kumaon (Uttarakhand, India), charged with the duty to determine exploitative metallic resources.[13] Then there was the veterinary surgeon William Moorcroft, who had crossed the mountain range and reached Tibet in 1812, keenly noting Russian ambitions in the region.[14] Moorcroft resumed his travelling activities in 1819, and in 1820 he had reached Leh, the capital of Ladakh, as the first European to do so for over a century. He would also spend some months in Kashmir, at the end of 1822, subsequently setting off for his original destination, Bukhara, in Afghanistan, which he sought to reach via the Hindu Kush. Along the way, the British traveller accumulated, in addition to his scientific observations, critical political and trade knowledge, which he hoped would increase British influence in and facilitate the commercial penetration of central Asia.[15] Yet, Moorcroft's journey came to an unplanned end in 1825. While passing through Afghanistan, exhausted from the considerable exertions of travel, he died from a fever.[16] To many contemporaries his death symbolised the dangers and personal sacrifices involved in British exploratory missions beyond the Company frontier. Generally, it was scientific interests and the commercial prospects of the partly

---

[11] See Wessels, *Travellers*, for their and others' travelogues and reports.
[12] An almost exclusively British phalanx of pioneers is given in Markham, *Memoir*, esp. 342–8.
[13] Webb, 'Memoir'.
[14] Waller, *Pundits*, 11.
[15] Alder, *Bokhara*.
[16] Baigent, 'Moorcroft'.

untapped markets and natural resources of 'high Asia'[17] – and, before the Opium Wars, the search for an alternative entry into the Chinese markets beyond the restricted bottleneck of Canton (Guangzhou) – that drove these expeditions, especially into Tibet.[18]

In the wake of Moorcroft's death, British travellers continued the exploration of the trans-Himalayan region. Among the early British officers in that great mountain range was James D. Herbert, Deputy Surveyor-General of India between 1829 and 1831, who surveyed parts of it between 1816 and 1828, attempting to provide the first coherent geological picture of the Himalayas. Another itinerant military man was Sir Richard Strachey (1817–1908), who carried out scientific investigations of the western Himalayas in the mid 1840s and joined the botanist J. E. Winterbottom on a trip into virtually inaccessible Tibet in 1848. While amassing a botanical collection of more than 2,000 samples, Strachey and his companion also focused on other scientific fields, which resulted in well-received treatises marked by the interdisciplinary thrust of Humboldtian science.[19] Some British Residents (political agents), such as Brian Houghton Hodgson, also became scientific collectors and prolific authors in their own right.[20] Hodgson, first stationed in Nepal, was despatched to Darjeeling in 1845, ten years after the British had seized the region. Despite a lack of any formal scientific training, the self-taught naturalist drew on his firsthand observations and continuous exchanges with local peoples to become arguably the greatest British authority on the natural history of the Himalayas in the 1840s – although Hodgson was acknowledged as such more in India than in Europe. He combined his naturalist activities with notable ethnographic studies.[21] Yet, despite his credentials, Hodgson struggled to receive due acknowledgement and material patronage from metropolitan experts and institutions.[22]

As these few examples of a much longer list of Schlagintweit predecessors demonstrate, British interests and modern explorations in the Himalayas by no means started with their mission. On the contrary, British scientific and commercial engagement with the northern mountain regions looked back on almost a century-long history of increasing

---

[17] The Schlagintweits coined the term 'High Asia' with their travelogue *Results of a Scientific Mission to India and High Asia*. While its precise meaning and geographical scope was never clearly defined, this 'extended' region encompassed 'the Himalayas, the Karakorum and the Kuenlun'; R. Schlagintweit, 'Besteigungs-Versuch', 314.

[18] Waller, *Pundits*.

[19] Strachey, 'Geography'.

[20] On the breadth of Hodgson's interests, Waterhouse, 'Hodgson', 7.

[21] Gaenszle, 'Hodgson'.

[22] Arnold, 'Frontier', 189–90.

exploration, intelligence gathering and systematic knowledge production.[23] Viewed from the British side, the German brothers were seen as merely contributing to this tradition of research. Yet, as the following chapters demonstrate, the brothers themselves were at times keen to fashion an image of themselves in front of popular audiences that could suggest that they were the only modern scientific explorers of these areas – even if their work built heavily on the accomplishments of others and must be placed within this longer continuum of imperial exploration. Their repeated neglect of this fact drew, unsurprisingly, heavy polemic in mid Victorian geographic circles, where questions over first achievements, scientific merit and social etiquette clouded numerous explorers' careers in stark controversy.[24]

Yet, it was not only British Company servants who engaged in scientific pursuits at the roof of the world. In addition to French interests in the region, embodied in the travels of Victor Jacquemont (1801–32), there were also German missionaries with natural-historical interests in the western Himalayas. These included members of the Moravian religious order 'Herrnhuter Gemeinde', with whom the Schlagintweits later exchanged information. Subsequently, one missionary claimed that even though the brothers had received help taking measurements and gathering collections in the area, they had later 'boastfully taken all the glory for themselves'. The missionaries therefore suggested that 'one should rather call them "Schlagaufsmaul"'.[25] The brothers were thus by no means the first German naturalists in the Himalayas. Nor were they the first German-speaking scientific travellers: the Austrian Carl von Hügel (1796–1870) had explored different parts of Asia and Oceania between 1830 and 1836. Hügel had also investigated sections of the north Indian and central Asian mountain chains, producing some well-received learned works, parts of which were then translated into English on behalf of the East India Company.[26] The growing wealth of knowledge in the Company's imperial archive and the amount of information gained from expeditionary science kept by the Royal Geographical Society in London meant that the brothers entered an already crowded field in setting their sights on becoming great Himalayan explorers.

---

[23] Bishop, *Shangri-La*, 98.
[24] As the violent polemics between African explorers, among others, demonstrate; on the 'Nile Controversy', see Dritsas, *Zambesi*; on the contended means of European exploration, Driver, 'Stanley'.
[25] A play on the German idiom for 'to slap somebody in the face'. Friedl, 'Forscher', 80.
[26] Hügel, *Travels*; Hügel, *Gebirge*.

Joseph Hooker and the Ideals of Gentlemanly Science

British knowledge of Indian and Himalayan natural history reached new heights through the travels of Joseph Hooker. He was also to become the German naturalists' most ferocious critic, which requires a closer understanding of his initially troubled trajectory and difficult relationship with the imperial Company. Before Hooker embarked on his Himalayan travels between 1847 and 1851, he had already had a noteworthy, though not unhindered, career as a traveller and naturalist. As the younger son of the eminent botanist William Hooker (whose post as director of Kew Gardens he would take up in 1865), Joseph 'did not so much learn botany as grow up in it'.[27] The fact that his father maintained an extensive network with government officials, naturalists and colonial collectors from various corners of the globe undoubtedly paved the way for Joseph's own scientific career as one of many 'metropolitan-based naturalists', who engaged with but were also eager 'to rise above the various forms of "colonial science"'.[28]

Having trained as a surgeon and graduated in medicine in Glasgow, Hooker was eager to make his name, and for this purpose sought to emulate his role model Charles Darwin. Consequently, in the capacity of assistant surgeon and on-board botanist on the ship MHS *Erebus*, he embarked on a 'voyage of discovery and research in the southern and Antarctic regions during the years 1839–43' under Captain James Clark Ross.[29] Such extended overseas expeditions were then regarded as a formidable rite of passage for young and aspiring naturalists.[30] Hooker had no choice but to accept a minor paid position on board, yet he struggled to accept this fate, writing to his father: 'I am not independent, and must not be too proud; if I cannot be a naturalist with a fortune, I must not be too vain to take honourable compensation for my trouble.'[31] The reason was that during much of the nineteenth century in Britain, 'men of science remained uncertain as to the respectability of being paid to do science'.[32]

Going back to a long tradition associated with the Royal Society founded in 1660, it was still widely presumed in Victorian Britain that the practitioners of the different branches of science ought to

---

[27] Huxley, *Letters of Hooker* I, 37.
[28] Arnold, *Travelling Gaze*, 40.
[29] Ross, *Voyage*. This Antarctic expedition was closely linked to the launching of Britain's 'Magnetic Crusade'.
[30] Leask, 'Genesis', 13.
[31] Endersby, *Imperial Nature*, 20.
[32] Endersby, 'Botanist', 163.

be gentlemen in the first place. In other words they should possess a personal wealth that was regarded as the guarantee of their scientific 'disinterestedness and thus ... truthfulness'.[33] Although the very notion of a gentleman shifted during the Schlagintweits' and Joseph Hooker's lifetime, as the rising middling ranks of society increasingly challenged the established power of the landed aristocracy, still the notion persisted that science should above all be pursued out of disinterested motives, not out of necessity as a paid service to either the state or industry. Both the wealthy Baron Alexander von Humboldt, whose personal riches were spent on his American travels and the immensely costly publication of his American *opus*, and the eminent Joseph Banks (1743–1820), who – as scientific adviser to the British monarch – never received a salary, embodied this idealistic notion. The social ideal of the gentleman scholar posed great difficulties for non-independent men of science like Hooker who lived and worked under considerable material constraints, but nonetheless sought to live up to the established notion and to be accepted by their peers as scientific equals.[34] Only personal misfortune could drive men from wealthy backgrounds into reluctantly accepting salaried positions. As the painter Ford Madox Brown noted bitterly in his diary in 1855: 'Gave my first lesson for a guinea ... and am no longer a gentleman.'[35]

Joseph Hooker in fact pursued two different, albeit closely related, objectives during his southern expedition, and throughout much of his subsequent career. Besides making a name for himself, he also strove to raise the prestige of the very discipline he was engaged in, namely philosophical botany.[36] Botanical pursuits were then highly popular with amateurs, including a good many women, and were practised by lay personnel in Britain and its overseas colonies. Hence, to distinguish himself from this army of dilettanti, Hooker strove to raise the status of botany as a science that was to stand above the mere collecting and classifying exercises – as it was regarded, and practised, by many at the time – while *de facto* relying heavily on the constant supply of specimens from the very same amateur collectors.[37] At least in the view of many British scientists at the time, as Jim Endersby has shown, 'true sciences were concerned

---

[33] Ibid.
[34] Rudwick, *Controversy*.
[35] Allen, 'Professionals', 9; Allen, *Naturalists*.
[36] Endersby, *Imperial Nature*, 12–20.
[37] See on the great importance of 'amateurs', including women, merchants, missionaries and others, for global science in the nineteenth century, Habermas, 'Intermediaries'.

with mathematics, experimentation, accuracy, precision, and – most of all – with discovering causal laws'.[38]

Becoming a philosophical botanist meant to draw on empirical observations and the scrupulous classification of gathered specimens from around the globe, but with a view to formulating higher theories about the geographical distribution of plants on the earth's surface, and to find natural laws about such vexing questions as to why new species were created in nature, and why a certain crop would thrive in one climate, but not in others.[39] Far from being only of scientific interest, these questions had tangible commercial implications, as the profits of maintaining and extending the British Empire depended in this period to a significant degree on natural commodities, including cotton, coffee, timber, opium and tea.[40] Joseph Hooker's life indeed encapsulates the rise of botany up the scientific hierarchy from being 'a peripheral aspect of medical training [to] one of the great imperial sciences, playing a key part in exploring, cataloguing and exploiting the natural wealth of the empire'.[41] On a semantic level, however, being a 'philosophical botanist' had the further advantage that one was judged by the 'quality of [one's] work, *not* by the way' in which a practitioner of botany earned his or her living.[42] Ultimately, Hooker's southern expedition spanning Australia, New Zealand and Tasmania did secure him a growing international standing, and he soon became both personally acquainted with Darwin and a close correspondent of Alexander von Humboldt.

However, despite his father's support and Joseph's rising reputation, which resulted from the publication of his treatises on the Antarctic flora, the aspiring naturalist struggled to secure sufficient patronage to pursue a scientific career in Britain.[43] As his application for a university chair of botany in Edinburgh had failed, Hooker was now, *faute de mieux*, keen to complement his extensive study of the plant life of the southern hemisphere with a tropical expedition, intended to explore the flora of the equinoctial zone at the northern fringes of the British possessions in south Asia.[44] In a letter to Captain J. Ross, he expressed his ambition 'of going to India … under the auspices of the Garden [of Calcutta],

---

[38] Endersby, 'Local Knowledge', 345.
[39] On the fragile status of botany as a science in the nineteenth century, see Bonneuil, 'Standardisation'.
[40] On the advisory role of Kew for cultivation schemes across the empire and schemes of crop transplantation, see Cornish, 'Curating'.
[41] Endersby, *Imperial Nature*, 34.
[42] Endersby, 'Botanist', 164.
[43] Hooker, *Flora*.
[44] Humboldt to JH, 30 September 1847, RBG, JDH/2/1/12.

who want to take advantage of the communication now opening up with Thibet and the rich passes of the Himalaya, which promise an extraordinary harvest of novelty'. However, the benefit of the scheme was not to be limited to Calcutta's Botanical Gardens, as Hooker saw his travels as 'a project which is to open a direct botanical communication between Kew and Calcutta in the first place and to explore a new country in the second'.[45]

Although his planned itinerary would still frequently change, Hooker's original intention was to explore the frontier region between the British possessions and Chinese-controlled Tibet, above all the region of Sikkim at India's snowy northeastern border. Still cautiously optimistic that he would receive not only Humboldt's scientific backing and some financial support from his father, Hooker also noted that 'The E.I.C. too have promised me every facility and recommendation.' Yet, while recommendations and logistical support were all well and good, Hooker needed more substantial patronage. For this purpose, he addressed among others Lord Auckland, then the First Lord of the Admiralty and a former Governor-General of India. If all of the considerable travel expenses in India were to be secured, he wrote that 'I should be content with £200 a year from the Gardens' as an additional salary. But even this project, with its several sources of meagre payments pieced together, might be built on sand, as '[s]o much may happen to frustrate these views that I cannot yet speak with confidence of the future'.[46]

The absence of a secure, paid position in botany was to accompany Hooker for a number of years, which overlapped with the generous employment of the Schlagintweit brothers by British authorities. While Hooker could embark late in 1847 on his Indian and Himalayan travels, during which he also found local scientific support from the Governor-General and a range of other British naturalists and administrators, his mission nonetheless lacked the decisive material support of the Company. Although the scientific gains of his expedition were considerable, it proved costly to the traveller himself, signifying the still fragile state of scientists without deep private pockets who then gradually emerged 'from the chrysalis ranks of gentleman amateurs' in British society.[47] Hooker struggled to recover his travel expenses, which led Humboldt to state that the fact that 'your so powerful Government hesitates to reimburse the whole, forcing your father to add 800£ St ... greatly puzzles

[45] JH to Ross, 27 August 1847, RBG, JDH 2/4/4, 42–4; subsequent quotations, ibid.
[46] Ibid.
[47] Arnold, *Traveling Gaze*, 206.

me'.[48] What was more, after his return in 1851, Hooker found himself in a delicate position similar to that before his Indian journey. 'At present I have no permanent employment, nor other sources of income but my Navy pay.'[49] However, concluding an immediate contract with the Company to arrange and publish his Indian collection was urgent, as Humboldt specifically reminded Hooker that it was best 'to draft and publish' his Indian observations 'in all the freshness' of his memory, further rubbing salt into his wounds.[50] Hooker even revealed in a private letter that '[e]xcept they do give me some temporary employ I cannot even marry, for I have not the means' – despite his engagement to his fiancée.[51]

Hooker spent the years after his Indian travels introducing a more strict classificatory system into Kew's inflated stock of botanical species, which was intended as the preparatory work for a more philosophical treatise on plant distribution in India and the Himalayas – his projected *Flora Indica*. Not without slight exaggeration, he wrote of his endeavours: 'I am a *rara avis*, a man who makes his bread by specific Botany, and I feel the obstacles to my progress as obstacles on my way to the butcher's and baker's. What is all very pretty play to amateur Botanists is death to me.'[52] Given the enduring lack of Company patronage, the first volume of Hooker's Indian flora, published conjointly with his travel companion Thomas Thomson in 1855, had to be financed through a private inheritance of Thomson's, and was meant to secure material support for the remaining portions of the work. However, convincing the Company of the prestige and commercial value attached to the patronage of the *Flora Indica* required a number of strategic moves on the part of the disregarded naturalists.

The first step was to put considerable pressure on the EIC's Court of Directors, as Hooker and Thomson engaged in 'fairly public efforts to embarrass the Company into supporting their flora'.[53] They wrote in both private and public statements about the double-edged policies of the EIC. In the introduction to the first (and ultimately only) volume of their work, they remarked sarcastically, 'the Court declined promoting the object, but expressed a willingness to take its merits into consideration on its completion'.[54] An anonymous treatise, arguably written by

[48] This was more than a third of the entire travel expenses; Humboldt to JH, 16 July 1851, RBG, JDH/2/1/12, 22.
[49] JH to William Wilson, 23 May 1851, RBG, JDH/2/4/4, 101–4, 103.
[50] Humboldt to JH, 16 July 1851, RBG, JDH/2/1/12, 22.
[51] JH to William Wilson, 23 May 1851, 104.
[52] Huxley, *Letters of Hooker* I, 473.
[53] Endersby, 'Botanist', 168.
[54] Hooker and Thomson, *Flora Indica* I, 7; RBG, JDH 2/9, Folder no. 1, pages unnumbered.

Hooker himself, dwelled on the Company's perceived failures. It decried the 'extremely limited' support the Company had given to the 'study of botany', for 'the keeping up of the various Seminaries, Horticultural Gardens and other Establishments, which promise return, more or less profitable, cannot be regarded as a patronage of the Science'.[55] The attack on resource imperialism in general and Company science in particular was intended to stress the almost moral obligation that the EIC had – at least in Hooker's opinion – to promote scientific work that did not solely promise financial profit.[56]

Despite Hooker's critique, the role of science for the running of the Company Raj was always more ambiguous. As David Arnold has argued: 'From time to time, when its economic and political interests were aroused, or the case for some scientific endeavour was convincingly made, science could command the Company's active attention, but for much of the period it was of secondary importance compared with the more pressing concerns of revenue, diplomacy, law and order.'[57] Many scientific pursuits in the early decades of the nineteenth century were carried out as a leisurely activity by EIC servants, outside government structures, precisely because the Court of Directors gave only erratic backing. This was even the case for suggested enterprises that would have met the Company's material interests – including Hooker's proposal for a comprehensive *Flora Indica*. Building on 'materials [that] have occupied nearly a century to amass' by the Company's 'own officers, at an enormous expenditure of money & of not a few valuable lives', it would have been a work of significant natural historical insight, but also with wider implications for plant acclimatisation and Indian agriculture.[58] In the end, Hooker would only complete a multi-volume account of India's plant life after the Company's suppression.[59]

In a letter to East India House in 1855, in which Hooker sought to secure the EIC's patronage for his ambitious project, he also introduced for the first time the previous employment of the three Schlagintweit brothers as a rhetorical tool to gently pressure the Company into supporting his own – and other British researchers' – scientific pursuits.[60] The naturalist in distress openly alluded to 'the liberality of the Court to the Schlagintweits', adding that 'no one rejoices at it more than I do, for

---

[55] Ibid.
[56] Brockway, *Science*, 28. For a critical reflection on the Company's 'profit motive', Desmond, *Discovery*, v.
[57] Arnold, *Science*, 25.
[58] JH to the Company director, Sykes, 3 November 1855, RBG, DC 102, 303.
[59] Hooker, *Flora*.
[60] JH to Sykes, 4 November 1855, RBG, DC 102, 306.

the sake of science & their own, as I am personally attached to them'. However, despite his supposed close ties to the German brothers, he continued that 'it is not a little mortifying to those who have worked hard and well, & spent hundreds of pounds in the service of the Company, to be denied any countenance, whilst comparative strangers are travelling with a carte blanche for unlimited credit on the local treasuries'.[61] This description of the German travellers captures a number of tropes that played a crucial role in later debates over the Schlagintweits' employment – above all their frequently emphasised foreignness, their grasping, greedy characters and the overly generous allocation of Company means to 'strangers', whose support seemed to be carried out to the disadvantage of longstanding, and equally deserving, British subjects.[62]

In sum, mid Victorian men of science like Joseph Hooker fiercely debated and tried to actively guide the precise directions their scientific disciplines should take, leading them to engage in 'boundary work' to distinguish their practices from those of mere amateurs both at home and in overseas locales. At the same time, they also remained highly sceptical about the respectability of pursuing science not merely as a vocation, but for one's own financial gain. Against recent claims that British men of science were united in their 'drive towards the dignity of professional status'[63], most of them still sought to live up to the ideal of disinterested gentleman naturalists, and did not see themselves 'as scientific tradesmen, much less as servants of centralised government, as were their French colleagues'.[64] Hooker, even in his later role as Kew director and therefore as a salaried government employee, shared this commitment, since 'the way in which he ... had put his personal expertise and resources at the nation's service, allowed Hooker to remain a gentleman'.[65] By contrast, in his and other British scientists' perceptions, the amount of money the foreign Schlagintweits not only spent on their travels and instruments, etc., but which they also received as personal salary from the Company over several years, made them look less like devoted and disinterested servants of Britain's needs and interests, but rather gave them the taint of being above all selfish poachers and mercenary opportunists.

---

[61] Ibid., 306.

[62] See also Finkelstein, 'Headless', 7.

[63] Gilmour, *Victorian Period*, 78. For perceptive reflections on the 'teleology of professionalization', Lucier, 'Professional'; Porter, 'Scientific Naturalism'.

[64] Endersby, *Imperial Nature*, 2. On the early trends of professionalisation of science in France since the Revolutionary period, and the German system, in which professionalisation occurred around the mid nineteenth century via the foundation of a growing number of specialised university chairs, see Allen, 'Professionals'.

[65] Endersby, 'Botanist', 168; Bellon, 'Ideals', 53.

Transnational Imperial Recruiting: A Typology

Whereas Joseph Hooker, Thomas Thomson and a host of other British men of science maintained a troubled relationship with the East India Company, this 'state within a state' acted, at the same time, as an important employer and patron for a range of naturalists and other occupational groups from Britain and the European mainland.[66] To be sure, this eagerness to recruit foreign experts to scientific and administrative positions within the British Empire was never confined to India. London's Royal Society enlisted the services of Robert Hermann Schomburgk, 'a German traveller in the mould of Humboldt', to explore British Guiana in the 1830s.[67] His brother, Richard Moritz Schomburgk, had accompanied him to Guiana and Brazil between 1840 and 1844, only to become (in 1865) the second director of the Adelaide Botanic Gardens in Australia and Secretary to the Board of Governors.[68] In Australia, Richard Schomburgk was in close contact with the highly decorated German-born botanist Sir Ferdinand von Mueller, FRS, who, after emigrating to Australia, accepted in 1852 the office of government botanist for the colony of Victoria, also undertaking exploration into the country's unknown interior. Between 1857 and 1873, Mueller then acted as director of the Botanical Gardens of Melbourne and not only greatly raised the scientific repute of that institution but also successfully worked to introduce an ambitious range of rare and useful plants into the country.[69]

While outside expertise thus got recruited into British possessions around the globe in the nineteenth century, the vast social and natural landscapes of India in particular created a heightened British demand for seeking out trained personnel from beyond the bounds of the English population. Well known are the crucial Irish, Welsh and Scottish contributions to the making of the British Empire.[70] The strong continental European dimension in 'British' India and its scientific services and activities is, by contrast, still much less explored. India thus provides a particularly useful case for reflections on the, to a degree, transnational nature of British imperialism in the nineteenth century. When approached from the perspective of the history of science, technology and medicine,

---

[66] Chaudhuri, *Trading World*, 20; on British naturalists in India and their harsh critique of the Company's insufficient support for purely scientific studies, see the private papers by Dr George Buist at RGS/CB4/279, e.g., from 14 November 1854.

[67] Drayton, *Nature's Government*, 144; on Schomburgk the ultimate study is Burnett, *Masters*.

[68] R. M. Schomburgk published a multi-volume travelogue, see, e.g., *Versuch*.

[69] Morris, 'Mueller'.

[70] From a large literature see Crosbie, *Networks*; MacKenzie and Devine (eds.), *Scotland*.

India was indeed far more exposed to non-British influences and connections than previous scholarship has acknowledged, as this section demonstrates. Therefore, in these fields at least, 'India was globalized through its imperial institutions and connections, not just colonized'.[71]

The pattern of drawing on foreign scientific expertise had occurred throughout the entire period of Company rule in India, yet earlier to a lesser extent and less systematically than at the time of the Schlagintweit brothers. Early figures encapsulating this phenomenon included the German-born Benjamin Heyne/Heine (1770–1819), who was first educated in Dresden, Saxony, only to become a surgeon with the Moravian Mission at Tranquebar in south India – a central site for many non-Britons to engage with Indian natural history.[72] Heyne subsequently found employment with the EIC, owing his appointment to the proposal of William Roxburgh that he 'attend to the plantation at Samulcottah' in 1793.[73] When the British botanist Roxburgh moved up to Bengal, Heyne took his place in 1794. Later, he rose to the position of assistant surgeon to the Madras Medical Establishment in 1796, becoming a full surgeon in 1807, and subsequently holding the office of the EIC's 'Surgeon and Naturalist on the Establishment of Fort St. George'. Another eighteenth-century naturalist taken into Company service was the Baltic German and pupil of the Swedish naturalist Carl von Linné (Linnaeus), Johan König (1728–85). He too had first been involved in the Danish Tranquebar mission before entering the service of the EIC as a surgeon and naturalist in 1778 on the recommendation of Joseph Banks. Banks was convinced of König's ability 'to repay his Employers a thousand Fold in matters of investment, by the discovery of new Drugs and Dying materials fit for the European market'.[74] Overall, a definitive figure remains difficult to attain, yet, my research suggests that at least several hundred German (and other continental European) naturalists, surgeons, advisers and other scientific and technical experts had been recruited into Company service by the mid nineteenth century.

Yet, the terms of their employment and the level of institutionalisation of their service could differ significantly, which indicates a guiding typology. While some non-British naturalists and travellers received permission to visit the Indian Empire, they did so without any specific government commission or Company salary. They usually received only unsystematic local support and reception from interested officers and

---

[71] Arnold, 'Contingent Colonialism'.
[72] See on Heyne, Berg, 'Useful Knowledge'; National Herbarium Nederland, 'Heyne'.
[73] Steenis-Krusemann, 'Heyne', 51.
[74] Quoted from Baber, *Empire*, 166.

other naturalists. This was, for instance, the case of the Austrian scientific traveller and collector Carl von Hügel, who, however, later received the Patron's Medal from the RGS in 1849 for his study of Kashmir.

There is a great ambiguity about two further groups of foreigners in British India, who sometimes maintained only lose connections with (or even complete detachment from) the formal structures of imperial power, and who sometimes got closely and wilfully drawn into the fabrics of British power: missionary orders and international humanitarian and philanthropic organisations.

As concerns the first category, members of *some* Christian missionary organisations did not enter into official relations with the EIC or the Raj, despite their often prolonged stay in the country.[75] After all, the Company had only reluctantly opened up its possession to proselytising efforts in 1813. Many missionaries, whether British, German, Swiss or American (whose presence in India was significant), were also keen naturalists and linguists, and thus added to western scientific engagement with India's natural worlds and religious and cultural diversity.[76] Western missionaries could also be responsible for economic impulses and technical transfers into India. Members of the Basel mission, for instance, introduced the fly-shuttle loom to weaving establishments in Madras, an important device to increase output. What is more, under the direction of the missionary and agronomist Sam Higginbottom, members of the American Presbyterian Mission established, in 1912, an Agricultural Institute in Allahabad, as they saw agricultural training and research 'as the best remedy against the perceived rural decay and poverty'.[77]

As previously suggested, however, there were also cases of missionary organisations that became conscious co-operators in the colonial apparatus. One striking example was the Christian Salvation Army, which closely collaborated with British authorities for the control and 'reform' of so-called 'criminal tribes' in the early twentieth century. With the sanction and financial support of government, the Protestant missionary organisation established numerous 'industrial homes', hospitals, leper and vagrants colonies, village schools and other reformatories throughout the subcontinent, as did the American Baptist Mission and the London Missionary Society. The aim was to transform social life in the colony

---

[75] Arnold detects different trajectories of outsider careers in India, Arnold, 'Contingent Colonialism'.
[76] Bray, 'Moravian Church'; Sharma, *Empire's Garden*; the best account of German-speaking missionaries in south Asia is Panayi, *Germans*.
[77] Fischer-Tiné, 'Rural Development', 201.

and, through various forms of social disciplining, to integrate members of the so-called 'criminal castes' and 'tribes' into industrial, 'civilised' society.[78] The work of missionaries in British India can thus be placed on a long continuum from distance to close integration into colonial structures.

Similarly, businessmen and philanthropic enterprises either operated consciously detached from government structures or tried to collaborate instead.[79] One example are the agents of the Rockefeller Foundation active in south Asia since the early twentieth century, who promoted anti-malarial campaigns and other large schemes of public health.[80] While the Rockefeller Foundation was never formally part of any government structure in India, it nonetheless had close relations to British authorities, with its ideology being seen as fiercely 'imperial'.[81]

Then there were those scientific travellers, like the Schlagintweit brothers, or the agricultural chemist John Augustus Voelcker (more on him below), who were formally recruited by the EIC or, after 1858, Raj officials and instructed to carry out specific scientific missions. For that purpose, those recruits usually received substantial scientific, administrative and political support. The limited period of their Indian service could in many cases prove crucial for future career paths, often in their home or neighbouring countries. A good example is the Munich-born naturalist Wilhelm Heinrich Waagen (1841–1900), who was struggling to secure a permanent university post in his native city. He subsequently accepted the offer to serve as a palaeontologist with the Geological Survey of India in 1870. Leaving the survey after five years, Waagen had by then accumulated a precious stock of firsthand knowledge and experiences of Indian natural history. He was able to use it for his subsequent university career in Prague and Vienna.[82]

Another German scientist who used British India as a stepping stone in his career was Georg von Liebig, son of the eminent German chemist and pioneer in agricultural chemistry, Justus von Liebig (1803–73). Liebig junior had trained as a medical doctor in Giessen and London, and entered the East India Company's service in 1853 – to the dismay of the British press since, 'to say the least, his qualifications were unknown'.[83] He soon became professor of natural history at the Hindu College in Calcutta from 1856 to 1858, before he returned after five years of Indian

[78] Tolen, 'Transforming the Criminal Tribesman'; Fischer-Tiné, 'Reclaiming Savages'.
[79] Dejung, 'Commodity Trading'; see also Ruprecht, 'De-Centering Humanitarianism'.
[80] Deb Roy, *Malarial Subjects*, 292–3.
[81] Fisher, 'Rockefeller Philanthropy'; Brown, 'Public Health in Imperialism'.
[82] 'Obituary of Wilhelm Heinrich Waagen', *Geological Magazine*, 7 (1900), 432.
[83] *The Athenaeum*, 1379, 1 April 1854, 408.

service to pursue his medical-scientific career in Bavaria.[84] Often, these more short-term employees maintained no sustained formal connection with the EIC or later the Raj. While they usually did not directly manage the cultivation and modification of colonial resources, many recruits in this group still collected information and useful knowledge on these subjects for the growing British archive on India's natural riches and populations.[85]

We can further distinguish those non-British recruits whose service in India was highly institutionalised and who held formal and long-term scientific and advisory offices in government structures. Their influence on colonial science was often pronounced and long lasting. Indeed, their activities could become central pillars of the British exploration, exploitation and rule of India and its natural and human resources. One such authoritative figure was Nathaniel Wallich, who significantly shaped British colonial botany in India during the first half of the nineteenth century.[86] The career of Wallich also indicated that in the late eighteenth and early nineteenth century, the British often turned to Scandinavian countries to recruit scientific personnel for the empire. That way they could satisfy a demand that Britain alone could not meet. One key reason for this was the international prestige of Linnaeus and his method of systematic botany, which subsequent generations of his pupils continued.[87] Incidentally, the early opposition of Wallich to the Schlagintweits' employment serves to demonstrate that imperial outsiders in British service did not necessarily form connections of mutual loyalty and support. Wallich's life rather shows how initial foreigners could, over time, fully assimilate into British and colonial society – in his case facilitated through marriage in 1815 to Sophia Collins, an English woman born in Bengal with family ties to Yorkshire.

As regards the early decades of the nineteenth century, it is salient to ask why there was such a marked shift from the influx of scientific expertise from Scandinavia towards a much greater presence of German-born scientific and technical practitioners and advisers in British India for the rest of century. Above all, academic developments changed the patterns of imperial recruitment. This applied in particular to the new and innovative forms of university teaching and research that evolved within the politically fragmented and scientifically competing landscape

[84] On Liebig in India, see Harrison, 'Medicine', 305.
[85] See on the importance, real and imagined, of vast stores of empirical data for British colonial officials in India, Appadurai, 'Number'.
[86] Arnold, 'Plant Capitalism'.
[87] Koerner, Linnaeus.

of the German states.[88] These modernising agendas of many German universities included a heightened focus on scientific research, which found its widely perceived and emulated expression in the establishment of the modern teaching laboratory by the chemist Justus von Liebig in Giessen in 1826. 'Nineteenth-century German universities became secular state institutions, and science instruction fulfilled a service function for the state in helping to train secondary-school teachers, physicians, pharmacists, bureaucrats, and other professionals.'[89]

Newly founded or reformed universities in the German states, with their particular scientific curricula, thus trained a new generation of proficient specialists, who then sought adequate positions in which they could apply their skills. A considerable number of them readily found employment within the medical sector and the nascent technological industries in Germany in the second half of the nineteenth century, especially in the fields of industrial chemistry, electrical engineering and precision optics.[90] Others, by contrast, looked for opportunities abroad. Given that no formal German overseas colonies existed until 1884 to absorb this scientific workforce, it was not surprising that some of these specialists turned to the established overseas dominions of other European powers. The Dutch East Indies and the British Empire in particular offered numerous openings for non-British employees.[91] By comparison, the French dominions appear decidedly more exclusivist and saw much smaller numbers of outsiders entering into higher imperial offices. Yet, a number of German travellers used French expansionist campaigns to explore extra-European territories, as was the case with Moritz Wagner, who visited and studied Algeria in the 1830s.[92]

The shift of perception regarding German expertise, and its increased value in the eyes of British administrators, had two important consequences. Not only were a growing number of British naturalists and doctors educated at German universities in the nineteenth century, a good many of whom later pursued scientific or medical careers in the Indian Empire.[93] German scientists' improved international standing was also reflected in expanding recruitment, already noted at the time.

---

[88] Daum, '*Wissenschaft*', 146.
[89] Dorn and McClellan, *Science*, 309.
[90] Ibid., 310.
[91] Kirchberger, *Aspekte*. The Dutch colonial establishment in Java also recruited the expertise of German *Forstwissenschaftler*; Furnivall, *Netherlands*, 201; empirically rich on the Prussian-born chemist/naturalist Caspar G. C. Reinwardt is Weber, *Ambitions*. On Germans in the VOC, Gelder, *Het Oost-Indisch avontuur*.
[92] Wagner, *Reisen*; Nyhart, 'Emigrants'.
[93] Biographical information on several such scientists can be found in Buckland, *Dictionary*. The prestige of German education at the time is also explored in Blackbourn, 'Germany'.

The French physician-scholar Alfred Maury wrote in the *Bulletin de la Société de géographie* in 1859:

The British are not the only ones exploring India. Scholarly Germany (*La savante Allemagne*), which appropriates for itself the discoveries of the French and the British in order to contribute to them, also supplies contingents of travellers, whose explorations are characterised by the same profundity and wisdom that characterises all of Germany's endeavours. When some great geographical problem needs to be resolved and all branches of science need to be summoned within a single exploration, the Germans are the ones who are called upon.[94]

There was another crucial reason to specifically recruit German scientists at this time. That is, their supposedly detached political stance was often considered an important precondition for their employment. The history of African explorers like Heinrich Barth and Adolf Overweg is a strong case in point. In the negotiations leading up to their appointment to a British-led expedition into central Africa in 1849, German officials stressed that these German travellers would not act as political agents with any 'national' agenda of their own.[95] As Bradley Naranch has noted, 'the rhetoric of scholarly expertise and the disavowal of direct political ambitions were part of a tactical arsenal that enabled individual [German] explorers to pursue their own personal ambitions of discovery' within the established overseas infrastructure of other powers.[96] A comparison with their French colleagues is telling. While solitary naturalists such as the Himalayan traveller Jacquemont might have passed through British territories, there was, by contrast, not a single Frenchman who held an office in the colonial establishment comparable to those held by Nathaniel Wallich or Dietrich Brandis. It was also improbable that a Frenchman would be appointed to lead a major scientific expedition of the British Empire, as Heinrich Barth did into central Africa, or the Schlagintweits across India and into central Asia.[97] However, French scientific and technical advancements and innovations did enter EIC circles, even if they 'had only indirect or intermittent influence on science, technology and medicine in British India, as, for instance, in the adoption of Alphonse Bertillon's system of identifying criminals through their physical attributes'.[98]

---

[94] Maury, 'Rapport', 58.
[95] Naranch, 'Beyond the Fatherland', 240.
[96] Ibid., 226.
[97] On British fears of Jacquemont being 'a spy of the French Government', see the anecdotal evidence in 'Victor Jacquemont', *Saturday Review*, 25, 25 January 1868, 119–120.
[98] Arnold, 'Contingent Colonialism'. On French advances in the field of bacteriology, especially by Louis Pasteur and his pupils, see also Chakrabarti, *Bacteriology*.

While it is evident that German scholarly and scientific engagement with India was never limited to or essentially defined by imperial concerns, a number of German experts did take up government office in India and ultimately had an important impact on the very practices of colonial governmentality. The history of western science and institution building in 'British' India thus possessed a strong European component. For instance, as Ravi Rajan and others have shown, German specialists were at the forefront of Indian scientific forestry.[99] Since the management of forests impacted on the life conditions of millions of wood-using peoples in the colony, the Indian Forestry Department was one of the most influential government departments in the realm of colonial science and was under German leadership for almost four decades. This pattern started with Dietrich Brandis, who served from 1864 until Wilhelm Schlich took over in 1881 as Inspector-General of Forests, only to give way in 1885 to the German Berthold Ribbentrop, who retired after fifteen years in that position.[100] The significant effect Brandis had on colonial governance is captured in the decisive role he played in bringing about the 1878 Indian Forest Act, which entailed a considerable increase in forest areas that were protected and monopolised by the imperial government.[101] The longstanding German leadership of the Forestry Department was, however, only the tip of the iceberg, as many more *Forstwissenschaftler* also filled middle-rank positions in that and other government departments.

Regarding the chronology of German expert recruitment into India, the influx of foreign personnel thus actually continued, and to some extent even increased, when the Great Indian Rebellion ended Company rule and authority was transferred to the Crown in 1858. This major political development sustained significant imperial transformations, and led to the further enlargement of the administrative structure of the colony. Several new government departments were created, among them the Indian Forestry Department in 1864, the Indian Meteorological Department in 1875 and the Botanical Survey of India in 1890. For the establishment or operation of the first two, German scientists played facilitating or indeed essential roles. Another government survey with considerable German participation was the Linguistic Survey of India,

---

[99] Rajan, *Modernizing Nature*.

[100] Both Brandis and Schlich were knighted for their services by Queen Victoria; Kirchberger, 'Naturwissenschaftler'; also Panayi, *Germans*, 100–1.

[101] On the many regulatory interventions of Brandis's work in Indian forestry management, see Saikia, *Forests*. On the EIC's early nineteenth-century practical experiments with timber control on India's west coast, and the temporary establishment of a Forestry Department (1808–23) in the context of an ambitious attempt to enlarge the Royal Navy, see Mann, *Forstbetrieb*.

set up in 1894.[102] In other words, the establishment of a formal German overseas empire in the 1880s by no means ended the transversal mobility of German scientific talent and expertise into foreign powers. As was the case with the Dutch East Indies in the later nineteenth century, where several German botanists and plant ecologists went to receive training and found opportunities to examine tropical nature at first hand in the superb infrastructure of the Buitenzorg laboratories on Java, so too did India maintain its attraction for German field scientists to engage with its diverse natural worlds and disease landscapes.[103] Indeed, India attracted wide interest from outsiders, not least since it was perceived as a source of danger to the health of other countries, as exemplified through several great cholera pandemics that originated in India, or the outbreak of plague in the 1890s. This led to different foreign government missions being sent to India to study the diseases' origins and treatments. Among the more notable was a Prussian-backed expedition that allowed the German bacteriologist Robert Koch to discover the hitherto unknown cholera bacillus in a water basin in Calcutta.[104]

These examples of trans-imperial mobility and the attendant transfer of scientific expertise and practices should lead us to rethink our assumptions of European empires as 'internally homogenous' and 'self-sufficient' and outwardly competitive entities.[105] Indeed, the nature and inner workings of European empires seemed to change in the second half of the nineteenth and the early twentieth centuries, when a highly itinerant group of trained experts came to play an increasingly important role in the formulation and implementation of imperial policies in the fields of science, technology and medicine in different imperial systems.[106] Crucially, those experts who sought to improve imperial statecraft and saw colonialism as a 'science' frequently founded new international institutions through which to exchange their skills and ideas.[107]

In this context, it is important to further note that some experts in the Indian service were not directly recruited from continental Europe, but came from the second generation of migrant families to Britain. This was, for instance, the case with the London-born John Augustus Voelcker

---

[102] Manjapra, *Entanglement*, 29.
[103] Cittadino, *Nature*, 2–3. The German government even offered yearly research grants for studies at Buitenzorg.
[104] On Koch and the slow acceptance of the contagion theory by the colonial government, Prakash, *Another Reason*, 136.
[105] Arnold, 'Contingent Colonialism'.
[106] Ibid.; important works in that regard are Mitchell, *Rule*; Hodge, *Triumph*; Tilley, *Living Laboratory*; Fischer-Tiné, 'Rural Development'.
[107] Daviron, 'Institute'; see for a critical look at the institutionalisation of inter-imperial knowledge exchange, Coghe, 'Learning'.

(1854–1937), an agronomist who excelled in another field in which many Germans took the lead in the late nineteenth and early twentieth centuries: agricultural chemistry.[108] In assessing the impact of Voelcker, who toured India between 1889 and 1891, and later compiled the highly influential *Report on the Improvement of Indian Agriculture* (1893) for the Government of India, it has been stated that his report 'became the basis of colonial policy for a few decades to come, and thus the impact of his particular interpretation of new chemistry and its relevance for India cannot be overestimated'.[109] Voelcker was the son of the agricultural chemist Augustus Voelcker (1822–84), who, born in Frankfurt in 1822, had moved to Edinburgh in 1847 and was soon afterwards appointed professor of chemistry at the Royal Agricultural College, Cirencester in 1849.

Finally, some German naturalists had first assisted the foreign empire in other locales before taking up formal employment in the Indian service. They thus brought previously acquired experiences and skills into the Raj. One example of such intra-imperial sojourning is given by the German botanist Gustav Mann (1836–1916). In 1859, William Hooker employed Mann at the Royal Botanic Gardens, Kew. Soon he was sent to West Africa to replace the British naturalist Charles Barter as botanist on the ill-fated Niger Expedition under the leadership of Captain Baikie. Mann made frequent stopovers that he used to collect specimens from the Canary Islands and Sierra Leone. He then reached the Gulf of Guinea, where he further amassed specimens that were sent back on naval ships to the British metropolis.[110] He subsequently joined the Indian Forest Service in 1864, soon becoming 'Assistant Conservator for British Sikkim', then the first conservator of the Forest of Assam in 1868, and later deputy conservator of forests. In these influential positions, Gustav Mann put his botanical expertise into practice amidst the tea and cinchona plantations at Darjeeling and Assam, always aiming at improving the commercial cultivation of cash crops in northern India. Among other things, Mann was one of the first to predict that the natural rubber resources in Assam would soon be exhausted. Through his authority as conservator of Assam, he ordered the large-scale plantation of rubber trees in 1873, which ten years later already encompassed an area of some 900 acres, and would increase still further in the following years. This was one of the first 'attempts at plantation of a commercially useful tree', and Mann suggested the indefinite yearly extension of the

---

[108] Bosma, *Plantation*, 146.
[109] Voelcker, *Report*; Kumar, *Indigo*, 124–6, quotation, 126.
[110] Cheek, 'Mann'.

plantation.[111] As these biographical sketches make abundantly clear, the Schlagintweits were only part of a much broader migratory movement of German sojourners and scientific specialists to the global British Empire, yet with India as a privileged destination.

The resulting considerable presence of German-born scientists, merchants, consuls, Company servants and university professors matters for our story, since their stay in India offered spaces for manoeuvring for the Schlagintweits and provided them with valuable connections on the ground. In 1856, for instance, they met Georg von Liebig, then 'Assay Master of the Mint at Calcutta', who provided them with scientific instruments, and also helped the brothers to take scientific observations in the area.[112] The scientific travellers also relied on commercial and logistical networks maintained by their countrymen in India for other purposes. The brothers thus engaged the Hamburg merchant and consul in Bombay, August Heinrich Huschke, for shipping their vast accumulated collections from India to Europe.[113] These informal German networks across British possessions in south Asia indeed provided the itinerant Schlagintweits with an alternative, non-British infrastructure – and channels for private communication, which the brothers eagerly seized for the exchange of information and goods with the German lands. As we will see later, far from being marginal to the controversies surrounding the brothers, they effectively used these alternative channels to engage in veritable double games with their British and German financiers.

Going beyond the realm of science, it is important to note that other sectors of the imperial labour market in British India were effectively closed to continental outsiders. A case in point is the Indian Civil Service (ICS), besides the army a central pillar of colonial statecraft in south Asia. Only 'natural born Subjects of Her Majesty' could enter into this elite corps of civil servants, over whose appointment the Court of Directors exercised a monopoly. Second-generation migrants to Britain or colonial India, however, could serve in it, while the question whether naturalised subjects, not least from the German-speaking world, were equally allowed into its ranks remained debated throughout the nineteenth century. With the introduction of competitive entry examinations into the ICS in 1853, Indian candidates could, in theory at least, now compete for entrance, even if they were in practice heavily discriminated against in the selection process. In other words, whereas British India's scientific and medical establishments saw numerous openings for trained

[111] Handique, *Policy*, 137.
[112] H. Schlagintweit et al., *Results* I, 147; II, 25.
[113] A., H. and R. Schlagintweit, *Report* I, 4.

non-British staff, including Indian surveyors, translators, guides and other assistants in imperial exploratory missions, the ICS bureaucracy upheld racial hierarchies and remained effectively in British hands until after the Great War.[114]

In sum, although the transnational recruitment of German scientists and advisers by the British imperial state was a common practice throughout the long nineteenth century (*c.* 1780–1914), it nonetheless led to resentment and *personal* competition between a number of British subjects and such privileged foreigners. As the analysis of Joseph Hooker's position has shown, this rivalry was often fought out over the limited means of the Company to staff high and lucrative offices, to finance scientific ventures in India and other British territories in Asia and to subsequently pay for their costly analysis and publication in Europe. This competition led to more general debates over the justice of employing non-British subjects. Such debates acquired xenophobic undertones, and set the toxic tone for the peculiar and long-lasting Schlagintweit polemic.

National Discourses and Senses of Entitlement

One such case of a foreigner's high-ranking employment was discussed in one of Britain's leading cultural-scientific journals, *The Athenaeum*, in 1857–8. After an initial criticism of the employment of a German naturalist who had been put in charge of Burma's forests was anonymously printed in late 1857, an unknown reader replied at length in defence of this individual. It is most likely that the replying author was the German naturalist himself, namely Dietrich Brandis, who had been appointed Commissioner of the Woods and Forests of the Pegu division in Burma in 1856.[115] Crucially, the *Athenaeum* editors directly linked the anonymously fought polemic with the Schlagintweit controversy, which had started to become a public issue around the same time. The Brandis debate already captures many of the tropes that were to be repeated regarding the brothers in the years to come. At first, the anonymous author (Brandis) restated the initial criticism against this case of imperial recruitment: 'I notice in your [earlier] issue ... the following questions from a Correspondent: – "Are there no botanists in the Indian service or in England? and further, if it be true that a German gentleman has recently been appointed a sort of Commissioner of Woods and Forests in Burmah?"' Against this background, Brandis continued by arguing and

---

[114] Banerjee, *Becoming Imperial Citizens*, 150–60.
[115] The reply to the initial letter was sent by *The Athenaeum*'s supposed 'correspondent' from 'Rangoon' in Burma, *The Athenaeum*, 1591, 24 April 1858, 531; quotations, ibid.

providing a rather elaborate self-defence that stressed the importance of the work of versatile and experienced 'experts':

To this I reply, no doubt there are many accomplished botanists in the Indian service and in England. But is a knowledge of botany all that is required for a Commissioner of Woods and Forests? Certainly not. If forests are to be worked so as to be profitable, a knowledge of the best method of girdling, felling, and dragging the timber to market; of dealing with foresters and lumberers; of controlling men of that wild class, in tracts remote from civilized haunts, and, finally, a knowledge of all the intricacies of forest management, which are only to be acquired by experience, are quite as necessary for a real working Commissioner of Woods and Forests as a knowledge of botany.

In this description of the various capacities and skills required for the post, many tropes of the 'rule of the expert' were echoed. Far more than an acquaintance with the science of botany alone, a real 'forest management' that was to be sustainable and commercially successful demanded an individual who possessed this vital embodied knowledge of how to administer these natural resources over longer periods of time. Lastly, to implement policies for such long-term forest administration, a high degree of state authority over these resources had to be established by the commissioner to prevent the ruthless exploitation of timber by private traders.[116] Hence, the ability to assert one's claims in dealings with the colonial authorities in order to codify scientific forestry in legal arrangements was another crucial requirement. In view of this combined need for scientific and managerial capacities, Brandis further claimed:

I believe I may say, without disparagement to the many eminent botanists in the Indian service, that not one of them has had the opportunity of acquiring such knowledge as referred to. I doubt if any one in England has … In the mean time, a German gentleman of high attainments has been appointed to the charge of the forests in Burmah, because he was not only a man of eminence as a botanist, but thoroughly acquainted with the working and management of forests in Europe. I trust the editor of the *Athenaeum* is not, under these circumstances, disposed to object to the 'right man' being put in 'the right place' in a British possession, even though the right man be a foreigner.[117]

The reply that this letter provoked, however, was just as telling. Its anonymous author, who had started the argument in the first place, nonetheless adopted the 'we' form. This was intended to suggest that his individual position would represent those of all 'Englishmen', a supposedly homogenous national community to which he frequently referred, and as

---

[116] These 'foresters and lumberers' were the private merchants with whom Brandis had to engage in long-standing conflicts over the use of the Burma forests. Pinchot, 'Brandis'.
[117] *The Athenaeum*, 24 April 1858, 531.

whose defender he pretended to speak. This juxtaposition between 'we' and 'the foreigner' was further reinforced, as the replying critic made it clear that he suspected Brandis of having composed the anonymous defence.[118] The strong national sense of entitlement he felt British foresters had was reflected in the highly sarcastic language he employed:

It will be new to some hundred of land stewards and English country gentlemen to learn that there is no one in England who understands the best methods of girdling, felling, and dealing with timber; – new, also, to Anglo-Indians to find that only in the forests of Germany can be acquired the power of controlling men of the wild class.[119]

Driven by xenophobic sentiments, the critic again focused mainly on the national origin of the appointed Commissioner, and used a rhetoric that seemed to imply such foreign recruitments were a corruption that needed to be brought to public attention:

Thanks to the *Athenaeum*, such appointments are not likely to be repeated, and if they be, are sure to be exposed. *We* do not object to see the right man in the right place, even if he is a foreigner. But Englishmen have a right to complain when they see foreigners of no higher qualifications than their own appointed to offices, which are their *birthright as Englishmen*, and should be their rewards as men of science.[120]

Hence, whereas some British contemporaries regarded the co-optation of expertise from beyond the realms of the national population as justified, especially when it increased the profitability of the empire's possessions, others were adamant that it was not. Those relied on a language of exclusion, even if such rhetorical nationalisation of empire by no means succeeded in the long run. Rather, Dietrich Brandis, in this case, first proved his worth in Burma, only to become the most influential scientific forester in south Asia in the second half of the century. However, in the existing literature on Brandis's internationally significant career, this initial hostility has been entirely overlooked.[121] Yet, to better understand the later Schlagintweit controversy and its very different nature and trajectory, it is important to note that Brandis's employment was attacked solely on the grounds of his nationality. Once his contributions were considered for the management of Burma's and subsequently British India's forest resources, the hitherto outspoken critic became

---

[118] *The Athenaeum*, 1592, 1 May 1858, 564.
[119] Ibid.
[120] Ibid., emphasis mine.
[121] Generally, the focus is predominantly on his impact on shaping scientific forestry practices literally around the globe, from Asia to Europe and the United States of America; see, e.g., Schenck, *Birth*.

silent. Given the ultimate value of Brandis's work for colonial interests, he was later elevated to the highest echelons of the symbolic hierarchies of the British Empire. In 1875, he was elected Fellow of the Royal Society and received 'The Most Eminent Order of the Indian Empire', first as CIE in 1878, and later as KCIE in 1887. Brandis's career demonstrates that there always existed different degrees of marginality and outsider status for continental Europeans in British employ, while this status of being perceived as an intruding foreigner could change considerably over time.

In conclusion, when we view the fleeting Brandis polemic against the background of the many other German scientists recruited into British service around the time, who usually served without provoking such strong public disputes, certain recurrent patterns emerge. First, given an imperial labour market open to external influx, a great number of foreign scientists, surveyors and administrators were appointed to serve in rather inconspicuous positions, as captured in the many middle-rank positions filled by German scientific and technical practitioners in ongoing and narrowly defined Indian surveys and government departments. These offices were not highly regarded, and did not promise lucrative salaries or social prestige. Many of those foreign, often temporary, employments went largely unnoticed by the British public. While a degree of personal rivalry was ubiquitous between British men of science and such foreigners, relatively rarely were their modest recruitments followed by sustained public debate. Second, there nonetheless existed a widely shared attitude in Britain and the Indian Empire that *generally* objected to the favouring of non-British personnel over national subjects. Third, outspoken criticism of appointing foreign experts emerged especially with regard to prestigious offices. This is unsurprising, as such positions promised a greater amount of public authority and – crucially in view of the still fragile position of many scientific practitioners in British and Indian society – an increased remuneration that could be vital to secure existence for those without private wealth. Fourth, however passionately this resentment against foreign employment was expressed, it often came in a depersonalised form, and served to oppose the co-optation of outsiders to exalted and lucrative positions at large.

The Schlagintweits, by contrast, faced a particularly ferocious critique not only of their initial employment, portrayed – with some justification – as an act of 'great injustice to a class of officers' in India.[122] *The Athenaeum* indeed started to fire against their scheme even before

---

[122] 'Our Weekly Gossip', *The Athenaeum*, 1378, 25 March 1854, 376.

it had really started. While its editors 'hope[d] that there is yet time for reconsideration', they also claimed that 'we have a right to protest against foreign diplomatic influence being brought to bear for the purpose of forcing strangers over the heads of Englishmen more distinguished for their attainments than the new comers'. Yet the public debate over the Schlagintweit mission went further and would later also focus on their comportment as men of science, and the ultimate results they proffered to the empire. A detailed study of their contentious careers can thus provide us with more than the finding that strong patriotic feelings and a heightened sense of entitlement existed in Britain and India – and especially so after the Anglo-German antagonism in the Crimean War.[123] The various elements of the Schlagintweit controversy and its unfolding were not the simple outcome of British xenophobic tendencies – even though contemporaries including Humboldt assumed this to be the case.[124] Rather, the various layers of the fierce controversy require a careful contextual reconstruction that includes attention to the brothers' clever engagement with multiple sponsors and opponents and the latters' respective interests in either supporting or undermining this transnational scheme of expeditionary science.

---

[123] Humboldt to Bunsen, 20 February 1854, in Schwarz (ed.), *Briefe*, 170–8; also Finkelstein, 'Mission', 201.
[124] Humboldt to Bunsen, 19 August 1855, in Schwarz (ed.), *Briefe*, 191; also Päßler, 'Introduction', in *Briefwechsel*, 23.

3   An Ingenious Management of
Patronage Communities

This chapter demonstrates that although xenophobic discourses, personal competition and outright jealousy were at play in the mid-century reception of German itinerant naturalists in British overseas territories, it was the ambiguous behaviour of the Schlagintweit brothers themselves that fuelled the conflict over their employment. The scheming and self-serving Schlagintweits provide a gripping case to reveal the intricacies of science management from the 'field'. This concerns not only the specific modes in which the brothers, from early on, sought to represent themselves as dominant figures in the landscapes across which they moved, as reflected in their compilation of scientific reports assembled on the march and intended for international audiences. The at times cunning activities of the brothers in India also applied to the numerous double games the travellers engaged in to maximise the benefits from the co-financed Anglo-German expedition. Their enterprise thus testifies to both the opportunities and the fragile position 'in between' that imperial sojourners faced. Particular focus is given to the Schlagintweits' ingenious manipulation of the information orders of European patronage networks, especially in Paris, London and Prussia, which, once revealed, would become another significant line of critique. The increasingly violent polemic surrounding their employment was thus linked with, in British eyes, their perceived failure to conduct the mission according to the respectable norms of gentlemanly science. To better comprehend why their work became so ambiguously received across Britain and western Europe, a reception which would alternate between hostility and hagiography, we first need a more nuanced understanding of the plurality of expectations diverse groups developed before and during their Indian mission – expectations that were to a considerable extent spurred on by the brothers themselves.

Great Expectations

Even before the three brothers had set foot on Indian territory in late October 1854, the pan-European expectations about the scientific

90

outcome of their journey were considerable. Given Alexander von Humboldt's active encouragement and heightened belief in the abilities of his young protégés, the trio embarked on the Indian mission carrying with them not only Humboldt's letters of recommendation and scientific instructions, but also – and above all – high expectations from various quarters.[1] Since 1850, Humboldt had shared his appreciation of the Schlagintweits with numerous German, French and British benefactors and men of science, ranging from the Bavarian king Maximilian II and the British scientist Michael Faraday to the Scottish statesman, historian and governor of Bombay, Mountstuart Elphinstone. Yet, to thrive under Humboldt's patronage was both a blessing and a curse, since his personal expectations would prove difficult to fulfil. The legendary traveller hoped that the Schlagintweits would succeed in a delicate balancing act between a wide-ranging engagement with the natural worlds of India and the Himalayas, grounded in studies in a number of disciplines, and, at the same time, the achievement of a rigid scientific thoroughness. Humboldt expressed these contradictory expectations to the Prussian envoy Bunsen in London: 'Since Saussure, no scientific work has appeared that so generally reflected the progress in *all* the sciences. Much *thorough work* is to be expected from these industrious, well-educated and modest young scientific travellers, to whom you have given your patronage.'[2] Hence, in projecting his own hopes and his earlier praise of the Schlagintweits' interdisciplinary Alpine achievements onto the upcoming Himalayan expedition, Humboldt made clear that nothing less should now be accomplished regarding this fantastically complex mountain chain.

Nothing proves better that Humboldt drew close parallels between his own earlier overseas expeditions and the Schlagintweit voyage to Asia than a personal letter, sent two weeks before their embarkation. It made clear that while he did not expect to see them again, he still regarded their scientific mission as one of the greatest projects he had ever helped to initiate. Humboldt thus symbolically passed on the torch of German overseas exploration to his disciples:

I did not have time this night, during which I wrote four warm and ingenious letters for you, my dear, amicable friends, to give you a word of love, of remembrance, of inner regard, and of eternal farewell. Of all things, to which I have contributed, it is your expedition that has remained one of the most important. The latter will still delight me when I will die. You will enjoy what between the

---

[1] See, e.g., one of Humboldt's letters of introduction, *JASB*, 24 (1856), 184.
[2] Humboldt to Bunsen, 26 February 1854, GStAPK, FA Bunsen, Rep. 92 Dep. K. J. v. Bunsen, my emphases.

return from Mexico and the Siberian travel constantly also occupied my own imagination. May you fare well.[3]

Apart from Humboldt and Carl Ritter, whose works on Asia the Schlagintweits expressly set out to both verify and complement, other leading geographers placed similarly high hopes in the brothers' project.[4] With Berlin and Gotha acting as leading centres for the production of maps and geographic handbooks at the time, a number of cartographic experts expressed their desire to receive scientific findings of precision and quality that would contribute to greatly improving their depictions of this world region. Among them was Heinrich Kiepert (1818–99), one of the 'most distinguished' geographers of his age.[5] Kiepert worked in Berlin and collaborated with Carl Ritter and Humboldt as a cartographer, before taking up the chair in geography at the University of Berlin in 1859. He could draw on all sorts of geographic data for his work supplied to him by colleagues and informants across the European empires, and also by 'the British, Russian, and French War Offices'.[6] Kiepert's oeuvre, which showed how German cartographers were extremely well integrated into trans-imperial knowledge networks, reached international readers. Among his numerous and positively received works was the *New Hand-Atlas of all Parts of the Globe*, which included some of the most precise maps of Europe, Africa, Australia, Russia, Asia Minor and central Asia produced at the time.[7] This work was soon translated into other languages, and was celebrated for 'its accuracy, clearness, fullness, and cheapness'.[8]

To be sure, western geographical knowledge was then still patchy about the interiors of many non-European continents.[9] In Kiepert's explanatory section detailing the sources of his map of central Asia, he complained about the highly confusing spelling of many topographical names provided by British travellers. He further pointed to the fact that itinerant naturalists had only recently begun to pay closer

---

[3] These 'listige Briefe' were further letters of recommendation, which opened many doors in India's imperial and scientific establishment. Humboldt to HS and AS, Berlin, 4 September 1854, Stiftung Stadtmuseum, Berlin [SSMB], HU 99/62 QA.

[4] Adolph once wrote to Ritter that the more he studied his 'comprehensive and excellent work on Asia ... the more it commands one's sincere admiration. If our Indian travel should actually come about ... we will enjoy the priceless gift of having in your work a firm (though unattainable) model for our own researches.' SBB, NR, 26 November 1853, 1.

[5] 'Obituary for Heinrich Kiepert', *Geographical Journal*, 13 (1899), 667–8.

[6] Ibid.

[7] Kiepert, *Handatlas*.

[8] 'Chronicles of Science', *Quarterly Journal of Science*, 4 (1867), 407.

[9] Kennedy, *Spaces*.

attention to indigenous names and the correct labelling of geographical phenomena and human settlements in those regions.[10] In his widely circulated *Atlas*, which went through several editions, Kiepert's passage on Asia concluded with the following hopes: 'Further advancements on this matter [of Indian terminology], as well as regarding the more precise descriptions of physical-geographical facts, both for Tibetan and Indian territories … are soon expected from the publications of the brothers Schlagintweit.'[11] The brothers would indeed devote considerable attention to these terminological questions during their fieldwork, gathering, besides much locational data, a wealth of indigenous terms and their meanings.[12] The Gotha-based cartographer August Petermann similarly anticipated important results from them, whom he regarded as great scientific explorers, not mere surveyors in the Company service. He wrote to the EIC director Sykes two years into the mission:

These travellers have already, in so short a time, overlaid your Indian Empire with a net of their routes and lines of manifold observations, which will afford a new and more complete view of that country than we have hitherto possessed. I only hope that they may yet be enabled to push their investigations into that great and so little visited and known Central Region of the Himalayas of Nipal [sic], or the equally unknown Eastern wing … [that] interesting grand world East and West from [Sikkim] is as yet almost a sealed book to us.[13]

Yet only the most detailed observations allowed cartographers, who rarely visited overseas regions themselves, to authoritatively depict unseen landscapes on maps. A belief in the thoroughness, and hence trustworthiness, of the data provided by itinerant naturalists was therefore crucial in the constant decision-making of sedentary mapmakers, not least for the question of what sources to consider as erroneous and what as reliable.[14] Indeed, as many recent studies in the history and philosophy of science have forcefully shown, 'scientific knowledge is as secure as it is taken to be, and it is held massively on trust. The recognition of trustworthy persons', and hence their perceived authority, is, in Steven Shapin's words, 'a necessary component in building and maintaining systems of knowledge, while the bases of that trustworthiness are historically and contextually variable.'[15]

---

[10] Kiepert, *Handatlas*, 13.
[11] Ibid., 14.
[12] H. Schlagintweit et al., *Glossary*, in *Results* III, 133–293.
[13] Petermann to Sykes, 9 April 1856, Forschungsbibliothek Gotha [FBG], SPA, ARCH PGM 353/1, 48.
[14] Tammiksaar et al., 'Hypothesis'; on the struggle over scientific authority between scientific travellers and synergising geographers, Outram, 'Spaces'.
[15] Shapin, 'Sociology', 302–3.

Moreover, French naturalists had far-reaching expectations about the Schlagintweits' expeditionary project, since the brothers had generated considerable interest among scientific circles in Paris. Several visits to the Académie des sciences, the delivery of lectures and the maintenance of personal and scientific ties with a number of French geographers had forcefully established their names in the capital's academic circles.[16] Adolph also frequented Paris before their departure in order 'to discuss with the members of the Academy different aspects of our observations, and to have made some instruments for our travel'.[17] While the brothers met eminent naturalists and administrators, they also personally informed the French emperor, Napoleon III, about their upcoming expedition and presented several gifts to him to spark further interest – among them their Alpine treatises and atlases, plus the stunning cast mountain reliefs in a wooden box.[18]

The lofty expectations of French men of science are captured in a notification given by the influential geographer Alexandre de la Roquette to the Geographical Society of Paris in 1854.[19] After recounting the great value of Adolph and Hermann's Alpine investigations, he announced the upcoming eastern journey, stating that 'the society learns not without a lively interest that, on the pressing recommendation of our former illustrious president, Mr Baron Alexander von Humboldt, patriarch of the geographical sciences, these two German savants have recently been appointed, together with their third brother Robert, on a scientific mission to the oriental Indies, and particularly into the Himalayas', lasting 'three or four years'. In detailing their equipment, the financial arrangements and the ultimate scientific objectives of the expedition, Roquette continued that French academics would now expect significant results in 'the geological, meteorological and geographical sciences.'

Roquette made these grandiose statements because the brothers had informed their French peers about the precise objectives of their voyage only *after* the Company had granted their proposals for a significantly enlarged programme. From the start, their self-fashioning as great scientific explorers clearly fell on more fertile ground on the continent than in Britain. Roquette then singled out their prestigious patrons besides Humboldt: the East India Company, 'the enlightened protector and patron of scientific enterprises', and the EIC director Colonel William

---

[16] Adolph and Hermann's Académie lectures focused on their hypsometrical researches; 'Mémoires Présentés', *CRAS*, 35 (1852), 17 and 102.

[17] AS to the Graf von Hatzfeldt, Paris, 13 August 1854, GStAPK, I. HA Rep. 81, GuK, 5.

[18] Note by Mr Feuillet, Paris, 8 January 1855, ibid.

[19] Roquette, 'Note', 229–32. Subsequent quotations, ibid.

Sykes, who had offered the brothers 'his devoted assistance'. Roquette concluded that '[w]e must also place high hopes in the success of this enterprise, when we see by whom it will be executed, and who are acting as patrons and guides of the expedition'.[20]

Yet, what appears on the surface as an objective statement about the credentials of the Schlagintweit enterprise was in reality an orchestrated manoeuvre. That is, Roquette's public announcement had been the brothers' idea from the start. From London, on 16 September 1854, Adolph had specifically written: 'We would be very obliged, to tell the truth, if you would pay us the honour of devoting a little article about our voyage to the excellent journal of the Société géographique.'[21] He continued that it would be most helpful if not only 'the name of Colonel Will. Sykes, one of the East India Company directors who takes the greatest interest in all scientific observations about India', could be mentioned therein. Adolph also requested that Roquette personally write to Sykes with details about this commissioned notification. The reason was that 'it would be of some importance for us if Col. Sykes could see that people in France take some interest in our researches'. The whole manoeuvre was intended to titillate the Company's interest to further sponsor the Schlagintweits' researches, and thereby to promote its own image as an internationally acclaimed patron of the sciences. By thus carefully controlling what information about their employment reached whom, and when, the brothers were able to ignite considerable curiosity and anticipation about their Himalayan travels across Europe.

Yet, while many German and French metropolitan scientists considered the Schlagintweits as the right men in the right place, some of their British peers were, for more than one reason, sceptical. In particular, the perceptions of some of the most influential British geographers and naturalists, hence not of Company men, greatly differed from the brothers' self-representation as scientific polymaths. Joseph Hooker, the Schlagintweits' most eminent predecessor as an itinerant naturalist in the Himalayas and a central figure in the controversy, became wary about the mission's outcome. Arguably to avert its potential failure, and to prevent a considerable sum of Company grants from being misspent, Hooker sought to obtain more precise information about the exact terms of employment. For this purpose he addressed the geologist Roderick Murchison. Hooker made explicit that one more capable naturalist and/

[20] Ibid.
[21] AS to Roquette, Bibliothèque Sainte-Geneviève, Paris [BiSG], Ms. 3611, 263; subsequent quotations, ibid.

or geologist should accompany the brothers' Indian mission.[22] In his reply, Murchison agreed with Hooker's recommendation and stated that '[n]ow neither of the Schlagintweits (clever physicists as they are), are Naturalists or geologists & such a [companion?] would be desirable' – despite the fact that Adolph was, in fact, a university-trained geologist.[23]

To compensate for the brothers' perceived lack of natural-historical and geological prowess, Murchison recommended a personal acquaintance, 'Dr. Rutimeyer, Professor of Cpt. Anatomy[24] in Berne, whom I cite so much in my "Alps Apennines &c". He is a very clever, methodical good palaeontologist & a capital observer of rocks & stratification: in short of geological phenomena – moreover he is young and strong'.[25] However, given the fact that Rütimeyer was also of German-speaking origin, Murchison further stated: 'On the other hand I find a feeling beginning to prevail against employing Germans in which I do not participate' – especially when no better-qualified Britons were available.[26] Why Hooker and Murchison questioned the Schlagintweits' scientific qualifications as geologists and naturalists remains unclear. Murchison's shift of opinion is especially striking, since his earlier assessment of the brothers' Alpine studies had been largely positive, as in his 'Address at the Anniversary Meeting' of the RGS in 1852: 'The brothers Schlagintweit, who, belonging to the active and stirring school of Prussian geographers, are worthy pupils of Humboldt and Ritter, and have already distinguished themselves by their observations on the heights, climate, springs and glaciers of the Alps.'[27]

This striking sense of equivocality in Murchison's private versus public judgements about the Schlagintweits needs to be understood in the context of the polite scientific culture of the time. Whereas he confidentially questioned their abilities in several fields, his public position as RGS president required from him an open statement about Adolph's work, which he had received as a gift. What is more, he had long been engaged in a close correspondence with their mentor Humboldt, whom Murchison admired and had first met in Paris in 1830.[28] Yet, the fact

---

[22] Although Hooker's letter is lost, we can reconstruct its content using the immediate detailed reply from Murchison, 19 January 1854, RBG, DC 96.

[23] Ibid.

[24] That is: Ludwig Rütimeyer; comparative anatomy.

[25] Murchison to Hooker, 19 January 1854; see on Murchison's own interest in stratification, over which he himself engaged in a geological dispute lasting sixty years, Secord, *Controversy*.

[26] Murchison to Hooker, 19 January 1854.

[27] Murchison, RGS 'Address', 1852, xcviii.

[28] Humboldt, as Murchison knew, closely followed British statements about his protégés; Humboldt to Murchison, 16 August 1853, Edinburgh University Library [EUL], copy at BBAW; also Stafford, *Scientist*, 9.

that Murchison perceived of them as only 'clever physicists' proves that some of Britain's leading men of science regarded the Schlagintweits as capable only of effectively conducting India's geomagnetic survey. The brothers had indeed never been trained in a botanical garden, neither on the most advanced classificatory practices for new specimens nor on the adequate preparation of a herbarium. No specialist in zoological or ethnographic researches had ever instructed them, and the same applied to several other fields of enquiry they chose to engage with during the eastern mission. They thus embarked on the interdisciplinary enterprise as partly unskilled observers and amateur collectors – which in turn would later complicate their cooperation with European specialists charged to analyse their specimens and prepare them for publication. The importance of careful preparation, cataloguing and labelling of specimens in the field can hardly be overestimated in order to make such objects useful to science, and formed, according to the instructional literature for travellers, a core activity in the work of expeditionary science.[29]

Ultimately, Murchison and Hooker's plan to secure a scientific companion for the Schlagintweits failed. Hooker, in a further attempt at damage control, agreed to at least oversee the Schlagintweits' preparations in an official capacity. He thus sat on a 'sub-committee', formed by the Council of the Royal Society, which consisted of him (in his capacity as botanical adviser), the administrator, military engineer and magnetic crusader Edward Sabine and the great naturalist and evolutionary scientist Charles Darwin, 'to whom the consideration of Mr. de Schlagintweit's proposed operations [were] referred'.[30] Given their considerable enlargement of the mission, the Royal Society had set up this committee in March 1854, which the brothers could address when seeking scientific instructions on any of the 'departments of physical science'.[31]

When Hooker received Adolph Schlagintweit's 'list of proposed operations' and was asked by Sabine to comment on the feasibility of the scheme, his reaction was unambiguous: 'I have carefully gone over Schlagintweit's paper which contains a programme of at least 8 years work for himself, his brother & a staff of assistants, & which will require a much greater outlay than the E.I. Company will probably be prepared to allow, both for instruments & travelling schemes.'[32] Hooker was not only critical about the interdisciplinary scope of the programme, but

---

[29] See the contributions in Bossi and Greppi (eds.), *Viaggi*.
[30] Sabine, *The Athenaeum*, 1767, 7 September 1861, 319–20, 320.
[31] Ibid. Finkelstein, 'Mission', 189.
[32] JH to Sabine, NA, BJ 3/53, unknown date, but before 24 May 1854.

also cautioned against the planned itinerary. The brothers 'could no more than [wander?] over the country route checked out, had they no encumbrances of any kind'.[33] The difficulties of realising such a grand scheme 'would I should think require a camp of at least 300 persons'. Hooker then quoted Thomas Thomson, 'who has had 12 years Indian experience', saying that the latter would 'entirely [agree] with me, in considering the whole scheme much too comprehensive & costly to approve even provisionally, without much consideration'. To prevent an ill-devised spending spree, he lastly suggested seeing Schlagintweit 'at any rate as soon as possible', arguably to convince Adolph to greatly reduce the brothers' ambitions.[34] Before departure, the Schlagintweits, however, snubbed Hooker's invitation to meet, claiming that 'it was quite impossible for us to spare an hour for the last days'.[35] In the same letter, Hermann nonetheless did not refrain from informing the rebuffed colleague that 'I should be very much obliged to you if you might have the great kindness of sending us a few lines for Dr Campbell and for Mr Hodgson at Darjeeling' – to secure the vital support of these British naturalists in India.

While in London, the brothers also chose not to seek the guidance of the more experienced overseas traveller, Charles Darwin, on the careful amassing of natural historical specimens. Darwin did, in fact, ask the brothers to make observations on the yaks in India and central Asia. Later, he would, however, be bitterly disappointed by their observations and consequently write to the eminent geologist Charles Lyell: 'Do not trust Sclangenweit [sic] (the Indian Brothers or some such name) about Yaks, if you come across their statement.'[36] One of the greatest scientific minds of the nineteenth century clearly did not trust the brothers' ability to act as reliable observers for metropolitan science.

Only with regard to geomagnetic surveying did the Schlagintweits accept the British expertise on offer. Consequently, they were provided with the compulsory instruments, in the use of which the brothers were specially instructed at Kew Observatory.[37] Edward Sabine trained them personally in September 1854: 'I was glad to be able, on my part, to render them the same assistance in the preparation of their magnetical

[33] Ibid.
[34] Ibid.
[35] HS to JH, 16 September 1854, from East India House, London, University of Georgia Libraries [UGL].
[36] RS to Darwin, London, 25 September 1857, Darwin Correspondence Project, letter 2142; to Lyell, 3 October 1860, ibid., letter 2935.
[37] John Welsh, superintendent of Kew Observatory, also prepared the brothers; Welsh to HS, 12 September 1854, BSB, SLGA, II.1.5.

instruments that I had previously given to Capt. Elliot; and to assist in discussing with them the observations most important to be made.'[38] Elliot's former instruments were even sent from India to Britain and were tuned up and compared for the mission. The Schlagintweits also had 'the advantage of profiting repeatedly by the personal advice of ... Lloyd, and Lamont, so well known from their theoretical and practical labours in the science of terrestrial magnetism'.[39] Evidently, British men of science strove to ensure that at least their magnetic survey would be carried out correctly and conclusively.

When taking all the contradictory private and public statements into account, it is striking just how contested the Schlagintweits' scientific authority was even before the Indian mission. Humboldt and other German and French naturalists and cartographers, as well as Company men such as Sykes and other supporters in the Court of Directors (especially their supposed 'close friends ... Eastwick, Cautley, Mangles, Rawlinson'), were confident of the Schlagintweits' broad potential.[40] By contrast, British and German experts within *one* particular field of study, like Hooker in the realm of 'philosophical botany' or Weiss in the field of mineralogy, harboured a good deal of mistrust about the brothers' holistic aspirations. For them, the Humboldtian programme they set out to realise in India and central Asia was not a laudable attempt to grasp their natural worlds from an interdisciplinary point of view; rather, it seemed to be a harbinger of failure.

Science Management from Afar

As much as the Schlagintweits' multiple patrons and financiers had from the outset different and not easily reconciled expectations, their position 'in between', with affiliations both to British authorities and their Prussian benefactors, also opened up unexpected opportunities. Their at times fragile position as imperial outsiders within the colonial establishment of British India, with a foreign background and multiple loyalties, could also be turned to their advantage. The brothers regularly succeeded in playing both sides, and this during and after their stay in south and central Asia from October 1854 to the summer of 1857. The aim in this section is to unpeel yet another source of criticism that

---

[38] Sabine, *The Athenaeum*, 7 September 1861, 320.
[39] APS, PMP, v.1196, no. 5, 8. They refer to Johann von Lamont (1805–79) and Humphrey Lloyd (1808–81).
[40] The brothers inserted these names into a translation they had made about their travels for Roquette; BiSG, 3611, 26, 2.

originated in the Schlagintweits' ability to profit from the lack of *direct* communication between their multiple patrons. This increased their space for manoeuvring, especially when they asked different sponsors to increase their allowances to cover the costs of a greatly expanded mission.

Close attention will be given to the Schlagintweits' ingenious communication strategies. The intensity of public disagreement over their scheme can be partly explained by the systematic exclusion of British metropolitan scientists from the brothers' private and scientific correspondence. Adolph and Hermann always maintained close contacts with a small number of powerful advocates within the East India Company who were mobilised to act on their behalf in financial and organisational matters. In comparison, the lack of communication with their scientific peers, and especially rival travellers in India and the Himalayas, is remarkable. Much of their correspondence during and after their journeys – particularly concerning financial agreements – was only privately addressed to Company men, intended to keep critical voices out of the conversation. The quest for EIC assistance in matters of funding, career planning and collecting was almost exclusively arranged through such private channels. It is thus crucial to distinguish between those personal statements that often reminded the addressees of the secrecy of the content from those that were addressed to a broader audience. However, while the brothers sought to orchestrate precisely what information should be kept confidential and what was to be shared with the wider public, the decision of what information could be passed on to others ultimately lay in the hands of the recipients. Indeed, in some instances, the secret plans and negotiations of the Schlagintweits were deliberately subverted and revealed by their British correspondents, which in turn led to considerable friction in their relationship with the scientific establishment in London.

Despite such setbacks, it is worthwhile to ask how the brothers sought to use their correspondences to achieve specific aims. Doing so sheds fresh light on the social practices of science in this period. Large surviving sets of letters make clear who was, or became, eventually excluded from their inner circle of correspondents. Above all, Joseph Hooker, later the brothers' prime opponent, was gradually omitted from their circle and therefore only learnt about their publication schemes and financial allowances at second hand. Against Hooker's understanding of the code of honour in science, the brothers never asked for his advice on their findings from India. This can be revealingly contrasted with the brothers' earlier eagerness to discuss their *Alpine* findings with him, as at that time no competition over British means or scientific authority existed between them.[41]

Among several naturalists in Britain, the Schlagintweits soon acquired the unflattering reputation of being obtrusive, greedy and ungentlemanly. The art of communication (or a lack thereof) played a much greater role in the Schlagintweit controversy than even contemporary commentators realised. Letter writing was an important tool of self-fashioning for any man of science, and a crucial medium for the formation of alliances within and across the boundaries of national scientific communities. It was an art mastered by Humboldt, who relied on his correspondence skills to build up and use a wide-ranging system of patronage and informants while still being considered an archetype of the gentleman naturalist.[42] The pragmatism with which the Schlagintweits broke the conventions of this polite republic of letters was, by contrast, not well received.[43] Instead of consulting and thus acknowledging the expertise of those travellers who had crossed Indian and high Asian territories before them, they often addressed their British peers only if letters of recommendation or similarly mundane necessities were needed. What emerges from their correspondence, especially with British administrators and scientists, is a dysfunctional system of communication that was in many cases not based on mutual interest and trust, but on one-sided benefit.

Even though there is an increasingly rich historiography about the role and functioning of imperial information networks, little attention has been given to the inherent fractures between private and public channels of communication, and how knowledge transmission was often stunted by mechanisms of exclusion within and across scientific communities.[44] We can address these issues by exploring the gaps and obstacles in the transfer of information. Such an analysis equally sheds light on the ambivalent neutrality of the Schlagintweits in privileging German over British naturalists and editors in arranging for their publication.

Although German scientific institutions, such as geographical societies with their own means of funding, became increasingly important as employing bodies in later decades, the mid nineteenth century was a period when royal patronage could still decisively shape the trajectories of individual naturalists with aspirations to travel overseas but without

---

[41] AS to WH, RBG, DC 51, 10 January 1851, 549.
[42] See the excellent treatment in Werner, *Humboldt und sein Kosmos*.
[43] On the social conventions of this learned republic, Brockliss, *Calvet's Web*.
[44] A prime reference for communication channels (and their potential vulnerability) in a colonial context is, however, Bayly, *Information*.

the independent means for their realisation. The Berlin Geographical Society embodies this gradual shift from essentially a receiving institution, which first digested the travel accounts by foreign travellers, only to become an active funding body that sponsored German overseas expeditions from the late 1850s onwards. It had at first a focus on Africa, in consequence of the travels and influence of Heinrich Barth and other active members and prominent African travellers such as Heuglin, Rohlfs, Decken and Schweinfurth.[45] The society, however, widened its geographical interest under later presidents such as the China traveller Ferdinand von Richthofen, and enjoyed the participation of anthropologists like Adolf Bastian and travellers to various other continents, including South America by the explorers Wilhelm Reiss and Karl von den Steinen.[46]

In order to ignite, and later to maintain, a high level of royal and public interest in their explorations, the Schlagintweits continuously promoted their ongoing researches in Asia by regularly presenting accounts of their feats to large German readerships. While they were formally requested to compile reports for the British imperial authorities upon the march, all ten of which were also printed in the *Journal of the Asiatic Society of Bengal*,[47] the brothers presented more personalised accounts to their German mentors and patrons.[48] These included especially their scientific confrères, the envoy Bunsen, Carl Ritter and Alexander von Humboldt, who were all crucial in making their findings available to further groups of scholars and scientists, including the cartographic circle around August Petermann – and through the latter to even wider international audiences.

These close ties to German societies, geographers and mapmakers were not entirely matched with similar contacts in Britain, even if the logistics of travel meant that the Schlagintweits were in regular contact with the EIC director Sykes, who had helped to plan their itinerary and had provided useful travel advice, especially regarding questions of health in India's hostile climate.[49] The brothers also needed Sykes's support to get the Court of Directors to sanction their most-generous financial arrangements with the Indian government. After weeks of intense negotiations in Calcutta, in which it was ultimately conceded, with the support of the Governor-General, Lord Dalhousie, that all three brothers would receive 'the same treatment and advantages'[50] from the

---

[45] Lenz, 'Gesellschaft'; Schröder, *Wissen*, 27–28.
[46] Ibid.
[47] There were ten reports in total: A. Schlagintweit et al., *Report*, followed by the name of the respective region covered.
[48] AS and HS to Frederick William IV, 2 September 1854; Bombay, 14 November 1854; Calcutta, 4 April 1855, BSB, SLGA, II.1.43.
[49] See for the frequent communications with Sykes, FBG, SPA, PGM 353/1.
[50] Roquette, 'Prix', 228.

imperial government for the entire duration of the mission, Hermann wrote to Sykes in late March 1855 that 'we think we can make no better use of this liberal allowance, which we hope the Court of Directors will sanction, than by extending as much as possible the field of our works'.[51]

The Schlagintweits also sought to sustain the Prussian king's material support from the outset. On several occasions, they sent gifts from India that reflected their studies on the spot. Soon after the brothers had set foot in Bombay in late October 1854, an account of their passage from England was set down and addressed to the Prussian monarch, which also included their first impressions of the heterogeneous colonial port city. The brothers informed their royal patron that '[w]e have tried to take different ethnographic photographs with our beautiful camera, and shall have the honour to present copies of those with the next mail'.[52] The king thus received the first parcel of images for his private amusement: photographs of the peoples of India were then a great novelty in Prussia and therefore added considerable value to their gesture.[53] The travellers later offered Frederick a fine selection of their sketches and watercolour views produced in the Mediterranean, Egypt, the Red Sea and the Indian Gulf, as well as some painted views of little-known mountain ranges in high Asia.[54] This personalised orchestration of texts, photographs and drawings for specific individuals was an important element of the Schlagintweits' attempt to secure and strengthen their ties to important benefactors into the future, and their visual materials never failed to impress. Humboldt confirmed the success of these offerings, noting that '[t]he views of the Karakorum pass … [of] cloisters, the old cradle of the Buddhist civilisation, of Ladakh and Cashmere, have greatly delighted the king, and this to such a degree that he on many occasions praised himself for having initially entrusted them with this travelling project'.[55] It is not without symbolic significance that the brothers hardly ever sent any watercolour or photograph from India as gifts to patrons or administrators in Britain – unless a direct request for a favour was attached.[56]

Their field reports were, however, also a crucial tool of self-promotion and the establishment of scientific authority across Europe. The

---

[51] HS to Sykes, 31 March 1855, FBG, SPA, PGM 353; also AS to Sykes, from Kumaon, 17 May 1855, ibid.

[52] RS, HS and AS to the Prussian king, 14 November 1854, BSB, SLGA, II.1.43.

[53] See on the early development of photography in India, Jarvis, 'Porträt'.

[54] 'Ostindien', *Allgemeine Zeitung München* [*AZM*], 258, 15 September 1855, 4118–19.

[55] Humboldt to Bunsen, 11 March 1857, GStAPK, Rep. 92, B I d 59.

[56] See, e.g., their correspondence with Henry H. Montgomery, 9 November 1860, BSB, SLGA, IV.6.1.

recipients of personalised accounts were often specifically instructed on steps towards publication. This ensured full coverage of their travels in scientific journals and popular newspapers. The distribution networks for their field observations not only demonstrate how transnational patronage communities and larger audiences were actively recruited for the Schlagintweit programme, but also how their own credibility as field observers was bolstered through the presentation of their facts through the mediation of metropolitan authorities. A letter the Schlagintweits sent to Humboldt in 1855 reflects these overlapping mechanisms of royal reporting and the reliance on scientific intermediaries in their search for public recognition: while they informed Humboldt of their latest report to the king, they also suggested that the same would 'be adequate for the geogr[aphical] society, maybe with the exception of the first and last sentence'.[57] The brothers instructed Humboldt to also make use of parts of their 'official report to the Indian Government', which they likewise had earlier sent to the Prussian monarch. After stating that 'short accounts' of this formal report should also be 'presented to the Akademie [der Wissenschaften in Berlin], we would strongly urge you (please forgive this rapid succession of nearly too presumptuous wishes) to present the first [part] to Professor Poggendorf for the *Annalen*, the second to the Geological Society'.[58] In the same letter, the Schlagintweits further named those scientists they hoped would revise parts of their Indian reports for publication, including Gerhard vom Rath (1830–88), a respected German mineralogist and geologist. This process of recycling and diffusing their reports allowed for a much greater coverage of their observations in German journals, while their patrons were crucial in fitting and refitting their findings to suit different formats.

Thanks to their multifarious promotion strategies to disseminate the results of their travels, the Schlagintweit enterprise had by 1856 generated such a wide public interest that German journals sought to provide a comprehensive account of their individual itineraries in India, almost down to the day. Carl Ritter edited bits of information for regular publication in the Berlin Geographical Society's own organ. The detail of the brothers' travel routes also enabled Petermann to produce fairly up-to-date cartographic depictions of their movements (Figure 3.1).

Petermann's sketch depicts Ceylon, India, the Himalayas and parts of central Asia, and it traces the itineraries of the brothers throughout

---

[57] HS to Humboldt, 21 April 1855, from Darjeeling, BSB, SLGA, II.1.43.

[58] Ibid. Such requests vividly illustrate James Secord's point about science as a 'form of communication', Secord, 'Knowledge in Transit'. *Annalen = Annalen der Physik*, a prestigious journal in the field of physics, which had appeared since 1799.

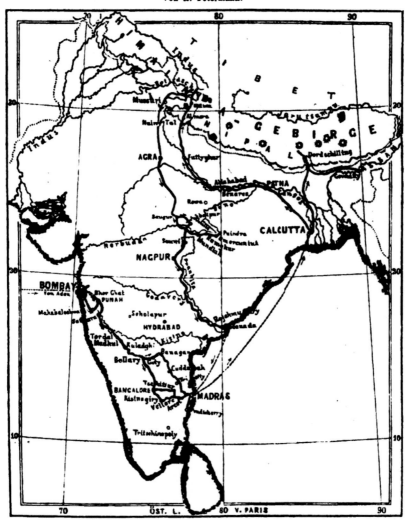

Figure 3.1 August Petermann, 'Overview Sketch of the Travel Routes of the Brothers Schlagintweit in India, from 5th of Nov. 1854 to 26th Feb. 1856', in Petermann, 'Reisen', 104.

the colonial territories, but also beyond the north Indian frontier, by inserting directional arrows onto the map. The editor's decision to leave large blank spaces where the Schlagintweits had not yet travelled stressed, but also greatly exaggerated, the pioneering character of their journey. By depicting only their travel routes instead of adding also those of numerous British, French and Russian travellers who had preceded them in several of the regions they had passed through, the image of the Schlagintweits as undaunted travellers was thus subtly reinforced. While it was Petermann's main purpose to illustrate the regions the brothers had by then already crossed, it is striking just how deceptive such a visual device was in forging a public image of the Schlagintweits as ostensibly self-sufficient travellers opening up uncharted regions to western science. With the help of the printing press, the Schlagintweits thus became celebrated explorers in the German states well before the extent of their activities and precise achievements was known in much detail. Based on the information available, a Munich-based newspaper even proudly pronounced that their 'scientific voyage of exploration … will add fresh glory to the old reputation of the German spirit of research'.[59]

As was integral to the art of exploration at the time, the Schlagintweits' pursuits clearly appealed to more general readerships hungry for stories about 'exotic' peoples and places, and accounts of human adventure and suffering, which the brothers' forays into the supposed unknown beyond the trans-Himalayan region seemed to provide in abundance.[60] The wide coverage of their scheme in hundreds of German learned journals, daily newspapers, popular magazines and papers serves to show how overseas expeditions were much more than the movement of individual travellers to the far corners of the globe. With a booming print culture, the opening up of supposedly unmapped territory in northern India and high Asia was staged, and avidly consumed by the public, as a 'media event'.[61] Indeed, 'from the point of view of metropolitan science and culture, exploration without writing and publication was no exploration at all'.[62]

While scholarship has only recently begun to more fully investigate this aspect, the reception and impact of travelogues have rarely been confined to singular national audiences. The consumption of such a popular work of exploration as David Livingstone's *Missionary Travels* (published in 1857 and hence simultaneous with the Schlagintweits' first reports) could become 'an international or even global literary

[59] 'Ostindien', *AZM*.
[60] Keighren et al., *Travels*.
[61] Laak, *Welt*, 31.
[62] Driver, 'Livingstone', 167.

event'.[63] While the brothers' early field reports and multi-volume oeuvre never achieved such broad appeal, their edited letters and first scientific accounts nonetheless succeeded in swiftly entering international circulation. For one thing, the famous cartographic journal *Petermanns Geographische Mitteilungen* multiplied the Schlagintweits' reader base. As Petermann proudly boasted to Sykes, 'my Monthly Journal ... has attained a new large circulation in and out of Europe (5000 copies every month)'.[64] Sykes acknowledged the importance of the *Mitteilungen*, writing to Petermann that this journal was, indeed, 'very instructive & gives an excellent summary of passing Geographical Knowledge'.[65] It thus served as an important forum in which claims could be made for new exploratory feats and scientific reputations forged. Here, as elsewhere, the brothers initiated the publications of their field reports themselves.[66]

However, while such an immediate publication of field observations was not uncommon, the question of which national journal could publish their findings *first* was a sensitive issue. The announcement of new exploratory feats was often intended to mirror a country's scientific achievements on the international stage. While the brothers formally complied with the official terms of their employment, and regularly sent reports to imperial officials and scientific institutions in India, there was nonetheless a tendency to undermine the rules of the game – the unwritten conventions of such foreign recruitment – by prominently releasing their 'discoveries' in German journals above all. Extracts from their first report to Humboldt were printed in Berlin as early as 1854.[67] A second and longer composition was published only a few months later.[68] The British version of the same account, by contrast, was first printed in Britain only in 1856.[69] This apparent favouring by the Schlagintweits of German publishing houses and scientific circles was critically noted and condemned by British men of science, who, like Joseph Hooker, even spoke of the brothers' supposedly wilful 'withholding of their Scientific results from our Societies'.[70]

Such criticism had already been articulated in the case of other German geographers appointed to scientific offices in the British Empire.

---

[63] Henderson, 'Historical Geographies', 228.
[64] Petermann to Sykes, 31 March 1860, FGB, SPA PGM 353/2, Sykes file, 55.
[65] Sykes to Petermann, 2 July 1860, FBG, SPA PGM 353/1.
[66] 'Notes générales sur la population du Yarkand (Extrait d'une communication faite à M. de La Roquette, par MM. Hermann et Robert Schlagintweit)', *BSG*, 18 (1859), 261–64.
[67] A. Schlagintweit, 'Schreiben'.
[68] Ritter, 'Reise'.
[69] A., H. and R. Schlagintweit, 'Mediterranean'.
[70] JH to Murchison, 19 July 1859, RBG, DC 96.

A telling example is the case of the traveller Heinrich Barth, who in conjunction with the Hamburg-born Adolf Overweg had joined a British expedition into central Africa in 1849 under the leadership of James Richardson. When the latter died *en route* in 1851, Barth took over as the head of the mission, which was of great political and trading interest to the British nation, as commercial treaties were to be signed. Soon, therefore, some RGS members voiced 'concern that the German scientists were providing some confidential reports from the expedition directly to Petermann and Bunsen instead of submitting them directly through British channels'.[71] Bunsen and Petermann, at that time still based in London, suddenly found themselves in the position of having to justify themselves. As Naranch put it, 'Petermann, himself under fire from his British RGS colleagues for allegedly withholding important information about the expedition and perhaps passing it on to German state officials, struggled to defend the roles of Overweg [and] Barth ... by accusing their detractors of discrimination against them for being German'.[72] Yet, even RGS president Murchison disapproved of the fact that printed accounts of the expedition's progress would first appear in the German press, thus denying his own society's journal the distinction of being the prime source of information on its achievements.[73]

The Schlagintweits' ostensible privileging of German newspapers and journals should thus be placed within the larger context of earlier British objections to such information double-dealing. Just as British men of science, frustrated that some of the scientific posts they eagerly sought were going to foreigners, had developed a sense of entitlement over these, so did British government officials, naturalists and the editors of learned journals claim a right of ownership over scientific observations. As the next section will show, the suspicion even prevailed that the Schlagintweits were providing information to one of the perceived rivals of the British in central Asia: the Russian Empire. Such public rumours reached the brothers whilst they were still engaged in exploration, and give an indication of the difficulties the German travellers would face upon their return.

The Debt of Exploration

Joseph Hooker's earlier fears became reality as the reorientation of the Schlagintweits' scientific commission did indeed entail much higher

---

[71] Naranch, 'Beyond the Fatherland', 243.
[72] Ibid.
[73] Ibid.

Figure 3.2 'Gold Leaf Electroscope Pair'. Materials: brass, glass, gold, leather, metal, sheet metal, wood; dimensions: 11.5 × 5.5 × 4.2 cm. © Harvard University [HU], inventory no. DW0780. There are altogether nineteen Schlagintweit instruments at Harvard.

expenses than initially envisioned. To properly equip them for engaging with the multiple scientific fields the brothers proposed to cover, the Company provided them with more than 200 of some of the most developed and hence expensive instruments available.[74] Humboldt, indefatigably acting as the brothers' intercessor, was aware of the high charges for their procurement, and the potential criticism this could attract in British circles – after all, the costs apparently ran to 'around 30,000 Francs' for the instruments' purchase alone.[75] Humboldt therefore proactively thanked Sykes, but also noted that the progress of the sciences would now require scientific travellers to be armed with such a wide range of precious apparatus.[76] These included such delicate instruments as an 'absolute electroscope' produced by M. Leyser in Leipzig, a 'Fortin type mercurial barometer' from Adie, London, and a precious 'gold leaf electroscope pair', on the reverse side of which the name Schlagintweit was specially engraved (Figure 3.2).[77]

---

[74] For a full listing of their scientific instruments, BSB, SLGA, II.1.5.

[75] Roquette, 'Prix', 228.

[76] Humboldt to Sykes, 18 November 1854; copy letter at Alexander-von-Humboldt-Forschungsstelle der Berlin-Brandenburgischen Akademie der Wissenschaften [BBAW].

[77] The Schlagintweits' planimeter is kept at the National Museum of American History, Washington, DC [NMAH], no. 87-4849. Only a few of their instruments remain in German holdings, as in the Deutsches Museum, Munich [DMM].

In addition to the increased expenses for the acquisition and climate-induced adjustment of their instruments in Europe, similarly high sums were necessary for transporting them. A small army of indigenous porters was employed to carry these bulky, yet fragile, objects. The extent of the brothers' demand for manpower can be understood by considering their photographic apparatus. One of their two cameras and its accompanying equipment alone weighed some 200 kilos. The size of their hardware meant that from the start they required the services of '20 camels (dromedaries) and six servants for the transport of our tents, collections, and our heavy luggage in general'.[78] In addition, '[a]ll the delicate instruments were carried by kúlis on long bamboo sticks', and these 'kúlis were changed every three or four marches'. Once the brothers started to pursue separate itineraries in order to 'spread our observations over a larger area', the size of the indigenous establishments of assistants and carriers multiplied accordingly.[79]

During their travels, the Schlagintweits organised the repair and replacement of their equipment whenever necessary and also repeatedly asked to be provided with up-to-date instruments from Europe. Despite the official agreement between the Court and the Prussian king that the Company would pay for the scientific apparatus, they also secured instruments from their Prussian benefactors. The way they proposed and carried out such schemes illustrates the brothers' sense of initiative and precaution in handling their multiple sponsors. In 1855, Adolph addressed Humboldt from the region of Jharkhand: 'It would be of great value, especially for our observations in the Himalayas in Tibet, if we could be provided with three little … theodolites from Berlin.'[80] Adolph drew on the informal networks that German expatriates had established in India to realise this scheme, urging Humboldt to send the objects to 'A. Huschke & Co, the Consul for Hamburg, at Bombay Fort'. Adolph openly stated the reasons for this double-dealing, claiming that 'it is impossible for us to be provided with new smaller theodolites by the East India Company, after the great expenditures it had accumulated for our instruments'.[81] Hence, the presence of numerous German scholars and naturalists, missionaries, merchants, a Prussian consul at Calcutta and a Hamburg merchant-consul at Bombay enabled the brothers to effectively use an alternative communication network *within* the British colony. This was the *sine qua*

[78] H. Schlagintweit et al., *Results* I, 12.
[79] HS's report to the Prussian king, from Calcutta, 4 April 1855, BSB, SLGA, II.1.43, 41.
[80] AS to Humboldt, Sangor, 15 December 1855, ibid., 210.
[81] Ibid.

non for allowing them to turn their position within the imperial establishment of British India to their considerable gain.

In 1856, the Schlagintweits asked Frederick William IV to greatly increase the monarch's financial support by doubling the amount the king had agreed to commit. They addressed the secret application to their intercessors Humboldt and Bunsen:

Our proposal, which we urge you to present to H[is] M[ajesty], would be the following. Namely, that through a letter of the ministry of commerce or the ministry of foreign affairs, Mr Kilbourn, the Prussian Consul at Calcutta, shall be authorised to pay a sum of 18,000 to 20,000 Thalers (which, of course, includes the yearly 3,000 Thalers already granted by HM) in order to pay for the expenses of our travels. We very much wish that the Prussian Consul shall be informed through direct communication from Berlin to Calcutta, and should not be notified through ... India House in London.[82]

Special precautions had to be taken for arrangements of this kind, since the 'Indian Government seems to be highly sensitive on this point, which we could see in the case of the £80 for instruments that had been sent to us by Mr Hebeler', then the Prussian general consul in London, on an earlier occasion. As Hermann further noted, British perceptions of the brothers had recently deteriorated to such a degree that open speculations had been made in India and Britain about their being engaged in acts of betrayal against the Empire: 'it has often been stressed in newspapers that we are not Englishmen. Only very recently, there was a lot of nonsense in all newspapers about "our supposed meetings with Russian agents in Turkistan"'.[83] Such allegations continued to flourish even after the brothers' departure. In 1858, William Howard Russell, a journalist on *The Times* sent to India in 1858 to cover the Indian Rebellion, wrote in his diary: 'There is a "sentiment" here that the Russians are pressing us dangerously close, and are moving down every year more surely towards India. One eccentric gentleman in Simla maintains poor [Adolph] Schlagintweit was a Russian spy.'[84] The fact that some journalists in Britain and India viewed the Schlagintweits as untrustworthy foreigners could swiftly lead to rumours and accusations of treachery, especially regarding Russia – Britain's alleged imperial rival.

---

[82] HS to Humboldt, 11 December 1856, BSB, SLGA, II.1.43, 371. For the almost identical letter to Bunsen, GStAPK, VI. HA Familienarchive und Nachlässe, Bunsen, Karl Josias von [FA], fifth folder, 11 December 1856, 275–6. Frederick had originally agreed to subsidise their travels for three years with a yearly grant of £350; 'royal order' 8 July 1854, 'Reisezuschuß'.

[83] HS to Humboldt, 11 December 1856, 371–2.

[84] Russell, *Diary* II, 136.

But just as striking is how the Schlagintweits immediately turned this suspicion to their own benefit, and used it to propose and justify covert financial arrangements with the Prussian monarch.

The brothers saw their requests for additional money as justified since circumstances forced them to guarantee their personal liquidity. Thus they needed this additional money to be able to complete their expedition at all. As they admitted in late 1856, 'we are writing only now, because we were only now able to compile a general list of our expenses'.[85] Those expenses 'averaged until now for each of us 1,000 Rupees (1 Rupee = ⅔ Thaler) per month, [or] from October 1854 until the beginning of 1857 for all three a little more than 70,000 Rupees in total'. Two years into the mission, the accumulation of costs came as a rude awakening. Their lofty scientific objectives had led them to greatly exceed their allowance. The brothers were now too embarrassed to even present these sums to the Indian government, instead choosing to call on the Prussian king to come to their rescue.

As they noted, with a trace of pride, the distance they had covered already amounted to '15,000 English miles', as if distance itself was a signifier of scientific achievement.[86] Parts of their collections had themselves travelled over vast distances throughout parts of the Himalayas and India, often sent in separate caravans from the interior to the coast for shipment to Europe. At one point, a hundred camels took more than 200 boxes of collectables from the foot of the Himalayas to Calcutta, a small expedition in itself, whose costs the brothers also had to meet. The Schlagintweits also specified those 'fortunate circumstances, which allowed us to almost always pursue separate routes':

The only thing that will delay our departure is the settling of the accounts with the Government [of India]. We have made it possible through private arrangements with our agents in Bombay and Calcutta, then through the Government's official advancing of money against later repayment, to be able to travel without time loss through all parts of India, and to temporarily defray all the necessary expenses for the collections. What is more, the biggest share of [the costs] for the inland transport [of the collections] has not as yet been paid to our agent in Calcutta.[87]

Their own initiatives with private merchant-consuls and the colonial government had thus enabled the German explorers to travel for years on unsecured and infinite credit. The accumulation of significant debts suggests that the Indian authorities never consulted the Court in London

---

[85] HS to Humboldt, 11 December 1856.
[86] Ibid.
[87] HS to Humboldt, 11 December 1856, 373.

on the question of up to what sum the brothers' financial demands were actually secured. These liberal arrangements were precisely what Joseph Hooker had criticised as their unusual 'carte blanche'.[88] Hooker must have been informed by a critical observer from India on the Schlagintweits' privileges – which only added insult to injury in view of the fact that the powerful Court of Directors had rudely ignored Hooker's early quest for adequately support for his own Indian travels and the publication of his *Flora Indica*.

While the brothers acknowledged that they had lost control of their spending, the travellers were still not prepared to compromise on the bold scale of their expedition.[89] On the contrary, Hermann explained in detail their ambitious travel plans for the future: 'Our plans for the cold season are that Adolph will travel to Peshawar, and then to follow the Indus to Kurachee [Karachi] and Bombay.' Robert, by contrast, was supposed to travel on a more northern route towards Bombay. Thanks to the persistent assistance of the Indian Governor-General, 'I myself will go to Lahore and Patra, and from there, which can now finally be arranged, I will visit Kathmandu. After a short stay in Nepal, I will come to Calcutta.' For only then, the brothers insisted, 'will we have completed our observations in India'.[90] (In 1857, to be sure, Adolph would make the fatal decision to return to the great mountain systems of the Karakoram and the Kunlun in central Asia.) Their projected final itineraries thus meant again three separate routes, and hence three times the cost of an indigenous establishment. Nothing should stand in the way of finishing their major scientific investigation of this world region, which in its vast scope was aimed at elevating European knowledge of India and high Asia to new heights.

The brothers linked their secret financial proposal to the Prussian king with yet another double game they played with their benefactors. This concerned the complex question of ownership over their vast collection of artefacts. As they informed Humboldt, and via him the Prussian monarch: 'Another fact that might be alluded to in order to excuse the sum named by us [20,000 thalers] is that it will be much easier for us to obtain a large part of our collections for Berlin if not all of the expenses have to presented to the Indian Government.'[91] While the issue of ownership requires a more thorough analysis later (Chapter 7), it is important to note that the brothers introduced this matter in their financial

---

[88] JH to Sykes, 4 November 1855, RBG, DC 102.
[89] HS to Humboldt, 11 December 1856.
[90] Ibid.
[91] Ibid.

negotiations from the field. Their promise to secure considerable chunks of artefacts for Prussian state collections was taken to heart by officials in Berlin – a fact that only promised further tensions in the years ahead.

The brothers' secret proposal for the 20,000 thalers never seemed to have materialised. However, through the mediation of their younger sibling, Emil Schlagintweit, the brothers soon settled on another scheme. Before their return to Europe, Emil addressed Bunsen yet again. After reminding the former consul of his earlier petition 'that the agreed Prussian grant shall be raised to more than double the amount', Emil suddenly claimed that the Prussian king and his adviser, E. E. Illaire had, in fact, confounded the amount of the promised royal contributions.[92] While the king assumed only £350, the brothers (and their ally Humboldt) now claimed that the three yearly grants would amount to 3,000 thalers, or £445, each.[93] Humboldt's name and reputation could thus also be used to lend higher legitimacy to their demands. Consequently, Emil proposed to Bunsen and via him to Frederick William IV that 'it would be sufficient' if the brothers were to receive 'a further subsidy for a fourth year'.[94] The brothers were by this time (March 1857), however, already set to return to Europe.[95] If the request was sanctioned, the additional sum 'should then immediately be sent to Mr Kilbourne', the Prussian consul at Calcutta.[96] Their persistence was their charm: in the end, Frederick William agreed to provide funding for a fourth year of an expedition that, for the brothers at least, lasted merely three.[97]

In contrast to other travelling naturalists, the brothers were not willing to draw upon their own salaries to meet the substantial debts they had accumulated. On the contrary, they even made a considerable financial gain whilst travelling in India. As they informed the Bavarian king Maximilian II: 'In India, the three of us had a monthly salary of 1,200 Rupees and 300 Rupees travelling allowances … Of this sum, we could put aside 21,000 fl. [gulden] between the years 1854 and 1857, which was made even easier for us, since the acquisition and the transport of the collections were officially paid for by England.'[98]

---

[92] Emil Schlagintweit to Bunsen, 11 March 1857, GStAPK, FA Bunsen, 273.
[93] These (incorrect) figures were also stated by Humboldt in a letter to Bunsen, 11 March 1857.
[94] Emil Schlagintweit to Bunsen, 11 March 1857.
[95] HS to Bunsen, 11 December 1856. Adolph intended to complement their observations in the western Himalayas and central Asia, but did not plan to stay for a whole year; AS to Sykes, 25 April 1857, FBG, 353/1.
[96] Emil Schlagintweit to Bunsen, 11 March 1857.
[97] Illaire to Manteuffel, 6 September 1857, GStAPK, III. HA Ministerium der auswärtigen Angelegenheiten [MdA], HEBS.
[98] HS and RS to the king, Munich, 1 June 1859, BayHStA, Adelsmatrikel, S 156.

Upon return to Europe, Robert and Hermann Schlagintweit finally claimed supposedly outstanding sums from the Court of Directors. East India House could merely remark that the only references to the mentioned sums are 'in their letters on the subject, in which the Court from the high character of those Gentlemen have every reason to place entire confidence'.[99] The Court's trusting view of the brothers thus stood in marked contrast to the dislike and even suspicion about the Schlagintweits in scientific circles and, increasingly, some sections of the wider publics in Britain and India. The still unclouded relationship with the Company was to prove crucial in order to strike a renewed contract with the imperial patron in the future. This second employment, negotiated between 1857 and 1858, was intended to allow the brothers to embark on the last, and no less expensive, journey in such an overseas expedition: the time-consuming preparation and careful analysis of their gathered materials in Europe, and the costly publication of their travelogues and illustrations. The brothers knew that their grand scheme would only be accomplished once an adequate *magnum opus* had been compiled and disseminated to European and colonial readerships – a printed monument to the size and ambition behind the brothers' Asian expedition.

Before we turn to the legacies the brothers produced over their Indian and Himalayan travels, it is first important to look at how the brothers actually realised their large-scale expedition: what colonial resources and institutions they mobilised, what sources of knowledge – both European and indigenous – they tapped and how their scientific practices shaped their perception of both the unfamiliar natural world and the indigenous peoples they encountered.

---

[99] James Melvill, East India House secretary, 'unadjusted amount of salary of H & R Schlagintweit', 1 October 1857, BL, E/2/25.

4 Making Science in the Field: A Eurasian Expedition on the Move

> A European, who understands travelling in the H[imalayas], whose preliminary arrangements and preparations necessary for his journey have been made with the requisite care and discretion, who surrounds himself with young attendants … may be likened by no means inaptly to *a thoroughly independent sovereign, who governs absolutely over an immense kingdom adorned with the rarest charms of nature,* which he can traverse in every direction, just as his humour and will may lead him. … *Every European, whom he may meet is a friend, every other person an obedient subject.*[1]

In retrospectively describing the experiences he and his brothers had gained as scientific travellers in the mountain systems of high Asia in 1855–7, these were the words that Robert Schlagintweit used in the 1860s and 70s to capture his seemingly unbounded sense of independence – a feeling he seeks to convince his audience any European could enjoy in those regions. One of the many striking aspects of this passage is that it testifies to a curious shift of perception on the part of the traveller. His actual role as a surveyor in the service of the East India Company is transformed by the powerful idea that he could turn into a 'thoroughly independent sovereign' of the land *himself*. His reign, he suggests, would not only encompass high Asia's natural kingdom. Rather, it would also extend to the rule over indigenous peoples and the south Asian assistants that made such exploratory journeys possible in the first place. The extract is a highly subjective and imaginary statement that formed part of the fantasies of domination the Schlagintweit brothers developed *after* they had travelled in and beyond British possessions in India. The quotation is also characterised by a number of striking omissions about the realities of such trans-cultural overseas explorations, including the fact that their voyage was carried out under the direction of and in accordance with the interests of the imperial Government of India.[2]

---

[1] Robert Schlagintweit, BSB, SLGA, Lecture notes, V.2.2.1, 50–51, emphases mine.

[2] All their travel and research plans were presented 'in full detail to the Government of India' and subsequently 'sanctioned'; the imperial authorities at Calcutta also provided

Here, however, the fantasies of the German naturalists were precisely not linked to British concerns. Their purpose was not simply to imagine how Britain could further explore, 'improve' and profit from the lands the brothers had studied in its service. Rather, the Schlagintweits developed an imagination that transcended their subservience to the empire's cause, and in which they were the main protagonists in the conquest and rule of those Himalayan regions and their inhabitants. It is precisely this intriguing self-perception, but also the blatant contradictions of this imaginary inherent in their own writings, that stands at the heart of two following interlinked chapters. The actual conditions of travel, the fear, dependence and often ignorance on the part of the European travellers are juxtaposed with the ideas, scientific practices and circumstances that later fuelled their feeling of mastery towards the lands they had crossed, explored and measured.

After the three Schlagintweits had prepared their Indian mission in the scientific hubs of Berlin, London and Paris, they travelled by steamship to Alexandria, from where they traversed a stretch of desert before arriving at Suez. From there, the brothers continued by steamer via Aden to Bombay, where, on 26 October 1854, they touched Indian ground for the first time. Already during the passage, they began their regular measurements of various natural phenomena such as Mediterranean and Indian Ocean sea currents, took their first water samples (later sent to the Court of Directors for chemical analysis), and also began their large series of watercolours with coastal views.[3]

During the following two and a half years, and in the case of Adolph until late August 1857, the brothers then travelled partly together, but mostly separately, on meandering routes through southern, central and northern India. They also undertook various excursions into different parts of the Himalayas, which demarcated the northern frontier of the Company Raj. They accomplished important explorations in those countries that lay within and beyond this partially explored mountain chain, including parts of Tibet and Nepal, and undertook a secret penetration 300 miles into Chinese Turkestan in central Asia.[4]

'verbal instructions' the brothers subsequently carried out; A. Schlagintweit et al., *Report* II, 106, as printed in *JASB*, 25. HS to Melvill, secretary of East India House, 31 March 1855, from Calcutta, *Report* I, 4. There exist different versions of each *Report*; those printed in the *JASB* tend to include further information.

[3] Finkelstein, 'Mission', 190.

[4] A wealth of new geographical information collected by the Schlagintweits in central Asia is inscribed on the map 'Turkestan: Comparisons of Positions in Turkistan, 1858–61', NAI, Cartography section, Historical Maps, F 93/34.

During the expedition, in which the Schlagintweits and their indigenous assistants and establishments traversed over 29,000 km, including long stretches of difficult terrain in high mountain regions, the travelling party made a number of pioneering achievements. Among them was the first crossing of the Karakoram chain from north to south, and the journey over the Kunlun range in central Asia by Hermann and Robert in 1855 as the first Europeans. As a result of these experiences, they developed new geographical concepts that lastingly shaped western understandings of the mountain systems north of India. They introduced the concept of 'high Asia', a geographical region that they believed was 'formed by the chains of the Himalayas, the Karakoram and the Kunlun in a constant mutual connection'.[5] They also established that the Karakoram was the 'watershed between central Asia and India'.[6] During their high mountain ascents, Adolph and Robert set a new altitude record of 6,785 m on Ibi Gamin (Kamet, Garhwal) in August 1855. The brothers also ranked among the first to employ the still rather novel technology of photography for the purpose of overseas exploration, experimenting with the medium in mountainous areas (over 4,000 m), as well as in badly lit caves – such as on the island of Elephanta.[7]

This enumeration of the brothers' 'heroic' achievements long dominated the popular, and to some extent also the scholarly, perception of their Asian travels.[8] The particular focus on the attainments of these geographical 'trailblazers' is further reinforced by the fact that their expedition was largely depicted as a single-minded, scientific undertaking in the tradition of the European enlightenment.[9] In the self-perception of European states at the time and, later, of some historians too, the age of enlightenment had inaugurated a new 'age of exploration', in which colossal state-backed undertakings had cast light into the dark interiors of unknown foreign continents. For contemporaries, exploration was indeed more than just a matter of advancing scientific knowledge: it came to symbolise the cultural superiority of exploring western nations in their encounter with extra-European societies.[10]

However, as a critical scholarship on the culture of European exploration and its popular perception has shown, European discoverers not only destroyed myths about foreign lands, but also created new ones.[11] A

---

[5] Defined in Hermann Schlagintweit, 'Bericht', 137; H. Schlagintweit, *Reisen* II, 3–18.
[6] H. Schlagintweit et al., *Results* IV, 524; Finkelstein, 'Mission', 195.
[7] H. Schlagintweit, *Reisen* II, 271; Körner, 'Photographien', 314.
[8] This is particularly true for the German press and historiography; e.g., Schlager, 'Helden'.
[9] 'Neue englische Expedition nach Inner-Asien', *Globus*, 1 (1862), 94.
[10] This 'second age of exploration' is sometimes juxtaposed with an earlier 'age of discovery' in the wake of Columbian exploits, Kennedy, 'Introduction'; Robinson, 'Exploration'.
[11] Driver, 'Livingstone', 166.

striking example of such myth making is the characterisation of Africa as the 'Dark Continent', which increasingly influenced European imaginings of this continent in the nineteenth century. Yet, as Patrick Brantlinger has argued, 'Africa grew "dark"', the more 'Victorian explorers, missionaries and scientists flooded it with light.'[12] The published works of European overseas travellers were also able to create new myths that often stood in sharp contrast to actual experiences on the ground.[13] The popular image of the Schlagintweit expeditions, which presents the brothers as solitary travellers, who despite all obstacles lifted the veil of ignorance from countries in high and central Asia, is ironically less a result of the brothers' own publications. Rather, the popular account of their exploits is just as much the result of a strikingly selective and glorifying reception history, which was already set in motion during their time abroad, but continues into the present.

The exploration of the Schlagintweit expeditions in this chapter, and their deep embeddedness in the colonial infrastructure of the British Empire in south Asia, seeks to offer a corrective to such enduring myths. The analysis in Chapter 5 is then concerned with exploring this under-taking not merely from the perspective of the European explorers, but also to enquire about the personal motives and interests of the numerous indigenous helpers and assistants. Together, these two chapters argue that it was only the indigenous employees, and the Schlagintweits' constant cooperation with colonial scientific, military and medical institutions, which made possible the realisation of this large-scale undertaking in its ultimate form. Since the brothers themselves acknowledged their at times complete dependence upon their indigenous assistants in their writings, it is all the more striking that so little attention has as yet been devoted to their extensive establishments, which certainly did not only encompass local 'native' helpers.[14] Indeed the very categories of 'local' or 'native' intermediaries are problematic as they imply a certain terri-torial rootedness and therefore a rather limited expertise on the part of the individuals analysed. Local expertise did, of course, play a role in the decisions of the brothers to hire ever-new guides along the way. Yet in general the distinction between mobile Europeans and local followers is a legacy of narrative conventions surrounding overseas exploration which recent research has begun to correct.[15]

---

[12] Brantlinger, 'Victorians', 166.
[13] See the groundbreaking analysis by Fabian, *Out of Our Minds*.
[14] A recent exception is Driver, 'Intermediaries', 11–13.
[15] Driver, 'Hidden Histories', 428.

In fact, a number of the indigenous leaders of the 'Schlagintweit expeditions' turned into veritable explorers themselves. This was especially the case when they led the moving party through territories that were also unknown to them. In acknowledging this vital role of Indian scholars, collectors and guides, the brothers differed in their publication strategies from the literary-scientific conventions adopted by many contemporary travellers who repeatedly whitewashed the contributions of hired assistants in their publications.[16] However, the brothers were at the same time also eager to re-inscribe in their writings hierarchies between these indigenous assistants and themselves to safeguard their own scientific credentials and authority in front of European audiences. While relying on numerous Hindu assistants, Robert, for instance, wrote of their linked religious views and understanding of worldly phenomena as 'sad and mournful errors of the human mind'.[17] This literary strategy should, however, be read more as a response to the unsettling experiences the German travellers had undergone during their expedition, when the position of power could considerably shift from the Schlagintweits to their guides and partners. A number of these Indian and central Asian assistants are identified, and we can also recover the German naturalists' appreciation of (in Hermann's words) the 'actual leaders' of some of their most important advances into and beyond the frontier regions of British India.[18] Of course, the relationships between the brothers and their various guides, porters, interpreters and protectors were not without their tensions, misunderstandings or mistrust. We therefore need to strike a balance between presenting the collaborative aspects of this encounter without silencing the asymmetric power relations between the brothers and their establishments.

Often it was the considerable time lag between event and narrative that helped transform existing hierarchies into imagined ones. This is true for the realms of knowledge production as much as for the representation of the role and authority of individuals. What should thus become apparent is how misleading it is to regard what is problematically called

---

[16] This practice had started at least in the eighteenth century, see Harrison, 'Networks of Knowledge'. Humboldt's American journey has also been critically revisited in that regard: Cañizares-Esguerra, 'Origins'. The long-running analysis of his diaries by a research team at the BBAW under the leadership of Ottmar Ette promises further important insights.
[17] BSB, SLGA, V.2.2.1, 62.
[18] H. Schlagintweit, *Reisen* II, 329 for Tibet; *Reisen* IV, 22 for Turkestan.

the 'Schlagintweit expedition' as simply a European undertaking. In fact, it entailed an ongoing intercultural collaboration, in which the Indian and central Asian travel companions exercised a decisive influence on the routes, the scientific results and the pioneering achievements of the enterprise. An expeditionary group was thus clearly more than a random collection of individuals. Martin Thomas rightly suggests we think of such a project of expeditionary science as a 'distinct socio-cultural formation' with a strong collaborative dimension as regards its internal workings and external links.[19]

The following analysis addresses a number of significant questions: how were the Schlagintweits' scientific practices and observations exemplary of wider concerns in the field sciences at the time? Were their diverse surveys and examinations 'simply instruments in the imperial ordering of India', addressing the East India Company's quest to better know and exploit the subcontinent's natural and human resources?[20] Or did their orientations also 'express a wider vision of scientific needs and opportunities' – beyond utilitarian concerns?[21] Given the Schlagintweits' wide-ranging interests and data requirements to fulfil their ambitious programme of climatic physiognomy and physical geography of vast expanses of land and sea, we can further ask: *how* could such a large-scale expedition be executed, not least because it went into politically sensitive and, from a European perspective at least, partly uncharted territories? This requires outlining the kind of logistical, political and financial support the brothers received through various channels from the British colonial administration. The following treatment will thus reveal the considerable extent to which the infrastructure of the British Empire made their travels, anthropological research and collecting practices feasible. The analysis will also shed light on the vexed relationship between observational-instrumental science in the 'field' and interpretative humanistic scholarship at the time.[22] While natural scientific researches dominated the brothers' activities in Asia, the expedition also generated a loose framework for the study of eastern religions and cultures through the accumulation of large amounts of texts, artefacts and visual depictions.

---

[19] Thomas, 'What is an Expedition?', 7.

[20] Arnold, *Science*, 22ff.

[21] Ibid.; these questions are increasingly asked about other large-scale imperial surveys undertaken in the earlier Company period; Robb, 'MacKenzie's Survey'.

[22] On 'fieldwork' and its material and cultural logics, Kuklick and Kohler (eds.), *Science in the Field*.

To See, to Know and to Remember: Concerns and Practices of Expeditionary Science

To address the question of the extent to which both personal and imperial interests shaped the Schlagintweits' programme, we must set their enterprise in the context of other ongoing surveys in India and compare the activities of the brothers and their Indian workforce with contemporary practices of field observations. While much of the existing literature has presented their expeditions as exceptional – considering their multiple sponsors, their accumulation of vast collections of data and artefacts and the drama of the killing of Adolph beyond the Indian frontier – little attention has so far been paid to the fact that their mission was in many respects characteristic of a wider culture of exploration.[23] As Lawrence Dritsas and others have shown, expeditionary ventures always had a strong geographical dimension. This is for instance expressed in their efforts to survey and map maritime and continental spaces while also looking for differences and similarities between encountered natural and human phenomena with those of other parts of the world.[24] The Schlagintweits, who came to India as established authorities on the physical geography of the Alps, were clearly inspired for their Indian and high Asian researches by models of Humboldtian global physics and trans-continental comparisons of local geographical phenomena. Indeed, they often employed the analytic triangle of seeking correlations and variances between 'the principal features of India and high Asia with those of the Andes and the Alps', an epistemic project facilitated by the surge in global travels since the eighteenth century.[25]

However, while Humboldt's enterprise acted as both model and challenge for the Schlagintweits' field observations, it is important to place their activities also within a larger continuum of earlier, simultaneous and future British imperial surveys and expeditions that were dispatched to the same or neighbouring regions.[26] One way to approach these connections is to look at the brothers' diverse scientific practices and fields of interests. Those can be compared with the requirements of Company surveys as expressed in a contemporary manual to travellers

---

[23] Driver, 'Face to Face', 441, 450.

[24] Dritsas, *Zambesi*. A similar concern lies behind the valuable collection of essays, Livingstone and Withers (eds.), *Geographies*.

[25] Crucially, many such comparisons were drawn whilst still in the field; e.g., A. Schlagintweit et al., *Report* II, 19. H. Schlagintweit et al., *Results* II, vii–xi; and esp. their *General Hypsometrical Tableau*, ibid., 473–505.

[26] On this logic of seriality of imperial expeditions, see the incisive reflections by Thomas, 'Expedition', 81.

and military personnel issued by the Office of the Surveyor-General. The brothers' close exchange with Henry Thuillier (1813–1906), Deputy Indian Surveyor-General at the time and author of this widely read manual, makes it clear that the brothers had knowledge of the prevalent practices in field surveying and military reconnaissance.[27] To be sure, to establish connections between the Schlagintweits' and other scientific missions under the official sponsorship of the Company is perfectly in line with their own positioning: they constantly referred to their predecessors and contemporaries as their guides and inspirations, on whose results they shaped their own itineraries, research objectives and publications. Hence, they opened the first volume of their *Results* by stating: 'In working out our observations, the well-known labours of numerous scientific predecessors have proved of the greatest importance to us' – then singling out 'Buist, Cautley, Cunningham, Eastwick, Elliot, Everest, Falconer, Gerard, Griffith, Hodgson, Hooker, Latham, Oldham, Prinsep, Thomson, the Stracheys, Sykes, Thuillier, Waugh and Wilson'.[28] Ongoing scientific activities and information exchanges with Company officials in India and beyond provided them with up-to-date data, maps and intelligence while they were travelling and clearly influenced their itineraries.[29]

In consequence of their transnational sponsorship, the Schlagintweits, in a sense, served more than one master, and therefore sought to fulfil various expectations through their extensive fieldwork. The brothers performed several roles at once, acting as officers and surveyors in Company employ, as ethnographic observers, empirical naturalists and scientific collectors, photographers and landscape artists.[30] However, by analysing their letters, publications, and the ten official *Reports on the Proceedings of the Officers Engaged in the Magnetic Survey of India*, which the brothers sent 'upon the march' to British imperial authorities in Dehra Dun, Calcutta and London, it is possible to identify a core set of practices and observations. These combined the material considerations of the Company with more disinterested scientific enquiries. All the brothers engaged in these core activities on conjoint or separate

---

[27] As elaborated in Thuillier and Smith (eds.), *Manual*. H. Schlagintweit, *Reisen* I, 230–3.
[28] H. Schlagintweit et al., *Results* I, 8.
[29] H. Schlagintweit, *Reisen* I, 231–2; A. Schlagintweit et al., *Report* IV, 23; V, 6 and 16; VI, 32; X, 7, on contact with British officers in military stations; also *Report* II, 106–11, as printed in *JASB*, 25, on intelligence from political Residents. See also the letters from HS to Brian Hodgson in 1855, Royal Asiatic Society Archives, London [RASA], GB 891, BHH/1/94; and Elphinstone to Humboldt, 31 December 1854, BSB, SLGA, II.1.43, 27–8; Waugh to Colonel Birch, 15 September 1854, NAI, Military Department, Consultation, no. 445.
[30] Driver, *Geography*, 28.

routes, to which they added further observations on the characteristics of distinct regions, thus exemplifying their mutable programme shaped by scientific opportunities and local advice and guidance from Company Residents and indigenous informants alike.

Principally, and symptomatic of exploratory missions at the time, their programme was based on the systematic and large-scale collection of different kinds of measurements, observations and material collections. However, in echoing Livingstone's confession that 'it is far easier to travel than to write about it', the Schlagintweits initially struggled to structure the constant flood of new impressions:

> Great is the variety of details preying upon a traveller's mind simultaneously whilst crossing large tracts of land novel to him and partly unexplored by previous observers. Finally, when he arrives at thinking over his impressions and at the heavier task of working up his manuscripts, he would be bewildered by the multiplicity of facts, had he not followed a certain plan in his books.[31]

Hence, in contrast to a mere jotting down of facts, scientific observance required a skilled eye and mind that immediately had to group the travellers' perceptions into useful categories in order to avoid being overwhelmed. This reflects the wider point that a naturalist in the field was never merely a passive observer, but 'more importantly one who sees within a prescribed set of possibilities, one who is embedded in a system of conventions and limitations' – while these 'conventions' and scientific practices were then anything but fixed.[32] Each Schlagintweit made use of 'one general book at a time', in which 'a detailed preliminary division of space had been made for the various branches' of the sciences.[33] This division of their fieldnotes was not final, however, since the pages (written on one side only) could later be reassembled 'into any systematic form' depending on the ultimate plans for publication. The Schlagintweits later bound together over forty volumes from their 'manuscripts of observations', starting with four volumes on 'Itineraries', 'Routes and Route-Books' (nos. 1–4) and ending with a thick exemplar (no. 42) on 'Botanical and Colonial Details in General'.[34]

While the original organisation of the Schlagintweit notes is lost, it is clear that ever-new categories were added in the course of exploration, owing to the transgressive nature of their programme. In other words, their initial 'List of proposed operations' only very partially accounts for their actual activities in Asia. Their mutable agenda also reflected the

---

[31] H. Schlagintweit et al., *Results* IV, 5.
[32] Driver, *Geography*, 52; Crary, *Techniques*, 6.
[33] H. Schlagintweit et al., *Results* IV, 5.
[34] Ibid., 6. To these were added three volumes of Adolph's recovered notes.

fact that the boundaries between different scientific fields were then by
no means static. Indeed, what constituted, for instance, accepted and
trustworthy practices in geographical fieldwork was hotly debated among
colonial and metropolitan scientists and travellers alike.[35] Some scholars
have suggested that the increasing publication of manuals and 'hints to
travellers' since the late seventeenth century helped to discipline scientific observers. However, by the mid nineteenth century there still existed
no consensus but rather much controversy about the right practices, and
even the epistemic status, of fieldwork – seen by many metropolitan men
of science as a potentially doubtful, at times dangerously flawed source
of information.[36] The assumed truthfulness of accounts from the field
depended on the one hand on the deployment of the right set of ever
more refined instruments and means of computation, which promised to
reduce subjective interference on the part of the traveller in the results
of expeditionary science. On the other hand, the question of credibility
always had a strong social dimension that was attentive to the gender,
'race' and class of the observer or informant in question.[37]

Equipped with large-scale instrumentation whose use they meticulously explained and which, the brothers believed, ensured the reliability of their data, the Schlagintweits overlaid each traversed region
with a dense matrix of observations and measurements.[38] The following
account does not attempt to give an exhaustive treatment of their myriad
scientific activities. Rather, the analysis will focus on a few but significant fields of enquiry to provide general insights into the kind of work
they undertook. Against much of the existing literature, this means
squarely placing their mission into the history of European science in
south Asia and the official structures that underpinned such large-scale
enterprises. It will thus become clear, as Felix Driver and others have
suggested, that '[e]ven the purest aspirations of the scientific avant-
garde', as expressed in the notion of Humboldtian science and contemporary interests in terrestrial magnetism, were 'necessarily implicated' in
imperial infrastructures.[39] In allowing the Schlagintweits and their contemporaries to pursue magnetic, meteorological and other studies on a
potentially global scale, it furthermore becomes evident to what sizeable
extent 'empire' (in the form of Europe's possession of world-spanning

---

[35] Cooper, 'Voyaging', 44.
[36] Driver, *Geography*, 49–52; Urry, 'Notes'; on the disciplinary controversies in the field of
anthropology, Sera-Shriar, *Making*.
[37] Bourguet and Licoppe, 'Voyages'; Withers, 'Voyages et crédibilité'; McCook,
'Legitimation'.
[38] For example, H. Schlagintweit et al., *Results* II, 11–38.
[39] Driver, *Geography*, 36–8; Fara, *Attractions*.

territories with corresponding observational stations) influenced the operations and outlooks of European science itself.[40]

As the initial trigger for their mission, the Schlagintweits began each *Report* to the EIC with 'Magnetic Observations'. Whenever possible, they punctiliously measured four magnetic forces: the declination, the inclination and the horizontal and vertical intensity, while paying attention to their 'daily variations'.[41] They took magnetic measurements at 112 stations across India and high Asia, twenty of which were situated in high mountain landscapes.[42] The highest point of recording was on the Karakoram pass, 5,575 m above sea level. From the data acquired, they were able to produce intricate visualisations of physical forces across the subcontinent, including a 'Map of Isoclinal Lines' (with the same inclination) and a 'Map of Isodynamic Lines' (with the same intensity; Figure 4.1), which marked a number of irregular modifications, such as a region with significant increase of the total intensity in central India.[43] In exemplifying their search for causal relationships between various forces in nature, the brothers also sought out explanations for any irregularities, for instance that 'the powerful action of a tropical insolation considerably modifies the physical and magnetic conditions of the soil'.[44]

In consequence of their magnetic observations in the highest mountain ranges of the globe, the brothers could disprove Humboldt's earlier hypothesis that the magnetic intensity would decrease with height. While others had already doubted his assumption, the Schlagintweits sided with his critics such as Lamont, and stated that hardly any influence of height could be shown on the total intensity.[45] Their results were, however, less clear regarding the debate over whether the earth possessed two or four magnetic poles. While the Schlagintweits quoted Gauß's *Allgemeine Theorie des Erdmagnetismus* and thus suggested the existence of only two, they also made references to differing opinions, including the works of Halley and their patron Edward Sabine, without taking a clear position.[46]

Meteorology was the brothers' next central field of enquiry in every region. While they and many of their assistants took regular measurements of humidity and air pressure, and studied 'the transparency and blueness

[40] Vetter 'Introduction', 9; Reidy, 'Spatial Science'.
[41] A. Schlagintweit et al., *Report* IV, 6.
[42] H., A. and R. Schlagintweit, 'Latitudes'.
[43] Ibid., 171; for a useful contextualisation of their magnetic surveys, Reich and Roussanova, 'Magnetometer'.
[44] H. Schlagintweit et al., 'Latitudes', 171–2; H. Schlagintweit et al., *Results* IV, 444.
[45] Brewster, *Magnetism*, 275.
[46] H. Schlagintweit et al., *Results* I, 478–82; Reich and Roussanova, 'Magnetometer', 207.

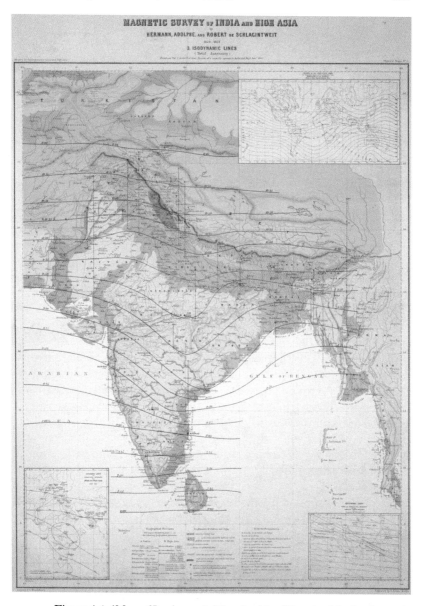

Figure 4.1 'Map of Isodynamic Lines', with a 'Sketch of the Isodynamic Lines of Relative Total Intensity for the Surface of the Earth, Reduced from General Sabine's Map', a product of Britain's global empire, depicted in the top right corner; Physical Maps 3, Hermann Schlagintweit et al., *Atlas*.
© DAV.

of the atmosphere',[47] for them *temperature* was the ultimate key to unlock a whole range of natural and other phenomena. In revealing both the lasting influences of theories of geographic determinism on their thought and an acute awareness of imperial concerns relating to the field of climatic variation, Hermann stated:

Temperature is the active power, the *force vive* of the meteorological phenomena, and the tables introduced by a *descriptive characteristic of climate* can be connected, at the same time, with the modifications of cultivation and settlement – with the habits and manners, even the character of the inhabitants. By its influence upon vegetation temperature is not less important for the types of foreign scenery.[48]

The brothers also studied specific cloud formations and their effects on insolation, trying to capture in aesthetic paintings the resulting 'impression' of various Indian landscapes at different seasons.[49] Yet, the bulk of their routine work was the acquisition of immense volumes of data for later calculation, taken in search of larger patterns and periodic variations.[50]

However, given the short period of time the brothers spent in crossing vast tracts of land on the Indian subcontinent and beyond, they could only realise their meteorological researches through close collaboration with imperial institutions and a large number of recruited observers stationed across India and high Asia. They installed a range of temporary 'observatories' for taking meteorological measurements to complement their own corresponding data sets.[51] Either British officers, or members of the brothers' group of trained scientific assistants, supervised these observatories during short and long periods, even up to a few years.[52] Yet, only the extensive institutional presence of the British allowed the realisation of their ambitious programme, and the brothers actively involved different government departments in their data collecting. In consequence, throughout and also after the expedition, a number of existing government branches – including the officers of the 'Medical Department' – supplied them with comprehensive and invaluable sets of climate recordings. These significantly shaped the scientific results of the whole undertaking in the field of Indian meteorology. The volumes were

---

[47] A. Schlagintweit et al., *Report* II, 108, as printed in *JASB*, 25.
[48] H. Schlagintweit et al., *Results* IV, 3, original emphasis. On their meteorological work in conjunction with imperial concerns, see also Chapter 8.
[49] Ibid., 118, 468.
[50] Finkelstein, 'Mission'.
[51] See for an overview of such cooperation, H. Schlagintweit, *Reisen* I, 229–35.
[52] H. Schlagintweit, *Reisen* II, 334. A. Schlagintweit et al., *Report* V, 559, *JASB*, 25.

received through the support of Dr J. Macpherson, 'first secretary of the Medical Department of the Indian Army'.[53] Hermann, who organised the material for scientific purposes in Europe, later refused to return these unique official records bound together in thirty-nine volumes, to the Indian government – unless he received £300 for his labour.[54] They became the cornerstone of the fourth volume of *Results: Meteorology of India: An Analysis of the Physical Conditions of India, the Himálaya, Western Tibet, and Turkistan* (1866), which is thus essentially a synthesising work of various colonial recordings.

Cooperation with Company men abounded on land and sea. While the brothers had started to make records on board the ship that brought them to India, they later subcontracted the same task. The political Resident at Darjiling, Archibald Campbell, the former travel companion of Joseph Hooker in Sikkim, was 'requested by H. Hermann Schlagintweit to keep a register of the temperature of the surface of the ocean on the voyage round the Cape of Good Hope, as such a register was a great desideratum to him in connection with his other Meteorological researches in the East'.[55] Thuillier, Deputy Surveyor-General of India, personally furnished Campbell with the necessary instruments for the purpose, who then completed his observations from the 'Sandheads of the Hooghly till we entered the Thames'.

What thus emerges from the fields of magnetism and meteorology alone is a dense matrix of cooperation between the Schlagintweits, their indigenous assistants and multiple agents and institutions of imperial science. It would thus be extremely misleading to understand their mission as the linear movement of one exploratory party. Rather, the expedition – which was itself in reality a series of distinct itineraries by different groups of travellers – constituted a complex undertaking whose external boundaries are not easily established. Much observational and recording work was outsourced and independently undertaken by Company servants at numerous localities, even if the data acquired was inconspicuously subsumed under the expeditionary records.[56] The 'Schlagintweit expeditions' were a deeply collaborative project that involved constant reciprocal exchanges between the travelling parties and sedentary scientists, officers and long-existing government

---

[53] H. Schlagintweit, *Reisen* I, 234–35. See also Finkelstein, 'Mission', 198.
[54] NAI, Department of Revenue, Agriculture and Commerce, December 1877; Meteorology, Proc. 2.
[55] Quotations from Campbell, 'Register', 170.
[56] See, e.g., A. Schlagintweit et al., *Report* IV, 27.

institutions from the foothills of the Himalayas to Ceylon, and from Bombay over Madras to Calcutta, which all fed the expedition's chest of measurements.[57] Even if much of the existing literature has precisely portrayed it as such, it was by no means a detached 'German expedition' to India and the Himalayas, independent of Company interests and infrastructures.

In consequence of the Schlagintweits' considerable attention to topography and mountain morphology (an interest already evident in their earlier Alpine studies), their travel records also included a mass of hypsometrical data and observations. Their published volume 2, *General Hypsometry of India, the Himalaya, and Western Tibet* (1862), ultimately contained 'the heights of 3,495 points, of which 1,615 belong to India, and 1,880 to High Asia'. These locations were widely distributed 'from the southern parts of Ceylon to the environs of Kashgar in Turkestan, and from the eastern boundaries of Assam to Singh', yet the Schlagintweits themselves determined only 1,113 of the total.[58] This again exemplifies how much their expedition must not be considered in isolation, but should be seen to form part of a series of earlier and contemporary imperial surveys whose results the brothers only complemented.

Even so, the Schlagintweits showed ingenious skills in transforming the available hypsometrical data, and their own sketches, into new kinds of mountain profiles – thus effectively widening the visual language of Humboldtian science. Their *Panoramic Profiles* (Figure 4.2) were a piece of scientific ingenuity and quite different from their beautifully executed landscape panoramas. With them, the brothers depicted the spatial succession of summits and mountain formations in the Himalayas and Tibet: 'In the profiles, the principal objects, the snowy peaks, are reproduced in full detail and with their various modifications of form. In the middle ground, as far as we found it necessary for completing the general geographical character of the picture, we distinguish distances by modifications of shading.'[59]

The brothers' various visualisation techniques deserve closer attention, as they achieved several goals at once. Their landscape sketches captured important topographical information and provided botanical and meteorological knowledge on traversed regions; while they reflected Romantic sensibilities, they were also consciously used to purvey strategic knowledge to imperial institutions. Hermann Schlagintweit, for

---

[57] A. Schlagintweit et al., *Report* II, 20.

[58] H. Schlagintweit et al., *Results* II, ix.

[59] H. Schlagintweit et al., *Results* II, 264. They resemble Humboldt's hypsometric profiles, e.g., of the Iberian peninsula, as discussed by Dettelbach, 'Global Physics', 264–5.

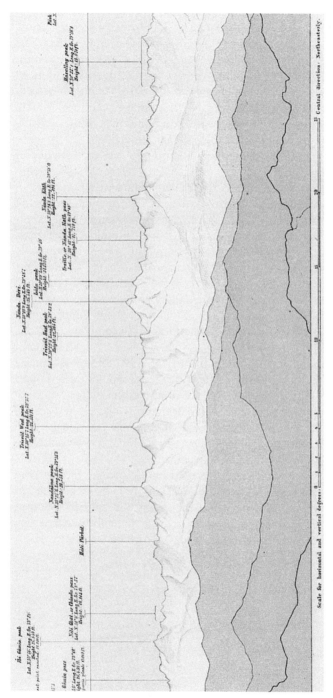

Figure 4.2 'Profile of the Chiner Panorama', drawn by Adolph Schlagintweit, surveyed by Robert Schlagintweit, May 1855 (original *aquarelle* no. 417). Inserted are precise geographical coordinates (height, latitude, longitude) of the peaks and important passes. Additional information denotes the highest elevation where 'phanerogamic plants' were found (19,809 ft), and the highest point the Schlagintweits reached (22,259 ft), on the Ibi Gamin pass and peak, respectively; Hermann Schlagintweit et al., *Atlas*, Part I, *Panoramic Profiles of the Snowy Ranges of High Asia*, II. 'The Himalaya of Kamaon and Garhval', 3.
© DAV.

131

instance, once sent a series of '100 to 120' views of a hitherto imperfectly known mountain range in Sikkim to the Surveyor-General's Office. The German traveller described the particular value of the 'panorama of 360°' at some length: 'The drawings of the same range of mountains having been made from different points of known position, they form pictures complementary to each other, like stereoscopic pictures.'[60] Such a pioneering depiction of a little-studied mountain chain in a politically still sensitive area was of both scientific and strategic value.

To capture characteristic mountain and jungle scenes, varieties of human architecture, holy sites and other *sujets*, Hermann and Adolph produced a few charcoal sketches, but more significantly hundreds of watercolours. They were predominantly begun *in situ*, such as during high mountain ascents, and 'taken from points commanding an extensive view', while the best spot for the drawings was carefully chosen from many potential sites.[61] The sketches and watercolours were usually only completed once the brothers had returned to basecamp. To aid their memory and record topographical features, the brothers also produced accompanying 'explanatory sheets' for many drawings, with written comments on the particular sceneries. Whether these explanatories were produced after the *aquarelle*, or served as their template, remains still unclear.[62] Robert Schlagintweit, and to a lesser extent Hermann and one of his indigenous assistants, also took photographic images of landscapes, buildings, ethnographic 'types' (around 400 alone) and rock formations.[63] Since black and white photographs did not capture the aesthetic quality of natural scenes, Hermann, while relying on his observations on the specific colour range of different landscapes, also made watercolours based on Robert's photographs, thus revealing the aesthetic dimension of their scientific agenda.[64]

One of the earliest field sciences pursued by Company employees in south Asia had been geology (itself an increasingly fashionable pursuit among metropolitan men of science), while many surveys had followed directly in the footsteps of military marches. Geological discoveries by Company servants, such as the hundreds of Siwalik fossils in 1837,

[60] A. Schlagintweit et al., *Report* IV, 3.
[61] H. Schlagintweit et al., *Results* II, 262; A. Schlagintweit et al., *Report* VIII, 8.
[62] Cf. Trentin-Meyer, 'Hochasienreise'.
[63] A. Schlagintweit et al., *Report* II, 109, as printed in *JASB*, 25; Ritter, 'Reise', 164–5; H. Schlagintweit, *Reisen* I, xxxvi, 41. The Schlagintweits' photographs of different Indian 'castes' and occupations were not as standardised as they would become later in the nineteenth century; Ryan, *Picturing Empire*, 142; on the introduction of photography in Indian anthropology, Pinney, 'Colonial Anthropology', 252.
[64] H. Schlagintweit, *Reisen* I, 397–8.

exemplified the fact that some remarkable scientific findings in British India were accomplished without the formal support of government structures.[65] The brothers, in bringing their expertise in this branch of science to India, faced a wealth of existing geological studies. When the 'Geological Survey of India' was formally launched in 1856, there existed, besides the Schlagintweits, dozens of Company officers and naturalists eager to explore the precise nature of sedimentary strata and to detect exploitable deposits of various minerals and resources across British possessions and beyond. The profitable mining of coal at Raniganj and earlier accounts of Company officers 'had invited the attention of the government to the mineral resources of the country', which in turn had led to a resolution in 1835 to found in India a Museum of Economic Geology.[66] Following the orders of the London Court of Directors, the museum was established by the Indian government at Calcutta in conjunction with the Asiatic Society of Bengal, while the society's journal published long lists of locations in India that contained 'ores of iron, copper, tin, lead, silver, gold' and numerous precious stones.[67]

Hence, when the Schlagintweits examined 'different coal localities in the Bhootan Himalayas, and in the Naga Hills' on the border with Burma, which they noted on into maps sent to the Surveyor-General's Office at Calcutta or directly to the Court of Directors in London; when they visited and studied several salt and 'diamond mines' as well as the 'very large quarries and mines, from which is dug the Yashem [Jade] stone' in Chinese Turkestan of which they 'procure[d] for future analysis a good supply of the different varieties ... much valued throughout Central Asia', then the brothers' pursuit of economic geology only *continued* long-established activities by Company servants.[68] In fact, the brothers drew heavily on existing knowledge of minerals and other valuable materials, which turned parts of their field *Reports* and published works into synthesising inventories of profitable resources.

Yet, even if the Schlagintweits' fieldwork incorporated the material interests of the Indian government, imperial concerns by no means exhausted their enquiries. While amassing large collections of 'shells and petrified wood', they were, for instance, eager 'to ascertain with accuracy the age of the [geological] formation' of different regions, taking numerous mineralogical samples along the way.[69] They also constantly

[65] Stafford, 'Annexing', 83; Arnold, *Science*, 26.
[66] Sen, 'Sciences', 47.
[67] Ibid.
[68] A. Schlagintweit et al., *Report* V, 19. A. Schlagintweit et al., *Report* I, 17; 23; A. Schlagintweit et al., *Report* VIII, 8. Hannay, 'Economic Geology'.
[69] A. Schlagintweit et al., *Report* X, 15.

used geo-thermometers 'some of which had a total length of nine feet, the bulb reaching two metres below the surface', and examined over a hundred sites of 'hot-springs' and filled 'a considerable number of fine glass bottles with water'.[70] This was done in the 'hope that the chemical analysis of these waters' may provide insights into sedimentary layers.[71] Being closely attentive to indigenous uses, they further noted on hot springs in Sukki and Mugger Pir: 'In both localities the quantity of water which issues, is pretty considerable, and extensively used by the natives for medical purposes.'[72]

Closely connected with their geological work, and another field of enquiry in which disinterested science and imperial concerns colluded, were the Schlagintweits' detailed studies in Indian and Himalayan hydrography: the surveying and charting of bodies of water – including seas, (mountain) lakes and especially rivers. Here too they entered a crowded field of previous exploration. According to the Victorian geographer and writer Clements Markham: 'The rivers flowing from the Himalaya, and forming the two great systems of the Indus and Ganges, have been studied with minute attention. Upon the water supply brought down by these streams from the Himalayan snows the very existence of millions of people inhabiting the plains of India depends.'[73] Indeed, the 'physical laws regulating the direction and volume of the rivers are of such practical importance that they have formed the subject of close investigation for many years'. There had been concerted efforts on behalf of the Surveyor General's-Office to instruct scientific travellers and officers in the art of river studies. A prime example is the popular *Manual of Surveying for India*, 'prepared for the use of the Survey Department', compiled by Thuillier and Smyth in 1851. A significant part of its chapter 'On Route Surveying and Military Reconnaissance' was devoted to rivers. Since so much of Indian agriculture, irrigation systems, internal commerce and transport depended on them, the manual specifically instructed travellers to study:

The sources of rivers, and the direction of their course – whether they are rapid or otherwise; their breadth and depth, and what variations they are subject to, at different seasons of the year – the nature of their channels and of their banks – whether rocky, gravelly, sandy or muddy – of easy or of difficult access.[74]

---

[70] H. Schlagintweit, 'Remarks on Some Physical Observations Made in India by the Brothers Schlagintweit', *Literary Gazette*, 19 September 1857, 909.
[71] A. Schlagintweit et al., *Report* II, 109, *JASB*, 25.
[72] A. Schlagintweit et al., *Report* X, 15.
[73] Markham, *Memoir*, 255.
[74] Thuillier and Smythe, *Manual*, 356.

Besides further observations on the quality of bridges and the existence of fords, Company employees were further required to establish: 'What rivers are navigable, and from and to what points', which included determining their 'velocity', 'banks' and 'soil of bed', all of which 'may be reported on with advantage'.[75]

After the brothers had met Thuillier at Calcutta at the relative beginning of their mission, it is not surprising to find a considerable overlap between the surveyor's programme of river surveys and the Schlagintweits' own observations. After all, Thuillier had personally provided them with instructions, literature and instruments, and the brothers directly referred to his *Manual* in their *Results*.[76] One of their field *Reports* furthermore noted that: 'The observations on the temperature, velocity and quantity of water, &c., of various rivers, have been continued throughout the journey.'[77] Besides 'ascertaining the discharge of rivers', the brothers also took great pains to establish 'the form and height of their banks etc.'.[78] In addition, 'Observations of the breadth were made by long base lines and triangulations.' The Schlagintweits likewise prepared 'a detailed map' of the Brahmaputra valley, 'showing the different soils, with observations on former levels and beds of the river, and depths from the surface deposits to the sub-soil'.[79] They immediately dispatched this map to the Court of Directors in London, so that the Company headquarter could also receive the results of their ongoing labour.

To trace where the Schlagintweits materials went first and who immediately worked with, modified and completed them can indeed tell us much about the place of their mission in the longer chronology of Company science in India. The example of their hydrographical studies shows that there is a striking combination of moving fieldworkers (the brothers and their assistants) accumulating observations and 'raw materials' in the form of data and sketches, which were then sent to fixed centres of knowledge production and cartography in India. This applied above all to the Surveyor-General's Office – an important centre of accumulation, constantly processing data and visual materials from the Schlagintweits and dozens of other Company employees. Its staff at Calcutta then transformed the information and sketches into more refined products. This mechanism is exemplified through the collaborative production of a visual 'Section across the Brahmaputra River' (Figure 4.3). It was first

[75] Ibid., part IV, ch. VI, 'On the Orthography of Native Names, General Statistics, Geographical, Revenue and Agricultural Reports', 636–7.
[76] See, e.g., H. Schlagintweit et al., *Results* II, 11; 68.
[77] A. Schlagintweit et al., *Report* VI, 102, *JASB*, 26.
[78] A. Schlagintweit et al., *Report* X, 215, ibid.
[79] A. Schlagintweit et al., *Report* V, 568, *JASB*, 25.

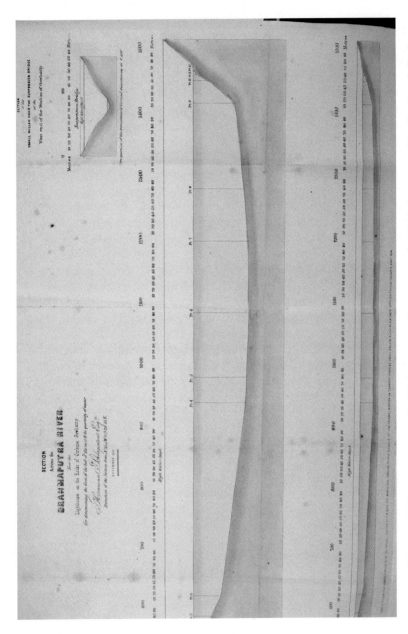

Figure 4.3 'Section across the Brahmaputra River, near the Lighthouse on the Rocks of Oorbosee, Gowhatty'. It includes a smaller 'Section of the small Nullah', i.e. ravine, 'near the Suspension Bridge at the West end of the Station of Gowhatty', with further information on the 'High Water Mark'; *JASB*, 25, 1.

surveyed and sketched by Hermann Schlagintweit in November 1855, but subsequently 'reduced to half the scale of the original & drawn on transfer paper by Abdul Hallim & lith. by H. M. Smith, Surv. Gen. Office Calcutta' in April 1856.[80] The purpose of this work was to 'determin[e] the form of the bed of the river & the quantity of water'. The finished section, whose making thus involved multiple hands of British and Indian artists and lithographers, later adorned Hermann Schlagintweit's next field *Report* and was elaborately printed in colour in the *Journal of the Asiatic Society*. The fact that not only were the river observations and sketches made by the brothers and their indigenous assistants, but that a larger workforce at permanent offices also worked out their combined results, again demonstrates the deeply communal nature of scientific fieldwork.[81]

It seems, however, as if the navigability of rivers and their role in facilitating trade within India, between the Himalayas and the plains, and even between India and China via Burma, became an ever-greater concern to the Schlagintweits especially after their return.[82] While they had taken numerous measurements relating to issues of navigation (such as on river beds at high and low water level), they fully analysed the consequences for commercial interactions only in their publications.[83] The brothers' contiguous engagement with rivers as arteries of trade is relevant, since it demonstrates that their proposed programme of resource exploitation in India and the Himalayas clearly *evolved* over time, and was constantly readjusted to ongoing British expansion. One telling example concerns Hermann Schlagintweits' earlier 'discovery' of 'very valuable brown coal' seams 'in the hilly ranges at the northern end of the Tarai', samples of which Hermann immediately presented 'to the Chief of the provincial administration, Colonel Jenkins', with details about the localities.[84] Yet, as he reflected in hindsight,

the entire area from Bhutan towards Assam was so inaccessible to the Indian government that one could not think of a further exploitation of these workable seams. The last war with Buthan, which brought a considerable extent of the outliers under English rule, may provide the initiative to use such coal seams.

[80] Printed in *JASB*, 25, 1; 6.
[81] Camerini, 'Wallace', 45.
[82] Hermann considered the potential of a 'great commercial exchange between Chinese and European manufacturers' via the 'navigable part of Irávadi valley' [Irrawaddy River; in Burmese: Ayeyarwady] and the source of the river in a region called Lokathra by the Tibetans, H. Schlagintweit, *Reisen* I, 477.
[83] For example, H. Schlagintweit et al., *Results* II, 161, on the limited navigability of rivers in central India.
[84] H. Schlagintweit, *Reisen* II, 110–11.

They would have an even greater importance due to the fact that they are at no great distance from the Brahmaputra and its navigable tributaries, which would facilitate transport across wide distances.[85]

While some of the Schlagintweits' suggestions for resource exploitation thus emerged after a considerable time gap, some British officials, by contrast, were quicker to use their observations and materials in trying to improve the output of India's natural worlds. For instance, John Forbes Watson, the newly appointed 'Reporter on the Products of India' at the India Office, had the brothers' water samples from various rivers tested in connection with experiments on the irrigation of Indian cotton fields.[86] Forbes Watson also lost little time to have a selection of the Schlagintweits' 1,200 soil samples chemically tested – with the aim of gaining insights into the potential expansion of flax, rice, wheat, indigo, tea and cotton cultivation in India.[87] In winning a gold medal for his achievement, he subsequently presented his findings at the 1862 Colonial Exhibition in London.[88] In suggesting another continuity between the Schlagintweits' work and earlier researches by Company officials, it stands to reason that Forbes Watson had personally instructed them to take these soil samples when they met in Britain. Returning to London from India in 1853, Forbes Watson himself had collected and tested such samples 'with the view of furthering the agriculture and commerce of India', while serving in the EIC's medical service in Bengal and Sind between 1850 and 1853.[89]

The results Forbes Watson gained from the brothers' soil collections, were, he was certain, 'of importance', because they provided 'fresh grounds of encouragement to those who at length appear inclined to engage in private enterprise for the development in India of the raw textiles for which this country is at present dependent upon somewhat questionable and uncertain sources.'[90] The example of the Schlagintweits' water, sand and soil samples at the same time demonstrates how closely connected landscapes and labscapes were in their project of expeditionary science. That is, ultimate insights from the raw samples could only be gained

[85] Ibid.
[86] Forbes Watson to J. Danvers, 4 August 1859, BL, IOR, MSS EUR F195/5.
[87] 'Indian Soils. Analysed under the direction of the Reporter on the Products of India', extract from chart. Forbes Watson, *Classified and Descriptive Catalogue*; Class I: India. Mining, Quarrying, Metallurgical Operations, and Mineral Products. Subdivisions VI. Soils and Mineral Manures, 23–6.
[88] Ibid., 286. For the importance of soil testing, Kumar, *Indigo*, 87–8.
[89] 'Portrait and Biography of Dr. John Forbes Watson', *Journal of Indian Art*, 3 (1890), 8–11, 8.
[90] Forbes Watson to T. L. Seccombe, Secretary in the Financial Department, 13 June 1860, L/F/2/241,802; also quoted in Armitage, 'Collections', 70.

through careful industrial analysis in the confined space of laboratories, which in their case were mostly located in Europe.

Yet, beyond such utilitarian interests, the brothers were also deeply interested in the sources of various rivers, and 'the astonishing vastness of their erosions' that shaped, over large periods of time, the morphology of mountains and led to the 'gradual production of valleys'.[91] In 'the upper course of the Ganges, the Sutlej and the Indus', the brothers claimed that the erosion 'even attains the extraordinary magnitude of 3000 feet'. Typifying their mutable programme, the Schlagintweits discovered the spectacular effects of this force of nature only relatively late. It was 'so new a subject of investigation, that we had at first much difficulty in finding those data which might guide us in the definitive determination' of the various consequences of this phenomenon. The brothers had, to be sure, already studied the effects of erosion in the Alps, but were taken aback by its sheer magnitude in the Himalayas.[92]

As skilled and enduring mountaineers, it is not surprising that the Schlagintweits obtained some of their most important results at high altitudes. Not only did they establish, in several regions, the limits of vegetation and eternal snow; they were also the first to determine the mean elevation of high mountain plateaux in the different high Asian ranges.[93] In using their own bodies and those of their assistants as physical instruments, they further provided acute observations on the 'Influence of height upon Man'.[94] Yet, perhaps their most relevant and innovative studies concerned the investigation of over forty glaciers that 'according to the classification in the Alps, must be termed glaciers of the first order'.[95] Not only was this considerable number something new, but also the brothers' ability to compare Himalayan glaciers with those they had studied in the Alps. Indeed, in seeking out physical laws and intercontinental commonalities, they observed that the 'physical struc-ture of the glaciers of the Himalayas, the laws of motion, the distribu-tion of the moraines and of the crevasses, is precisely the same as in the glaciers of the Alps'.[96] The Schlagintweits contributed enduringly to the field of Himalayan glaciology by taking detailed measurements and pro-ducing a series of sketches and *aquarelles* of their present forms. These depictions, together with topographical maps of several glaciers in their

[91] H. and R. Schlagintweit, 'Erosion', 91. Subsequent quotations, ibid.
[92] A. Schlagintweit, *Thalbildung*.
[93] A. Schlagintweit et al., *Report* II, 109; 114, *JASB*, 25.
[94] Ibid., 115.
[95] Ibid., 123.
[96] Ibid.

larger environments, were astonishingly reliable, and are still used today for comparative, diachronous studies.[97]

As they progressed, the Schlagintweits also produced a number of outline sketches that added textual information to their *aquarelles* and named salient topographical features of the landscapes, then a standard practice of transit surveys. Added to this imagery was the brothers' gathering of route intelligence. As suggested to Company surveyors in Thuillier and Smythe's *Manual of Surveying*, they made a wealth of observations on what roads and mountain passes were accessible, for whom (men, horses, 'animals of burden'), and to what degrees during different seasons.[98] The Schlagintweits also meticulously calculated the distances between major trading and political centres 'in the western parts of the Himalaya, Tibet and Central Asia', and the number of (in a telling military rhetoric) 'daily marches' necessary to complete the itineraries.[99] While providing in their *Route-Book* for 241 different routes 'explanations as to the roads, nature of the passes, and heights of various places, supplies, fuel, and general information' for their successors,[100] the Schlagintweits also marked out those 'provinces which, for the present at least … are only accessible to Europeans under the frail protection of a disguise'.[101] By extending the range of their information, the brothers further inquired about unknown roads and commercial arteries through high and central Asia with 'intelligent native merchants' and the leaders of encountered caravans.[102] In signalling their mistrust of foreign testimony, they regularly compared, however, the accounts of different traders with one another and with their own observations and estimates.

Related to their practice of using caravan traders as informants, the brothers paid minute attention to indigenous designations of routes, lakes, peaks and other topographical features. To fight 'the wild confusion that still prevails respecting Indian orthography', imperial survey manuals always stressed that: 'Care must be taken that the names of towns, villages, rivers &c., are spelt in the same manner as by the natives of the country, and when the spelling and the pronunciation differ very much, the name should also be written (in a parenthesis) as it is pronounced.'[103] The Schlagintweits heeded this advice. They

[97] Mayer, 'Gletscherforschungen'; Nüsser, 'Natur'.
[98] H. Schlagintweit et al., *Results* III [*Route-Book*], 6. Thuillier and Smythe, *Manual*, 357.
[99] R. Schlagintweit, 'Angaben über die Entfernung zwischen den wichtigsten Städten', 393. On the relation between the internal hierarchies of expeditionary science and military organisation, Thomas, 'Expedition', 67.
[100] H. Schlagintweit et al., *Results* III, 4.
[101] Ibid., 4, also 17.
[102] Ibid., 3.
[103] Thuillier and Smythe, *Manual*, 359.

ultimately produced an entire *Geographical Glossary from the Languages of India and Tibet, including the Phonetic Transcriptions and Interpretation*, in which they explained the meaning of 1,200 indigenous names – an undertaking only made possible through the help of one of their hired munshis, Sayad Mohammad Said, from Calcutta.[104] This dictionary was complemented by detailed information on the precise pronunciation of various indigenous terms in multiple languages, together with a discussion of the best method of transliteration. While the brothers claimed that this endeavour led them 'to acquire a practical knowledge of Hindustani' for 'directly consulting the natives of the various regions, for dialectical forms, as well as for the meaning', the published work was the result of the linguistic expertise provided by their diverse indigenous companions.[105] Indian assistants, translators and guides were also responsible, as the next chapter will show, for acquiring a vast amount of cultural and religious artefacts and indigenous manuscripts, even from places the brothers were forbidden to enter, such as Tibetan monasteries. This collecting through mediators, together with the philological and religious interests the brothers pursued thanks to the service of several south Asian go-betweens, reveals that the diverse social composition of the expedition allowed the combination of natural scientific studies with humanistic and philological enquiries.

While the brothers approached Indian orthography and languages mainly through intermediaries from their establishments, their ethnographic and especially physical anthropological programme directly touched on their 'objects of study'. In keeping with their numerical approach to documenting natural historical phenomena, the Schlagintweits also trusted the supposed precision of numbers to identify 'ethnographic' variation. In this, they followed contemporary practices. Among the most influential British racial scientists at the time was the medical doctor Joseph Barnard Davis. By 1867, Davis had amassed 1,474 human skulls in his private study, 'which made his collection larger than those of all the public museums in Britain put together – in fact, the largest in the world'.[106] To understand why the Schlagintweits measured human bodies – and what parts of them – in Asia, it is imperative to understand Davis's and other contemporaries' quest 'to determine the physical proportions of different human races, and to ascertain their essential diversities'.[107] The physical anthropologist loathed the

---

[104] H. Schlagintweit et al., *Glossary*, in *Results* III, part 2, 133–293, on Said, ibid., 138; and Chapter 5.
[105] H. Schlagintweit et al., *Results* III, 137–40.
[106] MacDonald, *Remains*, 96.
[107] Davis, 'Measurements', 123; subsequent quotations, ibid.

fact that: 'Travellers have generally contended themselves with speaking in indefinitive comparative terms of the people with whom they have come into contact' without submitting 'any considerable number of these people to the test of measurement.' The human head was the main target, since:

[W]e are fully assured that this latter division of the body is the seat of those faculties which lie at the base of all the peculiarities of human races, bearing essentially and intimately upon their manners and customs, all their institutions, their religious impulses, their degree of civilizibility, and the developement [sic] to which it has attained.

Since Davis and his followers believed that fundamental distinctions 'prevail among human races', which were 'capable of metrical appreciation', it is no surprise to find that he showed significant interest in the Schlagintweits' anthropological work.[108] In this field of enquiry, too, we can see how deeply their activities in India were shaped by the imperial context of the whole enterprise.

By Grace of the Company: The Schlagintweit's Physical Anthropology in the Context of British Rule in India

Few examples better reflect the power structures that enabled a number of the studies undertaken by the brothers than the fact that the Schlagintweits received permission to produce facial masks of Indian prisoners. Hermann thanked Dr Mouatt in particular, 'since he, as Inspector of Jails ... offered me the opportunity to measure individuals of those Indian races, which otherwise would not have been approachable. I thereby managed, on several occasions, to even induce some members of the wildest tribes to have their facial structures copied by applying plaster directly' to their faces.[109] For these living 'objects of study' (Figure 4.4), the procedure – which lasted between thirty and forty-five minutes – was highly unpleasant and could result, depending on the quality of the plaster, in the serious irritation or even burning of the skin. In all cases, however, the men, women and children being cast had to endure the discomfort of breathing through moistened paper cones inserted into their nostrils.[110]

Since the brothers' series of plaster casts that were later displayed in various Indian and European museums marked, not least in the eyes of

---

[108] He lauded their 'many curious results, chiefly pointing to the different proportions of parts of the bodies and limbs of ... tribes of the Himalaya and of India', ibid., 124–5.

[109] H. Schlagintweit, *Reisen* I, 236. See also Finkelstein, 'Mission', 198.

[110] A vivid description of the procedure is given by Zimmerman, 'Gipsmasken'.

Figure 4.4 Example of a Schlagintweit facial cast, 'Dinanáth, Bráhman, Kalkútta, Bengál', plaster cast taken in Alipore jail, Calcutta, 15 April 1857; according to the date, it could only have been produced by Hermann Schlagintweit.
© Muséum national d'Histoire naturelle, Paris [MNHN], MNHN-HA-3160, former catalogue number, Schlagintweit 157.

contemporaries, an important episode in the development of physical anthropology, it is instructive to explore how and why they started to produce such masks in the first place.[111] Already at their port of entry to India, Bombay, the brothers mingled with the city's elite. One of their most important contacts here was the British scientist, museum curator and newspaper editor, Dr George Buist (1804–60), widely considered the most eminent man of science in town. In 1839, Buist had left Britain to become chief editor of the *Bombay Times*. Since then, he had combined his position with an active engagement in the Royal Geographical Society of Bombay, delivering numerous papers on various subjects of natural

---

[111] 'Government Central Museum', *Allen's Indian Mail*, 22 July 1859, 608.

history to that and other learned institutions. Buist also made extensive meteorological surveys in his capacity as honorary inspector of the Colaba Observatory.[112] Yet, his scientific ambitions went further: Buist had started in the 1840s to create a collection for a projected 'economic' museum in Bombay.[113] In 1851, for London's Great Exhibition, Buist was put in charge of gathering specimens from the presidency to be displayed in the Empire's capital, and a few years later, Lord Elphinstone formally appointed him as 'curator and secretary' of Bombay's Economic Museum, which was established in the mid 1850s.[114]

When the Schlagintweits met Buist, he had already 'made some dozen of [ethnographic] masks … of the most interesting description, as illustrative of the physiognomies of the races amongst whom we live'.[115] While it remains unknown if the Schlagintweits had seen such objects of physical anthropology in Europe before, it is certain that they acquired the technique in India.[116] Buist himself claimed that '[w]hen the brothers Schlagentweit visited Bombay in November 1854, they were unacquainted with the art of plaster-casting. I taught them. They seemed struck with the facilities it afforded for ethnographical enquiries, and wished me to provide them with stucco enough for their journey'.[117] For practice, the brothers took each other's casts in Bombay before parting ways.[118] With the Schlagintweits having (as we shall see later) their own 'India museum' in mind, they adapted the technology in order not only to cast indigenous faces and heads but also hands and feet, then regarded as the most important parts of the human body for racial studies. 'The hands and feet were unique because their functions differed more from each other in humans than in any other animal. Furthermore, the feet were associated with the erect posture of humans, and the hands represented human art and skill.'[119] While many Indian subjects resisted the procedure, prisons in particular guaranteed them enforced access to their 'living models'.

[112] Markham, *Memoir*, 215.
[113] Basa, 'Museums', 468.
[114] Ibid.
[115] Buist and George Birdwood, 'To the Editor of the "Bombay Times"', Österreichischer Alpenverein, Innsbruck [ÖAV], 'Collectanea Critica' [CC]. Buist had previously worked with plaster to produce replicas of archaeological objects, both in Scotland and India.
[116] Anthropological plaster casting was first used by French physical anthropologists between 1837 and 1840 during the Pacific expeditions of Jules Dumont D'Urville; Sysling, *Racial Science*, 74.
[117] Buist and Birdwood, 'To the Editor'.
[118] Ibid.
[119] Zimmerman, *Anthropology*, 165.

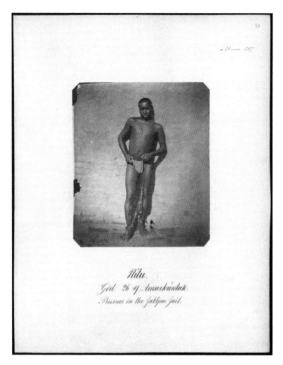

Figure 4.5 Enchained prisoner, 'Nítu, Gôd, 26 Y[ears], Amarkántak. Prisoner in the Jáblpur Jail', photograph by Robert Schlagintweit. © BSB, SLGA, IV.2.53.

The Schlagintweits' use of jails also allowed them to complement their anthropometric studies of different 'Indian races' in two other ways. First, they made photographic records of partly bound or shackled prisoners (Figures 4.5, 4.6). Second, they took extensive anthropological measurements of the entire bodies of these convicts, and had the data noted down on previously printed forms (Figure 4.7). It seems that a member of the prison staff (or one of Hermann Schlagintweit's companions) had assisted in this process for the study of female convicts in particular, since the written measurements do not correspond with either of the Schlagintweits' own handwriting, and are partly written in Hindi. In Calcutta's Alipore jail alone, some twenty-five individuals were thus methodically measured and twenty heads reproduced in plaster. Back in Europe, the brothers would seek 'to produce the heads in as great perfection as possible', which 'led to a different method, consisting in making strong metallic casts of zink the basis, coated with

Figure 4.6 'Group of Hindú-Women from Bengál in the Alipúr-Jail. 1 and 4 Bráhmans [sic], 2 Rajpút, 2 and 5 Súdras', photograph by Robert Schlagintweit.
© BSB, SLGA, IV.2.2.

a galvanoplastic deposit of copper, varied in [four different groups of] color according to the different degrees of color of the native tribes'.[120]

The brothers had sought expert advice on how to take such anthropometric measurements. 'Before leaving Europe, [in] 1854, we had drawn up a list of objects to be measured, in composing which we were chiefly guided by the advice of our venerated friend, M. Quételet.'[121] Adolphe Quetelet (1796–1874) was one of the foremost social statisticians of the time, and also a leading theorist in the field of anthropometry. The brothers used his suggested 'list of objects to be measured' throughout India and central Asia, merely 'modify[ing] it in some details'.[122] Like other

---

[120] Gilman, 'Ethnographical Collections', 235.
[121] H. and R. Schlagintweit, 'Statistics', 500.
[122] Ibid.

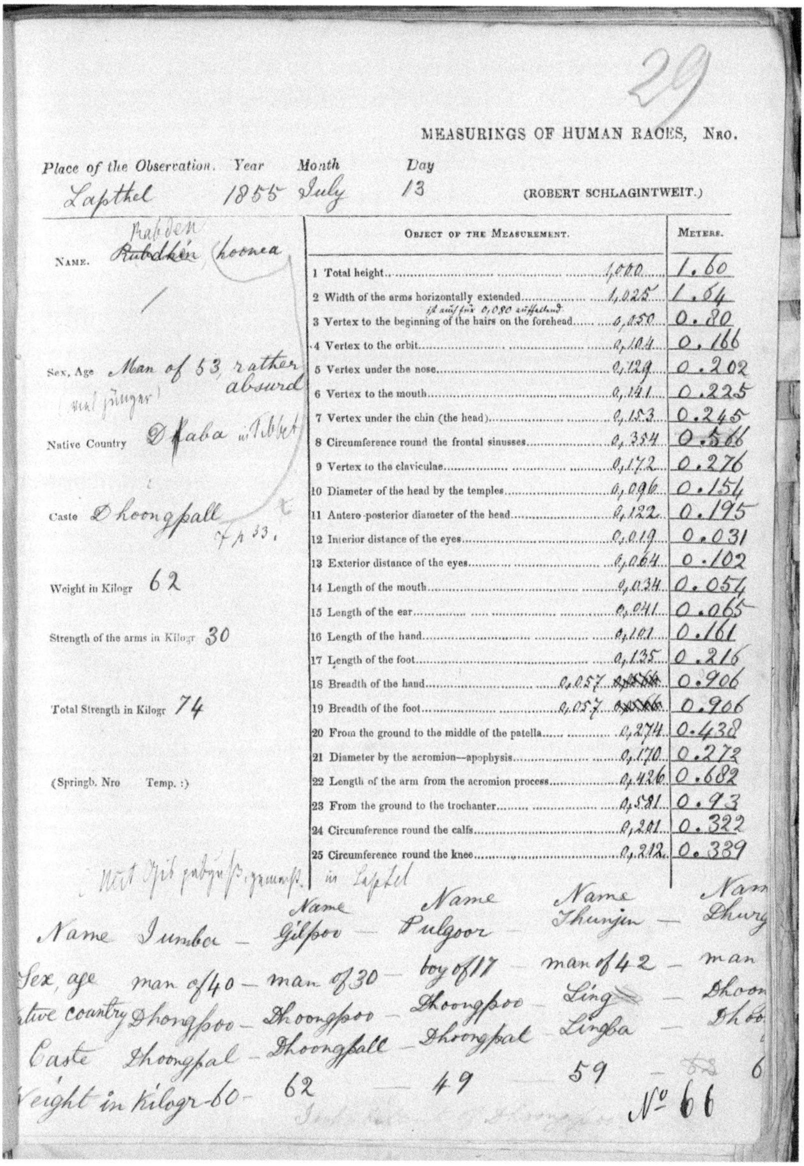

29

MEASURINGS OF HUMAN RACES, NRO.

Place of the Observation. Lapthel Year 1855 Month July Day 13 (ROBERT SCHLAGINTWEIT.)

NAME. Rabden Rubdhen Choonea

Sex, Age Man of 53, rather absurd

Native Country Dfaba in Tibet

Caste Dhoongpall

Weight in Kilogr 62

Strength of the arms in Kilogr 30

Total Strength in Kilogr 74

(Springb. Nro Temp. :)

Object of the Measurement.		Meters.
1 Total height	1,000	1.60
2 Width of the arms horizontally extended	1,025	1.64
3 Vertex to the beginning of the hairs on the forehead	0,050	0.80
4 Vertex to the orbit	0,104	0.166
5 Vertex under the nose	0,129	0.202
6 Vertex to the mouth	0,141	0.225
7 Vertex under the chin (the head)	0,153	0.245
8 Circumference round the frontal sinusses	0,354	0.566
9 Vertex to the claviculae	0,172	0.276
10 Diameter of the head by the temples	0,096	0.154
11 Antero-posterior diameter of the head	0,122	0.195
12 Interior distance of the eyes	0,019	0.031
13 Exterior distance of the eyes	0,064	0.102
14 Length of the mouth	0,034	0.054
15 Length of the ear	0,041	0.065
16 Length of the hand	0,101	0.161
17 Length of the foot	0,135	0.216
18 Breadth of the hand	0,057	0.906
19 Breadth of the foot	0,057	0.906
20 From the ground to the middle of the patella	0,274	0.438
21 Diameter by the acromion—apophysis	0,170	0.272
22 Length of the arm from the acromion process	0,426	0.682
23 From the ground to the trochanter	0,581	0.93
24 Circumference round the calfs	0,201	0.322
25 Circumference round the knee	0,212	0.339

Name	Jumba	Gilpoo	Pulgoor	Thunjin	Dhurg…
Sex, age	man of 40	man of 30	boy of 17	man of 42	man
Native country	Dhongpoo	Dhoongpoo	Dhoongpoo	Lingba	Dhoon…
Caste	Dhoongpal	Dhoongpall	Dhoongpal		Dho…
Weight in Kilogr	60	62	49	59	6…

No 66

Figure 4.7 Example of a preprinted and completed list by Robert Schlagintweit for racial studies of a 'man of 53' years, 13 July 1855, in Lapthel.

© BSB, SLGA, II.1.38.

contemporaries, they were convinced that only 'direct measurements of a sufficient number of individuals' would allow the 'physical definition of the various subdivisions of the human race' in Asia. In what may be dubbed salvage physical anthropology *avant la lettre*, the brothers also stressed 'the high importance of such observations [i.e. measurements], particularly in reference to the aboriginal tribes, many of which are now already very limited in number and disappearing fast'.[123] In total, the brothers took 'measurements of the head and different parts of the body' of '730 individuals', and produced 275 facial casts.[124]

Besides their access to the prisons, the brothers were also able to procure skulls or even entire skeletons from colonial hospitals, provided to them through their contacts with Indian and European doctors.[125] These complemented those human remains that the Schlagintweits obtained by single-handedly plundering tombs: 'In the territories of those nations which bury, such as the Mussulmen and the Buddhists in Tibet, it was possible to find, by opening not too old tombs, well-preserved skeletons.'[126] How effectively they used opportunities provided by their privileged status and rights of access to morbid sites in India is reflected in the fact that the brothers procured '83 skulls and 32 entire skeletons'.[127]

Company Power and Imperial Knowledge Networks

It was furthermore important for the execution of their expedition that the political influence of the Government of India reached further than its possessions formally extended. A number of British political Residents were often well informed about the political conditions and local conflicts that might affect the brothers' proposed itineraries – even in lands beyond formal British rule in high and central Asia. The securing

---

[123] Ibid.
[124] Ibid.; H. Schlagintweit *Reisen* I, 487. Vol. VIII of their *Results* was to be on 'Ethnography', yet was never completed. It was to be 'based on the critical examination and comparison of the following materials': their collected 'Skulls and skeletons', 'Numerous [anthropometric] measurements', '275 plastic facial casts' and a 'great number of photographs … of the different races and tribes'. H., A. and R. Schlagintweit, *Prospectus: Results of a Scientific Mission*, 10.
[125] *Reisen* I, 235.
[126] Ibid., 235–6. On the practices of collecting and looting in the context of British imperial surveys, and how this was linked to commodifying India, its resources and peoples, see Cohn, *Colonialism*, esp. ch. 4: 'The Transformation of Objects into Artifacts, Antiquities, and Art in Nineteenth-Century India', 76–105.
[127] H. Schlagintweit, *Reisen* I, 487.

of political intelligence was of fundamental interest to the British rulers and the brothers did profit immensely from the established information channels of colonial officials, who were stationed at the borders of, or even within, the at least formally still independent Indian princely states.[128]

Among these political informants – who proved important not only for the Schlagintweits' endeavours – were Colonel Ramsay, British Resident in Kathmandu, Brian Houghton Hodgson in Darjeeling and Lord William Hay in Shimla, later the elevated 'summer capital' of British India.[129] Many of these Residents were also engaged in scientific pursuits. They thus possessed precious sources of local and regional expertise for the travelling brothers. Hodgson alone compiled several hundred scientific papers, and was arguably the greatest European connoisseur of Himalayan natural history when Hermann met him in Darjeeling: 'In view of Hodgson's wide-ranging scientific investigations, it was highly beneficial for me during my first visit of the Himalayas to meet with him, the more so since he informed me with the greatest liberality of all his important experiences.'[130] Hodgson also compiled, in Nepali, a letter to the king of Nepal asking for access and support for the Schlagintweits' researches just as help had previously been given to Joseph Hooker.[131]

Former 'diplomat-geographers', British travellers with official political assignments such as Mountstuart Elphinstone as the past leader of a diplomatic mission of the EIC to Afghanistan in 1808, also possessed valuable travel experiences and insights into the countries beyond the north Indian frontier. The German brothers sought to make use of this store of knowledge and to tap Elphinstone's expertise during long days spent with this prominent informant.[132] In order for the brothers to be introduced to and accepted within the political and scientific circles of British India, they once again made ample use of Alexander von

---

[128] On the crucial role of information gathering for the establishment, expansion, maintenance and ultimately the demise of EIC rule in south Asia, see Bayly, *Information*. Given the brothers' constant cooperation with Company servants and south Asian assistants near and far for the exchange of information, skills, instruments, recordings, letters, charts or bills of exchange (hundis) with Indian merchants to finance their expedition in the northern episodes, it is useful to take seriously the importance of 'circulation' in their enterprise, a theme expertly explored in Markovits, Pouchepadass and Subrahmanyam (eds.), *Circulation*.

[129] For Ramsay, H. Schlagintweit, *Reisen* II, 235; for Hay, ibid., 393.

[130] H. Schlagintweit, *Reisen* II, 271; On Hodgson's far-ranging expertise, see Waterhouse, 'Hodgson', 7; Arnold, 'Frontier'.

[131] I thank Nancy Charley for this information.

[132] H. Schlagintweit, *Reisen* I, 45.

Humboldt's 'cunning letters'.[133] Humboldt's own reputation among the learned in India meant that his letters of introduction to people like Sir James William Colville, president of the Asiatic Society of Bengal (1848–59), had a snowball effect, unlocking government and personal support across the British possessions.[134]

While the brothers became swiftly integrated into British imperial knowledge networks, they could also profit from the political pressure that the colonial government could exert on indigenous rulers. Even before the Schlagintweits' travels, the Indian government had developed a particular diplomatic parlance in its interaction with other Indian or Himalayan rulers, in which the right of access into non-British territories could, ostensibly politely, be sought. Yet, through the evocation of specific terms, the Indian government could at the same time exert unmistakable diplomatic pressure. Such strategies are epitomised in the expedition of Joseph Hooker. When the British botanist, in 1849, sought to penetrate into Sikkim – a small, poor state flanked by Nepal and Bhutan, with borders to the Chinese-controlled Tibet in the north and the Company territories in the south – the raja of Sikkim, Chomphoe Namgye, was 'understandably anxious not to annoy any of his powerful neighbours so he and his chief minister, the Dewan, were suspicious of travellers like Hooker who surveyed and made maps'.[135] Consequently, the raja placed a number of restrictions in the path of the British travellers.

To resolve matters, the 'political superintendent' of the British in Darjeeling, Archibald Campbell, weighed in and reminded the raja of 'the duties of friendship' that he was supposed to fulfil in his relations with the British hegemonic power. The subtext of this reminder was clear, and Campbell's aim was effectively to blackmail the raja into granting Hooker's party free passage.[136] In Campbell's report to the secretary of the Governor-General, he stated explicitly in reference to this incident: 'It has always appeared to me that we owe it to the maintenance of our proper position towards the Raja to claim the privilege of free resort into Sikkim for all European and Indian British subjects.'[137]

The Sikkim raja, in turn, responded to British pressure by trying to marshal not political, but rather religious reasons in order to keep the

[133] Humboldt to HS and AS, 4 September 1854, SSMB; also Humboldt to Hodgson, 26 February 1854, RASA, BHH/1/81.
[134] See parts of Humboldt's letter reprinted in *JASB*, 24 (1855), 183–4; Colville's response to Humboldt, Calcutta, 16 March 1855, BSB, SLGA II.1.43, 71–2; H. Schlagintweit, *Reisen* II, 171.
[135] Endersby, 'Joseph Hooker and India'.
[136] See a number of letters, including the reply of the raja, RBG, JDH/1/11, 236ff.
[137] Campbell to Elliott, 'Sec. to Govt. of India', Darjeeling, 15 March 1849, ibid.

intruders at bay. Yet, since he was eager to avoid any open conflict with the Governor-General, he thus proposed the following compromise:

British Gentlemen are prohibited from travelling in my territories .... With the exception of this ... I can do anything else in virtue of my friendship. I have consulted the Lamas, as to whether it is good and proper that British Gentlemen should examine the Trees and Plants of my Country, the result is that it will not be proper. I cannot however quite refuse the requisitions from Calcutta; [will] ... the Governor General ... therefore be so good as let me know what Trees and Plants are required, and I shall send them by my own people.[138]

The British refused this proposal, and Hooker and Campbell ultimately forced their way into Sikkim. The political escalation following the imprisonment of the two British subjects by Sikkim officials entailed first a diplomatic and then nearly a military confrontation. In the end, the Indian government emerged victorious in the conflict, as they annexed parts of Sikkim (around 640 square miles), and stopped the payments of a yearly pension of Rs 6,000 to the raja.[139] This episode, which occurred shortly before the Schlagintweits' explorations, demonstrates how local power relations between the empire and its neighbours were negotiated through the right of passage and of unrestricted scientific scrutiny of (semi-)autonomous states. Apparently, British officials exerted similar pressure to secure the Schlagintweits' access into Nepal and other states. The brothers were aware of the seclusion of this little-explored country. It was therefore no surprise to them that it 'was only after two years' of 'diplomatic negotiations, very kindly entered into upon our behalf by the governor-general and Colonel Ramsay, British Resident in Kathmándu, that the Court of Nepal allowed' at least Hermann 'to visit a portion of its territories'.[140] Hermann, in turn, copied a 'Nepalese map of the country north of Phaywa Lake' from Jung Bahadoor, and inserted information on houses and shelters along different routes in the country, not all of which housed 'guards'.[141]

Alarmed by the experience of the still ongoing British northward expansion, many indigenous rulers perceived the brothers as potential vanguards of Company interests and schemes over their territories. Therefore, upon arriving at the frontier of Nepal, Hermann 'had quite an official reception, and a guard of sepoys constituted themselves my constant companions, partly in the capacity of guides, but more especially for keeping watch upon my operations'.[142] The fact that the German

---

[138] 'From the Sikim Raja to the Superint. at Darjeeling', ibid., Campbell's translation.
[139] H. Schlagintweit, Reisen II, 180. Bhattacharya, Contagion, 24.
[140] H. Schlagintweit et al., Results I, 29.
[141] See the map with HS's inscriptions at BSB, SLGA, IV.5.49.
[142] H. Schlagintweit et al., Results I, 29.

travellers were always equipped with travel passes and official letters of protection by the Indian government further reinforced the perception that they were British agents. The brothers gratefully acknowledged this important support:

[E]very official assistance was most kindly given to us, and we found ourselves liberally provided with the necessary orders to the respective civil and military authorities, and with diplomatic introductions to the Courts of the Native States. These documents were of the most essential importance in enabling us to extend our mission into countries, which, otherwise, we could never have hoped to reach, and which, indeed, were far beyond the limits of our original intention.[143]

Several excursions by the Schlagintweits into the kingdom of Kashmir reflect this indirect influence that the British had beyond formally ruled territories. During their first visit to Raja Gulab Singh, then reigning over Jammu and Kashmir, the brothers were successfully 'introduced ... through a letter by the governor-general during an official handing over of our papers'.[144] The formality of their meeting with the raja, during which Robert presented him with a photograph as a gift, eased the way for later negotiations with Gulab Singh's officials, when the Schlagintweits had to rely on their hospitality once more.[145] As the brothers noted, 'the influence of English power upon Kashmir was tangible enough' by the mid 1850s to ensure that their researches in Leh could not be forcefully prevented.[146]

Nevertheless, how unwelcome the European travellers were in Kashmir was well illustrated by the fact that the brothers, like Hermann later in Nepal, were presented with an unsolicited 'guard of honour' – a group of men was ordered to accompany them and watch their every step. Spying on these foreign naturalists was a serious assignment: when the brothers managed through deception to get rid of the company in order to secretly penetrate into Turkestan, Basti Ram, the governor of Leh, being 'highly exasperated with the members of our guards of honour' had – according to the brothers' account of events – the entire group of soldiers thrown into prison.[147] The brothers further claimed that he also 'immediately sent another group of soldiers after us with the strictest orders to bring us back again, even by force if it were necessary'.[148]

---

143 Ibid., 6.
144 H. Schlagintweit, *Reisen* II, 427–8.
145 Ibid., 428f.; III, 277–9.
146 H. Schlagintweit, *Reisen* IV, 201.
147 BSB, SLGA, V.2.2.2, 156–7.
148 Ibid.

The governor of Leh was under pressure to hinder the Schlagintweits from entering forbidden Chinese territories, as he could be held accountable for such intrusions. The reason was that the Qing Empire regarded the brothers as British spies. While they did not seek to prepare any military advancement, the Schlagintweits did, however, engage in applied natural history.[149] They studied thoroughly the supply and demand for a number of natural and manufactured goods across the different frontier regions they traversed in north India, the Himalayas and Turkestan. They explored, measured and mapped with refined instrumentation mountain passes, and noted the suitability of different regions within and beyond Company territory for future agricultural schemes and white settlement.[150] The brothers, in short, participated in imperial knowledge gathering in a manner that was also aimed at justifying the high expenses accumulated for their endeavours by British sponsors.

The Schlagintweits themselves noted that their extensive observations and measurements were perceived by non-European rulers as unwelcome or even hostile acts, and spurred rival powers with their own interests in high and central Asia to come up with countermeasures in order to keep those intrusive foreigners at bay. Robert even claimed that some officials and border guards – especially from the Chinese Empire – engaged in a sort of counter-espionage. Robert arguably dramatised this element of their journey as he spoke about it in front of popular audiences in Europe and the United States, whose appetite for adventure stories is likely to have sensationalised his narrative. Yet, he insisted on the point that every step they took along or across the borders of Chinese-controlled territory was closely registered in Peking. The Schlagintweits' penetration into the 'Chinese province of Gnari Khorsum' in the year 1855 had 'not a little disturbed the Chinese authorities, and had first directed their attention to us'.[151] The result was, he claimed, that '[b]undles of official papers, in which we are not spoken of in the most flattering terms, are to be met with in … the capital of China'.[152]

Since the brothers' first intrusion into regions under the suzerainty of Peking, the 'Chinese government, with whom we were now on anything but good terms … did not lose sight of us'.[153] It therefore did not take long until the Schlagintweits knew precisely what the purpose of the above-mentioned 'guards of honour' really was: 'namely faithful watchers

[149] Osborne, 'Applied Natural History'; Driver, 'Face to Face', 448.
[150] See the Schlagintweits' testimony to the Select Committee on Colonization and Settlement; BSB, SLGA, V.2.2.1, 106; H. Schlagintweit, 'Practical Objects', BL, IOR, MSS EUR F 195/5.
[151] BSB, SLGA, V.2.2.2, 19–20.
[152] Ibid.
[153] Ibid.

over all our acts and proceedings, official spies in fact'.[154] Whether this account of Chinese espionage was true or partly fabricated for European audiences is difficult to tell, since the evidence comes exclusively from the brothers themselves. What is beyond doubt, however, is that the brothers were seen as and behaved like British agents in regions outside India, and clearly profited from the nascent yet tangible power that their imperial employer exercised over various local rulers in the Himalayan borderlands.

'Voyage of Discovery' or Indian Tourism? Forms of Mobility in South Asia

The brothers' reliance on the colonial infrastructure was, of course, even more pronounced within British India itself and not only affected their collecting practices and ethnographic studies, but also modes of mobility. In contrast to the popular perception of their explorations as being marked by solitary travel and a certain heroism in their fight against extreme and hostile climates, much of their work on the subcontinent appears rather tame and sociable, as they followed in the footsteps of numerous predecessors: tourists, artists, merchants and Company officials. The different transport technologies used by the Schlagintweits and their establishment within British-controlled India – and partly also in the Himalayas – make clear to what great extent the brothers were travelling on well-trodden paths, for which they could rely on a convenient system of carriers stationed along major routes at fixed intervals.[155] These could even be pre-booked by the travelling naturalists through the Indian postal system.[156]

What was more, the brothers made use of a wide range of modern and traditional means of transport – from steamer and railway to elephant and palanquin. The brothers thus employed the means of the 'Camel caravan' and the 'Bhylie', a type of ox cart, to transport their luggage and scientific collections.[157] The same applied both to the 'Charry dawk' (a stagecoach) and the 'bullock train', which was utilised 'along the main line of communication between Calcutta and the Punjab'.[158] For their personal transport, the Schlagintweits also made occasional

---

[154] Ibid.
[155] A perceptive analysis of the question of access to infrastructure in colonial India is Ahuja, *Pathways*.
[156] H. Schlagintweit, *Reisen* I, 239–44.
[157] Ibid., 79. Note that the brothers' collections included entire tree trunks, heavy slabs of Indian marble and the corpses of large animals, such as that of a Bengal tiger.
[158] Ibid., 243.

use of trained elephants to ride through difficult-to-access jungles in north India and parts of the Himalayas. As a gesture of inclusion into their privileged circles, it was generally British colonial officials who personally provided these elephants to the brothers – and only to them, not their assistants.[159] Regionally specific means of transport – such as the Nepalese 'Dari, a type of portable hammock, which allows all kinds of observations of the traversed region' – were used and praised by the brothers for their comfort and efficiency.[160] At least within Company territory, the brothers thus usually travelled in an ostensibly more privileged manner than most of their helpers. While they were carried or drawn by south Asians, the great majority of their porters had to travel by foot; and even their personal assistants were confined to use simple palanquins. Some of the forms of travel the brothers enjoyed are summoned together in a contemporary illustration (Figure 4.8).

The image glorifies British rule in India as a story of technological progress. Starting in the upper left-hand corner, the most 'primitive' forms of travel are depicted, embodied by Indian subjects: the 'tramp', and the 'Hindoo pilgrim'. The composition then moves on to portray more 'civilised' forms of locomotion. The bottom image shows 'the East Indian Railway', thus portraying the most recent, and purportedly highest and most civilised stage of development in Indian transport history – an open celebration of Britain's modernising and 'civilising' impact on south Asian society.[161]

What further suggests that the Schlagintweit expedition was not exactly a pioneering 'voyage of discovery' within much of British India was the fact that the brothers hardly ever slept in tent camps, but rather enjoyed comfortable accommodation. This was possible because since the early nineteenth century, British India had developed the rudiments of a tourism industry, with fixed routes for curious travellers. In this context, Natasha Eaton has argued that Britons' experiences while visiting (often sacred) sites and landmarks in India were dominated 'by a concern with déjà vu, not so much a journey into the unknown as a confirmation of what was known about or desired from England thanks to travel capitalism'.[162] What is more, by mid-century, 'middle-class Europeans, women as well as men, continental Europeans as well as Britons, generally found few political obstacles to journeying from one part of India

---

[159] For instance in Sikkim and Assam, ibid., 245.
[160] H. Schlagintweit, *Reisen* II, 171.
[161] Aguiar, *Modernity*, 13; the only mode of travelling missing is the steamship, which the brothers used during inland river travels and along the sea coast to Ceylon. For how different forms of mobility interacted, see Huber, 'Multiple Mobilities'.
[162] Eaton, 'Tourism', 218.

Figure 4.8 'Modes of Travelling in India', *Illustrated London News*, 19 September 1863. 'Tramps, Hindoo pilgrim, Palky dawk, Camel caravan, Bhylie, Elephant, Charry dawk, Ekha, and the East Indian Railway'. © DAV.

to another'.[163] The flood of European tourists since the early decades of the century had led to the publication 'of hundreds, if not thousands' of works concerned with the colony: 'histories, biographies, political commentaries, economic analyses, evangelical tracts, chronicles of military campaigns, and tales of sport and hunting, and, above all, travel narratives'.[164] While these works were prime transmitters of impressions and information about India for metropolitan audiences, the sheer quantity of *Indiana* meant that Britain's reading classes soon showed 'expressions of exasperation' and 'tedium' with the topic – although this did not mean that the influx of visitors to India dwindled; indeed it increased over the following decades.[165]

Besides a functioning porterage system along many of British India's main roads, there also existed chains of 'bungalows' (even some 'Hôtels'), which had been explicitly 'erected for travellers', including merchants, sightseers and itinerant naturalists and officers.[166] The Schlagintweits frequently used such facilities, but also found suitable accommodation through numberless invitations from Company officials, and at British military or 'sanitary stations'.[167] These stations on elevated ground were designed for the needs of government officials and troops who sought to recover from the heat, tropical humidity and malignant miasmas of the Indian plains that seemingly arose from every jungle, swamp or graveyard.[168]

'Hill stations' ultimately fulfilled symbolic functions within the Company's system of rule in India. As Dane Kennedy has shown, they served as little 'homes-from-home', where English norms of civility played an important part of quotidian life.[169] Here, the children of the colonial elites could attend boarding schools, while the Anglican Church played a central part in the social life of these settlements. British visitors to hill stations took part in social and cultural practices that echoed the lifestyle of the elite at home, such as attending balls, concerts and hunting parties.[170] In a way, those mountain idylls represented an attempt to create a physical and ideological space for segregation between the white rulers and their Indian subjects through retreating from the supposedly

---

[163] Arnold, *Traveling Gaze*, 15.
[164] Ibid., 26.
[165] Ibid.
[166] H. Schlagintweit, *Reisen* I, 44; 98, 222; II, 170; IV, 439.
[167] H. Schlagintweit, *Reisen* I, 223.
[168] Kennedy, *Stations*, 1; H. Schlagintweit, *Reisen* II, 433, 468.
[169] Kennedy, *Stations*.
[170] The library at the Rashtrapati Niwas (formerly the Viceregal Lodge) at Shimla holds two unreferenced photographic albums with depictions of elite social events and formal gatherings at the summer capital, which may be consulted courtesy of the staff.

corrupting influences of the plains. The exclusion of Indian subjects from hill stations was, of course, never feasible, as their running essentially relied on an Indian workforce. Nonetheless, many hill stations sought a strict hierarchy of spatial differentiations that reflected the colonial order as a whole. In Shimla, arguably the most prominent of those sites, the finest houses on the ridge were owned by senior British officials and the slopes below reserved for English and Anglo-Indian clerks, while the Indian population was confined to the lowest elevations – 'out of sight and out of mind' – with these habitation patterns reflecting 'the symbolic significance of altitude'.[171]

The brothers frequently enjoyed hospitality at hill stations, and thus it becomes clear that the Schlagintweits did not dwell for long as 'solitary travellers' in supposedly unknown and unexplored territory.[172] In the early nineteenth century, a few indigenous surveyors had, indeed, anticipated parts of the routes taken by the brothers later.[173] The explorer Mir Izzet Ullah had crossed the Karakoram pass in the service of Company agent William Moorcroft in 1813. The indigenous surveyor had then reached Yarkand and Kashgar, only to return to India via Bokhara and Kabul, collecting a wealth of useful commercial information along the way.[174] To describe these frontier regions as entirely unknown and – before the Schlagintweits' excursions – never visited places simply ignores the history and knowledge contributions of British officials and indigenous travellers in the service of the EIC alike.

By contrast, we come much closer to the brothers' actual experiences during their explorations if we acknowledge that the Schlagintweits were fully integrated members of the colonial establishment in British India, even regularly participating in events of 'European sociability' among the social elites. Hermann Schlagintweits' description of his activities at Shimla captures how the brothers yielded freely to the pleasures offered by the presence of the ruling classes:

From March until September ... a circle of socially active Europeans is united in this place ... Balls and concerts, picnics and theatre for connoisseurs rapidly follow each other ... Since we encountered such a lively intercourse here for the first time since we had left Europe, we were able to judge it more impartially than if one indulges it continuously. I admit I rather welcomed it. Especially after a long deprivation of European sociability, one appreciates her provocative charms: the small chains of fashion and etiquette hardly constrain.[175]

[171] Kennedy, Stations, 197.
[172] Cf. Trentin-Meyer, 'Hochasienreise'.
[173] H. Schlagintweit et al., Results I, 33.
[174] Waller, Pundits, 22.
[175] H. Schlagintweit, Reisen II, 364–5.

Such remarks on the social life of exploration are significant, not least because they provide insights into the identification of the Schlagintweits as Europeans within British India's establishment. While their national affiliation with the German states certainly played a role within the unfolding controversy over the employment of these 'foreigners' among metropolitan circles in Britain, in the colonial realm the brothers could more easily move 'within and between multiple identities and networks in a seamless, almost effortless way'.[176] The Schlagintweits also identified with British colonialism as a supposed force of progress, as a 'civilising mission' spreading cultural and judicial norms among supposedly less cultivated and at times 'barbaric' peoples.[177]

The brothers' appropriation of a colonial mentality through their integration into administrative elites and British scientific circles in India also becomes apparent elsewhere in their writings. The established dichotomy in British thought between both physical and moral corruption associated with the Indian plains and a greater purity of the air and the mores of mountain dwellers was directly reflected in the Schlagintweits' concept of 'high Asia' and its inhabitants. This is captured in Robert Schlagintweit's later descriptions:

[T]he entry from the plains into the mountains is … enormously surprising. There everything appears to change at once, the temperature, the vegetation, the animal world, the current and flow of the rivers, yes, even the Indian dress. It is a splendid, dazzling, and magnificent contrast. We deemed ourselves fortunate in having exchanged the hazy air, as damp as it was hot, which we had hitherto breathed in the burning Indian plains and in the fever-generating Tarai, for the pure, clear, refreshing, invigorating atmosphere of the H[imalayas].[178]

Robert also commented on the moral virtues of 'all' Himalayan 'races', stating that 'the inhabitants of the H. know nothing of a number of barbarous, abominable customs, which up to recent times survived in India. Widow-burning, the Satis, which – it is incredible to say – took place openly in India as late as 1829 –, Infanticide, the killing and sacrificing of human victims, have never found an entrance into the H[imalayas].'[179] It was due to these glorifications of the greater 'purity' of the mountain range and its inhabitants that the Schlagintweits recommended to their imperial employers above all the 'colonisation and settlement

---

[176] MacKillop, 'Europeans', 35.
[177] See among many passages H. Schlagintweit, *Reisen* I, 424; II, 357–8; on the notion of the civilising mission, Osterhammel and Barth (eds.), *Zivilisierungsmissionen*; Osterhammel, *Europe*.
[178] BSB, SLGA, V.2.2.1, 29–30, 48f.
[179] Ibid., 55.

of Europeans' in these elevated regions, a proposal inspired by their prolonged stays in British hill stations.[180]

That the Schlagintweits could perceive of themselves as full members of the British imperial establishment was further reified by the fact that they enjoyed far-reaching privileges and rights otherwise held only by colonial elites. They enjoyed 'privileged mobility' in the form of special political and scientific resources liberally provided to them by the Company.[181] An example of this status was the right 'to officially press [helpers] into service, as in the case of military marches'.[182] From the point of view of those who had been forcibly recruited, the brothers did not travel as representatives of the Prussian Crown, or as detached scientific observers in British-administered territories. Rather, they travelled and recruited in British India as colonial officials.

In sum, as the study of their various surveys, measurements and observations across south and central Asia has shown, the Schlagintweits and their assistants contributed to both wider scientific concerns *and* the 'empire of knowledge' of the British in India.[183] To do so, they made full use of the established information networks and scientific and technical institutions of the ruling power – an effective mutual instrumentalisation between those imperial outsiders and the architecture and interests of Company rule on the subcontinent.

---

[180] Ibid., 106.
[181] A similar argument is developed in Nayar, 'Mobility'.
[182] H. Schlagintweit, *Reisen* I, 82.
[183] Ballantyne, 'Colonial Knowledge', 102; while knowledge was crucial for the British to eventually establish hegemony in south Asia, the complex administrative and survey arrangements of the imperial state nonetheless saw internal conflicts and multiple fault lines, see Appadurai, *Modernity*, ch. 6, 'Number in the Colonial Imagination'; Peabody, 'Knowledge Formation', 92.

5 The Inner Life of a 'European' Expedition: Cultural Encounters and Multiple Hierarchies

The nature of the Schlagintweits' scientific programme as much as their personal ambitions to explore places partly unknown to Europeans at the time called for a greater reliance on south Asian assistants and guides than most western travellers would have been comfortable acknowledging. This chapter is therefore concerned with the large group of travellers that accompanied the brothers on their expedition from northern India into high and central Asia. While the diverse group of indigenous servants, assistants and semi-independent followers performed a number of important functions within the configuration of the expedition party, it has not, as yet, been explored in its striking complexity.[1] Instead of imagining the 'moving colony' of the expedition party as a hierarchically organised group under the guidance of the European naturalists as their leaders, we can more fully appreciate the actual internal dynamics of the expedition if we acknowledge that in the course of travel, different and partly contradictory hierarchies emerged within – what the Schlagintweits themselves called – their 'establishment'. The existence of such multiple hierarchies significantly shaped both the dynamics of exploration and the ultimate scientific results of the mission. Ultimately, it is this lack of a firm hierarchy that stands in stark contrast to the popular perceptions of how an expedition was organised, and it prompts us to ask how 'European' this undertaking actually was – both on the ground, and later in its literary representation.[2]

Questions of precedence within the travelling party itself were, and could not be, established once and for all at the outset, but rather remained in constant flux. By focusing on a greater number of actors, their characters, mutual relationships and individual trajectories, it is

---

[1] Certain individual indigenous members of the brothers' entourage are mentioned, yet not explored, in Polter, 'Nadelschau', 92; Waller, *Pundits*, 34, 40; a first treatment was given in Brescius, 'Forscherdrang', and Driver, 'Intermediaries'; Driver, 'Face to Face'.
[2] Incisive reflections on the myth of European exploration are given in Thomas, 'What is an Expedition?'.

possible to recover the contingencies of scientific exploration and the crucial agency of those non-Europeans who facilitated the advancement of the party through difficult and, from the brothers' perspective, unfamiliar terrain. Although many works on the Schlagintweits' travels mention the existence of indigenous companions *en passant*, they often seem to be 'second class' travellers, pushed into the background and thus into oblivion, literally remaining in the shadow of the three German brothers.

The reasons for this misrepresentation are manifold. For one thing, the manner in which overseas experiences were recorded and archived retrospectively played a major part in fabricating an image – however inaccurate – of European self-sufficiency. Accounts of travels, both scientific and popular, were long-established literary genres by the nineteenth century, and carried with them a weight of conventions that successfully perpetuated the tale of the heroic, single-minded and solitary explorer. These conventions, in turn, evoked a set of specific images closely associated with the term 'expedition' itself: above all, the image of the 'heroic' European, who leads his expedition against all kinds of human and natural obstacles while risking his own life to fill in the last blank spaces on western geographical maps.[3]

In this vision, all travel companions are merely the means to this higher end, 'servants' to help accomplish the great cause determined by the Europeans. By drawing on the aesthetics of military experience and colonial portraiture, metropolitan painters and engravers helped sustain these tropes in their own way. As Johannes Fabian noted about African exploration, it was often in the illustrations of European travel accounts that 'those verbal flourishes' of the 'intrepid, heroic, courageous explorer … parallel pictorial ones: the traveler's quasi-military garb, his faraway gaze, his proud and determined posture. He rides or walks ahead of his caravan; a few porters and guards are recognizable, while the rest blend into a file that gets smaller and smaller until it disappears in the landscape.'[4] This attention to how popular perceptions of explorers were fabricated and visually expressed points at the same time to the wider networks of travellers' mediators and agents, who assisted in perpetrating a specific heroic image of the overseas traveller through particular, demand-driven representations for metropolitan audiences.[5]

---

[3] On the pre-existing images in the minds of their European readerships, Kennedy, 'Introduction'.
[4] Fabian, *Out of Our Minds*, 5.
[5] Riffenburgh, *Myth*.

To recover the histories and voices of those who were crowded together in the background of travel narratives is notoriously difficult, as the structure of European and colonial archives usually reflects the ideological focus on the European individual. As a result, relatively few documents have survived of those indigenous assistants who partook in the exploration of the Himalayas and central Asia as doctors, translators, guides and plant-hunters.[6] Yet the piecing together of information contained in private correspondence, official reports, field notes, sketches, photographs, lists of expenditure and published travelogues does provide us with an eclectic mix of sources that allow for more than a 'deconstructive literary analysis' of European published accounts.[7] The Schlagintweit case in particular provides a great stock of incidents and unique sources, in which the views and agendas of non-European travellers, scholars, collectors and merchants involved in the enterprise become strikingly visible. In pursuing an actor-centred approach, this chapter intends to contribute to a growing literature on the history and culture of European overseas exploration that significantly shifts the focus away from the 'heroic' explorer – for long the sole focus of attention. Rather, the history of scientific exploration can, and indeed ought to be, written from multiple points of view.[8] Only by doing so are we able to discern the diverging interests and motives that different groups of actors – both European and non-European – pursued during such expeditionary undertakings.[9]

We may begin by asking what different groupings made up the Schlagintweit 'establishment', how large it was in numbers and what its specific functions were. To start with, it seemed to be an all-male group of travellers; while the brothers provide two dozen indigenous names, there is not a single woman identified by name as companion. In this oddly womanless homosocial world of these travelling men, the composition of the travelling party was changing constantly. Some regulars remained most of the time with one of the brothers, while others joined the party only for certain legs of the journey. Porters made up the single largest group of the entourage. Arguably reflecting what little respect the brothers showed for this group, these porters have left the smallest trace in the archives and remain mostly anonymous in their travel accounts. Depending on the number of instruments, tents, food supplies and,

---

[6] In the case of Indo-British exploration of central Asia see, however, the writings of the pundit Sarat Chandra Das, *Pandits*; Das, *Journey*; as explored in Raj, *Relocating Modern Science*, 200.

[7] This 'reading against the grain' method, focused exclusively on European accounts, was applied by Fabian, *Out of Our Minds*.

[8] Incisive is Liebersohn, 'Narrative Perspectives'.

[9] Burnett, 'Exploration'; Fabian, *Out of Our Minds*; Jones and Driver, *Hidden Histories*.

above all, scientific collections taken along a specific route, the size of the group of carriers could swell enormously. Usually, there were between ten and fifty carriers and servants per European traveller. The reason why so many people were needed for rather mundane tasks greatly vexed the German explorers. Most of their companions were Hindus and followed a strict division of labour according to the caste system.[10] The long-established and seemingly immutable social hierarchies among their domestic servants, carriers and assistants entailed a considerable extension of the necessary 'personnel'. This, in turn, involved skyrocketing expenses. On the payroll of the brothers were, among others, washers and torchbearers, watchmen, water carriers and messengers, each of whom fulfilled specific tasks.[11]

For procuring natural historical objects for the Schlagintweit collections, which after the travels would amount to over 40,000 objects, a large group of plant collectors and huntsmen was engaged.[12] Although there were a few people who were permanently employed to identify, work on and accumulate artefacts, the workforce was temporarily enlarged if an area was particularly rich in wildlife, or particular skills were required. In such cases a number of additional paid 'hands' were taken on, sometimes even the populace of an entire village. The brothers were usually keen to recruit experienced men who, 'having been drilled to the work of collecting and preserving animals and plants', each received 'a salary of 9 rupees a Month'.[13]

The assistants' technical skills were of paramount importance for later assessments of the Schlagintweit collections, and required an ability to stuff quadrupeds and birds, to preserve insects, fish and other species in bottles of alcohol, but also to prepare fragile botanic samples for their transport to Europe. The hired companions also categorised specimens, for instance by distinguishing between plants with medical or hallucinatory qualities, those that could be used as dyestuff and trees of fine timber.[14] For the latter, the brothers enlisted the services of a 'woodcutter', and '18 men' to procure those valuable and polished tree sections

---

[10] H. Schlagintweit et al., *Results* I, 41.

[11] H. Schlagintweit, *Reisen* I, 85–6.

[12] While the number of objects collected is often given as a mere 14,777, this number emerged after the brothers had sorted through their collections and disposed of the many rotten or broken specimens; cf. Felsch, '14.777 Dinge'.

[13] Hermann to Capt. Atkinson, Secretary to the Military Department, 2 December 1855, BL, L/Mil/3/587.

[14] There survives a list of plant names written by an anonymous assistant in Assamese and included in the brothers' volume of travel observations on 'tree sections' and 'useful plants'; BSB, SLGA, II.1.42. On the intricacies of scientific taxonomy in the nineteenth century, Bonneuil, 'Standardisation'; Drayton, 'Knowledge and Empire', 238.

that later adorned the rooms of the Company's East India Museum in London.[15] Such a massive effort of collecting natural specimens was only feasible with a small legion of non-European collectors – even if their efforts in procuring these raw materials for British scientific and commercial interests were never acknowledged in the metropole.[16]

A smaller cohort of well-trained, often multilingual assistants represented the next group of companions that decisively shaped the inner workings of the expedition. They included indigenous cartographers, surveyors, scribes, munshis, merchants and caravan leaders, numerous translators for the different local languages and also a 'Native Doctor'.[17] Many of those assistants were widely travelled and learned men, who either joined the venture only for short intervals, or remained in the company of the brothers for several years – while some accompanied the brothers on every step of their journey.[18] What decided the size of the establishment was, however, the question of whether the expedition party moved through safe terrain or, rather, in disguise through politically sensitive areas. Especially during their secret excursions into Chinese Turkestan, the number of attendants was reduced to an absolute minimum. This was done partly out of fear of encountering shortages of supply in the barren mountain regions, but partly also because a smaller travelling party had a greater chance to escape from Chinese border guards and soldiers in case the intruders were discovered. The establishment was thus by no means a fixed group of people, but rather a social configuration that was profoundly and continuously shaped by its almost constant state of mobility through regions that greatly varied in their 'natural treasures' and political sensitivity.[19]

When the specific contexts of their exploration are acknowledged, it is not surprising to find that the expedition resorted to a way of travelling that greatly resembled the caravan trade that had crisscrossed high and central Asia for thousands of years.[20] For long stretches of the way, foreign pilgrims and merchants joined the exploratory party, who were arguably ignorant about the scientific agenda of the group, and yet appreciated the

---

[15] The enlistment of temporary collectors was sanctioned by the Governor-General in Council; the Secretary to the Government of India, Fort William, Mr Birch to Hermann, 14 January 1856, BL, IOR, L/Mil/3/587.

[16] Drayton, 'Knowledge and Empire', 238; Schaffer et al. 'Introduction', in Schaffer et al. (eds.), *Brokered World*, xxxiii.

[17] H. Schlagintweit et al., *Results* IV, 523.

[18] The dramatis personae of the Schlagintweit 'establishment' are given at the outset of H. Schlagintweit et al., *Results* I, 36–42.

[19] Brescius, 'Empires of Opportunity'; Brescius, 'Forscherdrang'.

[20] Robert Schlagintweit on the expedition party: 'We thus form a Caravan as varied and motley as it is stately'; BSB, SLGA, V.2.2.1, 91.

Figure 5.1 'The Chain of the Kuenluen, from Sumgal in Turkistan', showing the travelling party in Chinese territory; watercolour by Hermann Schlagintweit, August 1856, lithographed by Sabatier, printed in oil colours by Lemercier; Hermann Schlagintweit et al., *Atlas*, no. 29. Robert was a member of this expedition, his complete disguise, however, prevents his identification.
© DAV.

protection provided by becoming a temporary member of this caravan. The brothers employed similar methods for protecting themselves and their entourage. At various points they had to travel through central Asia covertly, often assuming a false identity, and the most convenient way to do so was to join a larger caravan, dressed as Indian merchants (Figures 5.1, 5.2). In doing so the brothers always stocked up on fine samples of textiles as potential trade goods for supplies.[21]

The ethnic, social and religious diversity of the travelling party was striking. It appeared as a conglomerate of individuals from highly distinct social ranks, with different belief systems, places of origin and divergent levels of literacy and education. The mixing of all these diverse people was a cultural encounter for every member of the establishment, not just for the Schlagintweits. In reflecting upon the sheer diversity of people present, the brother noted that: '[o]n one occasion our camp presented a most interesting variety of tribes and creeds, and for the time being might

---

[21] Ibid., V.2.2.2, 128. The caravan leaders sometimes also took charge of the financial arrangements for the expedition, H. Schlagintweit, *Reisen* I, 92–3.

Figure 5.2 Robert Schlagintweit, in 'native disguise' during their high Asian exploration, carrying a gun under his belt (*c.* 1855).
© BSB, SLGA, IV.2.90.

be almost said to form an ethnographical museum of living specimens'.[22] The camp then included members of six different religions. No fewer than twelve languages were in use: besides the brothers' native tongue, 'the languages spoken by these natives were, Hindostani, Bengali, Gujarati, Maharati, Panjabi, Kashmiri, Persian, Tibetan, Turkish, Portuguese and English'.[23]

At first glance, the establishment fulfilled a number of obvious roles. It managed the mobility of the travellers and guaranteed the safe transport of their fragile instruments and scientific collections. Yet, the changing internal composition also served as a rich source of information regarding such important matters as the availability of local food supplies,

---

[22] H. Schlagintweit et al., *Results* I, 42. For the importance of the establishment as a 'human laboratory' for the ethnographic studies of the brothers, see Chapter 7.
[23] Ibid.

knowledge about violent conflicts and recent robberies in specific areas and the course of regional routes and mountain passes. Some members of the establishment also had a deeper understanding of those lands that the Schlagintweits themselves could study only in passing, as the former had often lived in particular areas for decades. This treasure trove of experience could effectively be used with a view to addressing a number of scientific issues – including questions about climatic patterns in a long, comparative perspective.[24] It was partly only through the locally gained knowledge of their assistants and other informants they met that the brothers could formulate theories about then much-discussed topics among European geographers – such as the conundrum of the line of perpetual snow in the highest mountain chain of the world, or the causes of the frequent floods that endangered harvests in the Indian plains. The latter in particular were of great importance to the East India Company, as a series of Indus floods had taken many lives, devastated wide areas of agricultural land and even threatened the destruction of entire military cantonments.[25]

The changing composition of the Schlagintweit establishments also offered the required linguistic expertise for the different regions that were traversed. This expertise was not only important for intercourse with the inhabitants of various countries in order to secure news, geographical information and supplies, but also enabled the brothers' philological researches about the distribution of languages and dialects in the Himalayas. Equally important for the brothers and their British employers were the recording of indigenous denominations of plants and topographical landmarks, such as rivers, mountain peaks or valleys. To take but one example, the Schlagintweits' vast collection of seeds of useful and beautiful plants was accompanied by a comprehensive 'Index to Messrs. Schlagintweit's Collections: Seeds, Sent to the India House Museum December 1858'. This included the names of well over 400 specimens, featuring the local denominations besides their botanical and vulgar equivalents in several languages. It also contained original seed samples in small pouches, giving the small book the impression of a botanical inventory.[26]

Lastly, the linguistic skills of the brothers' assistants proved essential to secure geographical information and in negotiations with indigenous

---

[24] See for a wealth of such acknowledged indigenous 'testimony', H. Schlagintweit et al., *Results* IV, on Indian meteorology; on questions of marginalisation of indigenous testimony, Lässig, 'History of Knowledge', 53.

[25] The problem concerned many contemporaries, e.g., Drew, *Jummoo*, on 'Indus Floods', 414–21; also Kreutzmann, 'Habitat Conditions'.

[26] 'Index' at BL, IOR, T3787.

rulers and officials about the right of access into their territories.[27] Despite the later idealisation by Robert Schlagintweit that he had crossed the Himalayas like 'a thoroughly independent sovereign, who governs absolutely over an immense kingdom adorned with the rarest charms of nature, which he can traverse in every direction, just as his humour and will may lead him', in truth the brothers frequently needed to negotiate their passage with the actual rulers of those regions.[28] While the brothers consulted with British colonial authorities about the prospects of entering certain parts of high and central Asia, their assistants often mediated the actual encounters with local authorities. As we will see, they not only had the necessary language skills, but also the right contacts with influential merchants and officials in the trans-Himalayan regions to facilitate the progress of the expedition party.[29]

An episode that captures this important linguistic competence of their companions – and other vital functions of the establishment as a whole – is offered in relation to Tibet. At the time of the brothers' travels, Tibet was under the suzerainty of the Chinese Empire. At first, Robert and Adolph Schlagintweit had tried in 1855, 'under the guidance of Mani [Singh], the Patvari or head man' of the city of Johar in the central Himalayas, to secretly penetrate into Tibet. Mani Singh was not inexperienced in accompanying imperial explorations. Only a few years earlier he and his cousin Nain Singh had assisted the Company servants Richard and Henry Strachey during their excursions into the Himalayas, earning open praise for their services.[30] Mani came from a wealthy and respected family, several members of which the Schlagintweits would recruit over the course of their expedition. In order to guarantee the success of their intrusion into Tibet, Adolph and Robert had left behind the majority of their establishment in Milum, and were now 'accompanied by only 10, all well-armed Buthias'.[31]

After safely crossing the 'Kiungar Pass at the Border of Tibet' (Figure 5.3) and on the route to Gnari Khorsum (Ngari, Tibet/China), the small group ran into a Tibetan 'border guard of eight Hunias', who were under strict orders to impede the advance of the foreign travellers.[32] Since the brothers had mastered hardly any Tibetan themselves, it was their interpreter who informed the guards that they had no intention to travel into Tibet itself, and that the expedition would rather proceed

[27] H. Schlagintweit et al., *Results* I, 38.
[28] BSB, SLGA, V.2.2.1, 50.
[29] H. Schlagintweit, *Reisen* III, 72.
[30] Strachey, 'Note', 536.
[31] H. Schlagintweit, *Reisen* III, 65.
[32] Ibid., 67.

Figure 5.3 Adolph Schlagintweit, 'Kiungar Pass at the Border of Tibet', watercolour, 12 July 1855. Note how precisely the British officials and servants of the Great Trigonometrical Survey could follow the route over the pass through the carefully chosen viewpoint; gen. no. 471. © DAV.

from here to Niti, in a northwesterly direction.[33] This proposal seemingly appeased the guards. Now 'Mani, who was charged with the planning and execution of our Tibetan journey, suggested, in order to mislead the guard, to proceed a little in the direction towards Niti, and only then to further penetrate into Tibet by crossing one of the small passes ... by night.'[34] Following this suggestion by their indigenous leader, the expedition group, at first escorted by the still 'suspicious' guards, proceeded in a seemingly innocuous direction. Thence, the Schlagintweits and their companions attempted to secretly cross the border into Tibet once more, this time via the 'Sakh pass'.[35]

The group had pushed their horses forward throughout the night, and the following day – having reached the Sutlej river and starting to feel at ease – they were suddenly joined by their guards again. This led to a precarious encounter, whose later description, and the apparent bravery (or rather the recklessness) of the brothers' actions, should be taken with a pinch of salt. According to the brothers' matching narratives, the border guards surrounded the expedition party and when they attempted to capture them and their horses, the brothers suddenly hit them hard in the faces with 'long English riding crops'.[36] The brothers claimed to have been on the brink of firing their revolvers at the guards, who would have returned fire using their own rifles. In the end, it was apparently only Mani's astute intervention between both groups that prevented a deadly escalation.[37]

Since the brothers were now taken into custody by the border guards as 'Chinese state prisoners',[38] Mani Singh became once more the crucial intermediary and sent for a nearby 'Dzongpon' in Daba (i.e. the governor of a fort) to open negotiations between the brothers and the Chinese authorities. The latter only sent 'as his proxy an assistant (Dúik)'.[39] After an 'endless negotiation with the Dúik, who gradually had to be made obedient by means of Rupís, Brandy, Sherry, etc.', the brothers secured a written agreement on 19 July 1855. According to a copy of the treaty (Figure 5.4), the brothers received 'permission to travel up to the Sutlej, [and] to remain there for three days' – in the company of a group of local guards. At the same time, they 'committed' themselves 'to pay 600

[33] Ibid., 65.
[34] Ibid.
[35] Ibid., 66.
[36] Ibid.
[37] BSB, SLGA, V.2.2.1, 133.
[38] Ibid., 135.
[39] I.e. *drung yig*; H. Schlagintweit, *Reisen* III, 69; RS, BSB, SLGA, V.2.2.1, 137–8.

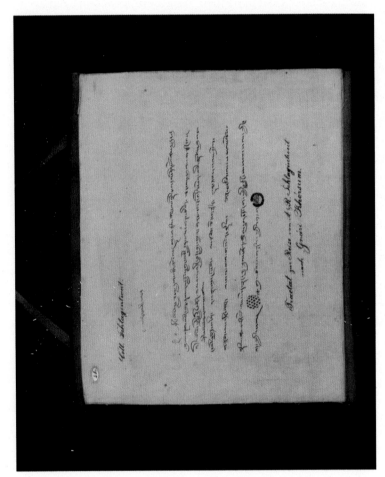

Figure 5.4 Copy of a treaty between Adolph and Robert Schlagintweit and Tibetan officials in July 1855. 'Tractat zur Reise von A. & R. Schlagintweit nach Gnari Khórsum', height: 37 cm, width: 41.5 cm.
© Museum Fünf Kontinente, Munich [MFK], Schl. 841.

Rupees as a penalty, if they were to cross the Satlej'.[40] Yet, the exclusive mention of the name of the Tibetan official and the brothers' signatures on the treaty belie the actual genesis of the document, which could not have been concluded without the crucial diplomatic and linguistic intercession of Mani Singh.[41]

The brothers never regarded the treaty as binding, however. They soon went well beyond the agreed end point of their excursion, a specific bridge over the Sutlej. After arriving at the river, 'a relative of Mani, their expedition leader, came to them, since he had heard that they [Adolph and Robert] were in some difficulty with regard to the continuation of their journey'.[42] This ability to draw on the kinship networks of their assistants represents another crucial element of collaboration between the brothers and their establishment. Mani Singh's relative was called 'Bara Mani', or 'the great Mani', and is introduced by the Schlagintweits as 'the wealthiest and most respected among the inhabitants of Johar. Due to his extensive trade and financial transactions' he had 'really a lot of influence even in Tibet.'[43] The powerful merchant now became the brothers' confidant, with whom they shared their ambition to travel deeper into Tibet. Consequently, Bara Mani took it upon himself to 'negotiate on our behalf with the Dzongpon'.[44] Thanks to the decisive intervention of this go-between, the previous terms of the written contract were nullified, and the rights of travel for the brothers greatly expanded. Now, the Schlagintweits had forced their access up to the 'Chakola pass': 'This is one of the few points of passage over the high mountain ridge, which here separates the Sutlej region from the Indus region.' The pass was therefore – also in the view of their British employers – a strategic route, which could now be measured and drawn for the first time.[45]

In the course of their remaining stay in Tibet, the brothers used Mani Singh's mediation, and candid bribery, to gain the 'confidence' of their Tibetan guards, who now allowed the foreign party to visit a number of other locations that again lay beyond the terms of their contract. For instance, the brothers and their assistants were accompanied to the monastery of Mangnang (Figure 5.5), where they purchased much of

---

[40] H. Schlagintweit, *Reisen* III, 72; I am most grateful to Christoph Cüppers, director of the LIRI, Nepal, for translating the treaty.
[41] BSB, SLGA, V.2.2.1, 138.
[42] H. Schlagintweit, *Reisen* III. 72–3.
[43] Ibid.
[44] Ibid.
[45] A. Schlagintweit, 'Mountains from Chakola to Indus', 26 July 1855, gen. no. 485, DAV. The brothers also broke the terms of the second formal agreement. For instance, they made secret excursions towards Gartok, the main city in western Tibet.

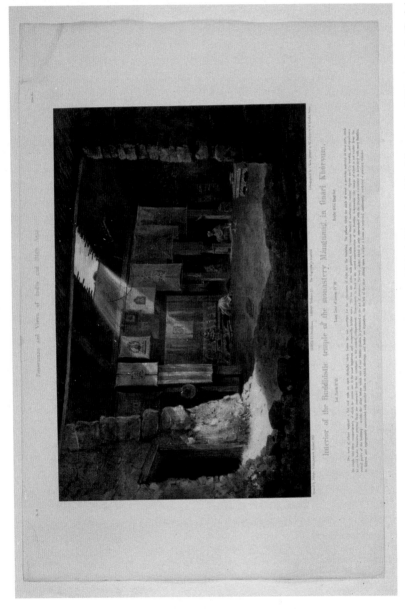

Figure 5.5 Adolph Schlagintweit, 'Interior of the Buddhistic Temple of the Monastery Mángnang, in Gnári Khórsum', August 1855, reproduced as a lithographic print in Hermann Schlagintweit et al., *Atlas*, no. 12. © DAV.

the temple's interior – including prayer flags and praying mills, carpets, Buddhist manuscripts and statues of the Buddha.[46]

Mani Singh and members of his family were also involved in many of the more risky expeditionary ventures the brothers undertook in central Asia. He even acted as a headhunter for the brothers, who were on the lookout for a knowledgeable and trustworthy leader for another difficult leg of the journey. From Leh, the capital of Ladakh, the brothers were determined to reach the forbidden Chinese Turkestan. To find someone who could guide them without attracting the attention of the Chinese border guards was crucial, but complicated. The Schlagintweits lacked the linguistic prowess to seek out a suitable candidate themselves, and therefore had to rely entirely on the choice made by Mani Singh.[47] The choice was significant, since, as Robert later acknowledged, when such a man has been found, 'one must entrust one's self [sic] unreservedly to him, yield one's self [sic] unconditionally to him, and give effect to his arrangements, even when these appear singular, peculiar, and surprising'.[48]

The responsibility placed upon the prospective leader could not have been greater, since the brothers claimed to know the content of 'a very peculiar command', which the Chinese government had supposedly issued to all of its officials in Turkestan: 'Should a European enter the district ruled over by you, his goods and chattels belong to you, but his head to Pekin.'[49] The martial rhetoric of this order reflects the geopolitical conflict between different powers in central Asia at the time – among them the Chinese Empire, Tsarist Russia and the British Empire.[50] The Tsarist Empire had launched a territorial expansion in a southeastern direction, sometimes cloaked in the terms of a civilising mission, yet with predominantly strategic and military aims.[51] China demonstrated its claim to supremacy in central Asia, also after the defeats in the Opium Wars, through such martial rhetoric of exclusive sovereignty over Turkestan, which was aimed at defending Chinese claims against foreign intruders and local oppositions by rebelling subjects.[52]

---

[46] H. Schlagintweit, *Reisen* III, 83.
[47] A further case of such headhunting of a 'leader and translator' was a man called 'Dávang Dórje', H. Schlagintweit, *Reisen* II, 102.
[48] BSB, SLGA, V.2.2.2, 42.
[49] BSB, SLGA, V.2.2.2, 59.
[50] Darwin, *Empire Project*, 30–1. On British and Chinese imperial ambitions, as played out over Tibet with its own internal conflicts of power, Stewart, *Journeys to Empire*.
[51] For the mostly military, not 'civilising' ambitions of the Russians in Turkestan in the second half of the nineteenth century, Morrison, *Russian Rule*, 36, 151.
[52] Blanchard, 'Silk Road'.

When Mani Singh returned with a potential candidate for their secret mission, the brothers arranged a meeting at night, concealed from the other members of their establishment. They had the man led into their tent. Two candles lit up the gloom, 'before one of them lay a purse, heavily filled with gold, before the other was a six-barrelled revolver, loaded with ball, in the middle between them was placed the Koran. The whole presented a serious and deeply solemn aspect.' After the brothers had convinced themselves that his 'frank, open demeanour inspired confidence' in the Turkistani, called Mohammad Amin, they let him swear 'that he would guide us faithfully, would never forsake us, would never betray us'. In return, they swore 'naturally likewise on the Koran ... that if he kept his promise he should receive the purse filled with gold, but if he broke it, we would shoot him like a dog, with the revolver lying before us'.[53]

The brothers' direct threat on Mohammad Amin's life reflected their own insecurity and fears, namely of becoming dependent upon the guidance of a complete stranger during the upcoming voyage into prohibited territory. This episode links up with Rachel Standfield's notion of 'intimate violence': 'for all the vulnerability Europeans may have experienced in the face of such intimate encounters of empire, such vulnerability was productive rather than inhibiting of violence'.[54] But how might the tense situation have shifted in the brothers' imagination and later through the process of narration? The way the episode was conveyed to European audiences, from a safe spatial and temporal distance, suggests a curious transformation in the balance of power. Looking back at this crucial moment of the expedition and on their relationship with Amin, the brothers later sought to turn their own position of vulnerability into a dominant one.[55] However, we should not be misled by their rhetoric in retrospectively describing this delicate situation. Even after the brothers had made a threat of vengeance against their guide, nothing had changed, in fact, about the internal hierarchies: the Schlagintweits were entirely reliant upon Amin's truthfulness, and his decisions regarding routes and food supplies. If Amin should want to betray or rob them, the Muslim merchant could have done so without any real threat to his life: even the Government of India could and would not have been able to prosecute and punish him (Figure 5.6). The ultimately peaceful collaboration between Amin and the brothers was, to be sure, only feasible through the mediation of the ingenious Mani Singh.

[53] All quotations in this paragraph from BSB, SLGA, V.2.2.2, 46.
[54] Standfield, 'Violence', 35. Ballantyne and Burton, 'Introduction', 13.
[55] BSB, SLGA, V.2.2.1, 50.

Figure 5.6 Mohammad Amin, a merchant of Turkish origin from Yarkand and guide to all three Schlagintweits into Turkestan. Portrait by an unknown photographer. Noteworthy are the later modifications: the addition of a column, a backrest and the blackening of his pupils were artistic interventions, aiming to show the individual portrayed as a civilised individual, not as a 'racial type'.
© BSB, SLGA, IV.2.94.

As his and Amin's case demonstrate, in a sense, the indigenous establishment thus recruited *itself* along the way.

The Schlagintweits' assistants played further roles during their secret penetration into Turkestan. With the help of Amin, the travelling party had adopted the appearance of Muslim traders from south India.[56] The fact that some of their assistants did actually originate from India, and spoke Hindustani fluently, now provided the brothers with greater protection against discovery. They noted that the 'company of Makshut', a 'Mussalman from Delhi' who had already accompanied the British

[56] BSB, SLGA, V.2.2.2, 26.

traveller William Moorcroft some decades earlier, was 'beneficial for us'.[57] The German brothers, who mastered only a few phrases in Hindustani, demonstratively engaged in conversations with Makshut, to increase their prospect of success in passing as natives of India themselves. This was particularly the case in front of those inhabitants of Turkestan, who 'did not know the language [Hindustani] at all'.[58]

The brothers did not trust all members of their entourage in equal measure. In fact, only a small circle of assistants was informed about their most important – and at the same time most secret – travel plans. The majority of companions instead were kept ignorant for strategic reasons. To illustrate this point, the travels by Hermann and Robert into Chinese territory are telling. On the recommendation of Mohammad Amin, the leader of the upcoming Turkestan mission, the brothers at first pretended to leave the city of Leh with a huge entourage in the direction of Kashmir. In July 1856, the brothers left 'with 30 servants, 20 horses, and 50 men carrying luggage, together with a number of tents, and as our people thought, a quantity of useless baggage of every kind'.[59] As the Schlagintweits knew only too well, it would have been 'sheer madness ... to attempt to penetrate into China and Turkestan with all these people and this motley caravan'. And yet, the appearance of the slow-moving camp was part of the calculation. It was precisely the demonstration of the inertia of this throng that served the brothers for their deceptive manoeuvre. In the eyes of the governor of Leh, their large establishment 'afforded a surer guarantee that we could not get across the frontiers, than a whole company of soldiers, which he perhaps might have posted there for our expulsion'.[60]

At the same time, Amin, of whose existence only Mani and the brothers knew, was planning a trick that would free the brothers from the opponents in their own establishment: the 'guard of honour' that was monitoring and reporting back on their movements. Amin's plan for the brothers was to lead the entire travelling party up onto the highly elevated Sasser pass at some 5410 m, and to remain there for several days under conditions of a constant shortage of supplies and icy temperatures, until a good part of the establishment were to suffer from severe bodily symptoms. Indeed, as one of the brothers noted in hindsight: 'Our

---

[57] He was taken into service by the brothers 'on the way to Ladakh', H. Schlagintweit, *Reisen* IV, 22–3.

[58] Ibid., 131.

[59] This and the following from BSB, SLGA, V.2.2.2, 53ff.

[60] Ibid., 55f.

entire encampment very soon became like a lazaretto; sighs and groans resounded from every tent and filled the air.'[61]

Meanwhile, Mohammad Amin waited nearby and, after Mani Singh had given him a secret sign, he ran into the group as if by accident with fresh supplies, servants and horses.[62] The brothers then struck a deal with their worn-out companions. It was agreed that the majority of the latter, including the watchful guards, should return to Leh, while the brothers would complete their observations on the pass with a small group of remaining assistants, before rejoining the main party. The Schlagintweits' fear that members of their own entourage could expose their plan for a surreptitious intrusion into Turkestan went so far that only Mani Singh was informed about their true motives. Except for him, it was only the servants of Mohammad Amin who made up the travelling party. This episode captures well how the brothers used members of the establishment as pawns in the game of misleading indigenous rulers and their spies.

Instead of perpetuating the myth that the Schlagintweits were marching at the head of a sworn-in establishment, whose members were all devoted to them and their 'higher goal' of scientific and exploratory advancement, it is thus more accurate to conceive of the expeditionary party as a 'society of strangers', in the sense that the expedition employees and patrons in south Asia were all initially strangers to the Schlagintweits, even if not necessarily to each other. Within the establishment, internal discords and shifting hierarchies between different groups and individuals was something the brothers could only partially control. To be sure, at times they themselves manipulated the establishment. Yet, while the perceived wisdom of western exploration may invoke the impression of the European travellers as omnipotent, larger-than-life-figures and unchallenged leaders, the Schlagintweits had limited authority over their fellow travellers. This was due to insurmountable cultural and linguistic barriers, personal fears and mistrust, which meant that the brothers often remained estranged from and yet 'entirely dependent' upon their entourage.[63]

Multiple Hierarchies and Unexpected Dependencies

The complicated internal dynamics of the expeditionary party had a great impact on the real and imagined hierarchies that coexisted – sometimes happily, sometimes fraught with tension – in the minds of the

---

[61] Ibid., 70–1.
[62] Ibid., 66ff.
[63] BSB, SLGA, V.2.2.1, 46, 80.

Schlagintweits. The volume of texts, objects and images they produced for posterity about the expedition's unfolding present an assemblage of conflicting facts and judgements about the members of their travelling party, as well as about their own role as leaders. To neatly disentangle the realities of travel from the imprint of the imagination is no simple task. In the end, many irresolvable contradictions remain. Nevertheless, it is worthwhile enquiring into the representation of power relations, as they reveal most clearly the fragile nature of the authority that the Schlagintweits possessed, struggled to keep – or sought to construct retrospectively.

Hierarchies changed not only over time and with distance from the events – when the explorers attempted to provide 'heroic' narratives in print and lectures – but also with the political context in which their excursions were executed. The brothers were in a relatively dominant position as far as *British*-administered territories were concerned. Equipped with a broad range of privileges by the colonial power, the Schlagintweits perceived of themselves as the unquestioned leaders of the enterprise: while drawing on the rich cartographic knowledge provided to them by the GTS and the Surveyor General, they set down the itineraries and also secured the provisions, accommodation and salaries of their large entourage. Moreover, several members of the establishment were formally assigned to assist them by different colonial institutions and governors. For instance, a number of draughtsmen from the Surveyor-General's Office or the Office of the Quarter Master General at Bombay and Madras joined the expedition by presenting their 'service record book' (*Dienstbuch*), thus executing an act of submission under the brothers' orders.[64]

By contrast, we are faced with a starkly different situation in those regions that were held to be *terra incognita* by Europeans. In addition to the dependence upon indigenous geographical knowledge, other factors were crucial too with regard to the shifting relations of power, with their assistants assuming ever-greater authority. Any *forced* recruitment of helpers and porters beyond Company-controlled territories was unthinkable. The brothers were rather forced to offer such lucrative terms of employment to their companions that the latter were willing to accept the hardships and personal risks that excursions into politically unstable regions entailed.[65] The accompanying assistants also had to be heavily armed, so that the risk the Schlagintweits could be robbed and murdered by their own people was far greater in high and central

---

[64] H. Schlagintweit et al., *Results* I, 36–7.
[65] H. Schlagintweit, *Reisen* I, 156.

Asia – where no kind of imperial protection was available to them. And while the brothers were able to freely communicate in English with officials in British India, only their assistants and guides were able to converse with the inhabitants, authorities and travelling merchants of the Tibetan, Nepali and Turkestani regions they entered. This helplessness, in turn, significantly increased the assistants' salary – during high mountain ascents and for the dangerous excursions into central Asia.[66]

This lack of cultural and linguistic skills represented a significant impediment for the Schlagintweits' exercise of authority. For the sake of their security, the brothers learned a set of formal Muslim salutations in word and gesture from their main expedition leader Amin. When they travelled on a caravan route towards Yarkand in Turkestan and passed a 'Chinese frontier guard', the Schlagintweits also 'received from Mohammad Amin precise rules as to how we should behave in case the guard should see and question us'. He even taught them specific phrases to be used during any interrogation by Chinese soldiers or civil officials.[67] Since the Schlagintweits were furthermore entirely reliant upon the Muslim merchant's decisions regarding itineraries and provisions, it was inevitable that their initially dominant position was totally undermined beyond the north Indian frontier. Even making the arrangements with Amin was in itself challenging, as the brothers had to find the right linguistic mediator for their establishment to avoid a 'double translation' in conversations with their own principal guide.[68]

While the Schlagintweits later maintained that they were able to freely exercise command over their establishment, this was only half the truth. In reality, they struggled a great deal with religious hierarchies and practices that time and again undermined their own position. The Munich-born travellers developed a particular dislike for the beliefs of their Hindu companions.[69] They despised the fact that the Hindu worshippers sacrificed food in time-consuming ceremonies to various deities on the steep mountain ascents in high Asia: 'the people are much addicted to superstitious ceremonies, upon the strict performance of which they insist, adding that '[r]emonstrances are of no avail'.[70] In fact, the Schlagintweits paint a picture of their explorations of mountain systems as if they had to fight as much against indigenous devotion to nature as against factors such as extremes of weather, lack of proper

---

[66] For example, BL, IOR, E/4/835, 48. H. and R. Schlagintweit, *Official Reports*, 3.

[67] BSB, SLGA, V.2.2.2, 113, 150.

[68] H. Schlagintweit, *Reisen* IV, 121.

[69] H. Schlagintweit, *Reisen* II, 323.

[70] H. Schlagintweit et al., *Results* III, 17; BSB, SLGA, V.2.2.1, 56–7.

infrastructure and reliable information or shortness of supplies.[71] Their own worldly claim to authority was subordinate to that of the religious leaders of the expedition, the Brahmins.

These conflicting hierarchies were captured during a critical moment on a high pass of the Nanda Devi in the Himalayas, as told by Adolph Schlagintweit:

> I was at once quite frightened by seeing three of my men, one after another, getting suddenly quite epileptic, they threw themselves down in the snow, turning their eyes and beating about with their arms and legs, evidently quite out of their senses, and all my people began to cry out 'Nanda Devi Aya' – 'Nanda Devi Aya'.[72] I was indeed rather a little frightened, since I feared that this nonsense might become contagious ... I therefore took aside my two ... Brahmins and told them ... that I had given to the Nanda Devi everything they had asked for ... I ordered them to calm the people at once, which they effected by Mantra, and snow applied to their temples.[73]

The situation confirmed that the German explorers were always forced to comply with the religious requirements of sacrifices. However, when a part of the group suffered from fits, Adolph *had* to rely on the mediation and authority of the Brahmins, since only they were able, by singing mantras and holding snow against the suffering men's heads, to end their inner turmoil. Tellingly, the passage on Adolph's loss of control was taken out in official reports on the expedition published by British patrons of the scheme.[74] Attesting to their contradictory nature, while the Schlagintweit books contained passages scornful about many of their companions as irrational individuals and religious fanatics, at the same time many of their collectors, servants and scientific assistants were clearly held in the highest esteem. Hermann, for once, remarked on the indispensable services of his 'butler' Dhamji, who 'tried successfully to maintain peace and friendship among my people', hence to settle erupting conflicts within the group.[75] The loyalty of some of their companions loomed so large in other parts of the brothers' writings that it was later used as cannon fodder by the British press in their critical campaigns against the German travellers. In the former's view, such extensive praise of 'natives', including their fine personal characters and vital support, was not appropriate conduct for a true European explorer.[76]

---

[71] BSB, SLGA, V.2.2.1, 98.
[72] Translated as 'The Nanda Devi has come', H. Schlagintweit, *Reisen* I, 326.
[73] 'Report of Adolph Schlagintweit and Robert Schlagintweit upon their journeys in the Himalayas of Kumaon in May and June 1855', FBG, ARCH PGM 353/1.
[74] A. and R. Schlagintweit, 'Notices', 152.
[75] H. Schlagintweit, *Reisen* I, 86.
[76] Driver and Jones, *Hidden Histories*, 45.

The Closest Companions

The variety of tasks carried out by indigenous assistants within the context of 'European' scientific expeditions has attracted growing attention. This revisionist strand of literature considers intermediaries and go-betweens as instrumental for both the scientific and political advancement of colonial regimes in Asia, Australia, Africa and the New World.[77] Yet, the scarcity of primary sources on indigenous companions and explorers has often prevented historians from going beyond praising their 'contributions' in abstract terms. The following account of a number of notable assistants to the Schlagintweits adds further dimensions to this line of enquiry through the detailed analysis of the intimate, fraught and decisive relations between the brothers' 'moving colony' and indigenous experts in the Asian highlands. Particular attention is paid to recovering information about their earlier lives, skills, local influence and alternative channels of information, which all significantly shaped the expedition's conduct. An important example were the trans-regional commercial networks of merchants, whose contacts with other traders at bazars and different nodal points along the former Silk Roads significantly influenced the brothers' itineraries in the trans-Himalayas. Clearly, their expedition meant different things for different people involved. Only if we consider the agendas and later trajectories of several members of the party can we appreciate the various legacies this undertaking would have in multiple contexts: it profoundly shaped not only the brothers' own biographies, but was also the starting point for a number of other imperial careers within British India, pursued by their former assistants in most remarkable ways.

Many of their indigenous and European helpers had been trained in scientific and military institutions. One of the central sites for recruiting trained assistants was Calcutta.[78] There, during March and April 1855, the Schlagintweits secured the services of the 'Indo-Portuguese Mr Monteiro', who became involved in the 'preparation and accurate packing of our collected items' to be sent to London. Monteiro, however, quickly distinguished himself to such a degree that he rose to a position of authority within the establishment, becoming the 'general superintendent of the collectors' of the entire expedition (Figure 5.7).[79]

---

[77] Safier, *Measuring*; Schaffer et al. (eds.), *Brokered World*; Raj, *Relocating Modern Science*; Kennedy, *Spaces*; Brescius, 'Cultural Brokers'; Driver, 'Face to Face'.

[78] British India consisted mainly of the three presidencies of Bengal, Madras and Bombay, each maintaining its own colonial administration and army. Some passages in the following account are based on Brescius, 'Forscherdrang'.

[79] H. Schlagintweit et al., *Results* I, 40.

Yámu,

Bhutia, 28 Y. Bhutan Tassi suden —

Mr Monteiro's Boneboiler.

(not measured)

Figure 5.7 Modified photograph of 'Yámu', a twenty-eight-year-old Bhutia from Buthan, 'Mr Monteiro's "Boneboiler"' for the collections of animal and human skeletons; note the remark '(not measured)' at the bottom, showing that members of the establishment were used for anthropological studies.
© BSB, SLGA, IV.2.63.

In carrying out his duties, Monteiro erected several temporary 'laboratories' along the travel routes, which he oversaw and managed for extended periods, whilst the expedition party moved on to new sites.[80] The 'Schlagintweit expedition' thus combined, in a sense, the ideal of Humboldtian studies outside in the field with the provisional use of closed-off workshops. Such a temporary station for recording various scientific observations was, for instance, erected at the 'hill station' in Darjeeling, where Monteiro also coordinated the collecting practices of a number of helpers – again out in the field.[81] During Monteiro's

[80] H. Schlagintweit, *Reisen* I, 239.
[81] H. Schlagintweit, *Reisen* II, 186.

sojourn in Darjeeling, Hermann Schlagintweit had trained him in the 'use' of a 'photographic apparatus', and both men produced a number of 'daguerreotypes on metal plates'.[82]

Monteiro's important position within the expeditionary party led him to develop great commitment to the cause of amassing scientific objects and anthropometric measurements for the brothers.[83] An episode in Kashmir in 1856 demonstrates his recklessness in the pursuit of that aim:

> Monteiro had obtained a cadaver for our collections in an unfitting manner. During the night, he had cut down a man who had been hanged a long time ago, who was displayed as a warning to the people and as plunder for the animals. Then, since Monteiro quickly aroused general suspicion, his belongings were searched. He only knew to help himself by hiding the anyway [dried-out corpse] in his own bed, where it was certainly the least expected.[84]

Monteiro's drive to operate independently from the brothers was further demonstrated by the fact that he 'continued his works for another year, after we had already left India'. The assistant accumulated numerous observations and sets of data between 1857 and 1858, and sent them to Europe.[85]

Other assistants, too, had close ties with British colonial institutions before they were co-opted. A Muslim 'writer and draughtsmen' called Abdul, born at Madras, was recruited, and from February 1855 onwards temporarily abandoned his service 'as draughtsman and assistant surveyor in the office of the Quarter Master General' of Madras.[86] His previous military training was to become highly beneficial for the brothers' explorations. This applied to Abdul's talent for quickly grasping the main features of an unknown landscape and fixing his impressions on paper. He put these talents to use during several independently executed excursions, which lasted up to six months. Hermann later integrated long passages into his travelogues entirely based upon the assistant's observations. Regarding the Tista river in the Eastern Himalayas, the German traveller noted that '[m]y draughtsman Abdul sketched a detailed plan of the river along its entire course, which was highly important to me for determining erosion conditions, the more recent changes of the riverbank, &c.'.[87] Abdul also managed to tap into the local knowledge of inhabitants of various regions, with whom the brothers could not

---

[82] Ibid., 271; cf. Körner, 'Photographien', p. 314.
[83] H. Schlagintweit, Reisen II, 31.
[84] Ibid., 428.
[85] H. Schlagintweit, Reisen I, 239.
[86] H. Schlagintweit et al., Results I, 21, 36–7.
[87] H. Schlagintweit, Reisen II, 153–4.

personally converse: 'Abdul was told about the Atri [Atrai] river that, circa fifty years ago (or perhaps periodically?), the latter had possessed a greater quantity of water than the Teesta' river. Such treasure troves of experience helped the brothers to develop their hydrographical theories about high Asia. The Muslim draughtsman also delivered information that was essential from a colonial perspective, such as insights into the accessibility and use of local river systems for the transport of resources such as timber.[88] In the brothers' perceptions and published accounts, the fact that Abdul and other assistants had been formally trained in British geographical and military institutions lent credibility to their observations. After all, the testimony of indigenous informants was highly debated in Victorian times.[89] By thus presenting their indigenous surveyors as well-calibrated 'instruments', the brothers legitimised the accuracy of their observations – and hence the trustworthiness of their entire mission.[90]

Using their assistants as 'satellites', the brothers garnered information from villages and regions they never personally visited. On one occasion, Abdul was able, 'while being disguised as a Lepcha and provided with merchandise …, to travel within the territories of the Raja of Sikkim [on the border with Nepal], and to take a number of observations according to previously carefully supplied instructions'.[91] In providing this data, the military draughtsman took great personal risks, 'since even natives of India are excluded from Sikkim during the entire summer' out of fear of spies.[92] Earlier European travellers Joseph Hooker and Archibald Campbell had feared for their lives after their capture in Sikkim. Thanks to his varied and proven skills, Abdul was soon promoted to act as Hermann's 'second assistant', working alongside the British lieutenant, later captain, Adams (who had 'received charge of the Deputy Assistant Quarter Master General's Office at Mooltan' in 1853 and thus also came from a military background).[93] Abdul remained in that position until almost the end of the brothers' stay in India. His personal commitment to the expedition could not have been greater, as he died in Calcutta, following several months of lingering illness and exhausted from his extensive travels with the brothers and independently.[94] While his early death was

[88] Ibid.
[89] Schröder, *Wissen*, 168ff.; Withers, 'Trust'.
[90] Their expeditionary project thus anticipated Montgomerie's use of pundits (surveyors) for the trans-Himalaya after the Indian Rebellion: Raj, *Relocating Modern Science*, ch. 6.
[91] H. Schlagintweit, *Reisen* II, 218–19.
[92] Ibid.
[93] NAI, Military Department, Branch, ref. no. 112, letter from Mooltan, 13 April 1853; H. Schlagintweit, *Reisen* I, 238.
[94] H. Schlagintweit et al., *Results* I, 36.

deeply regretted by the brothers, he was not the only casualty within the indigenous establishment, as one Tibetan helper fell to his death on the Boko-la pass, when the party was forced to travel in secret at night.[95]

The degree of independence that some assistants maintained throughout the expedition is also apparent in the case of 'a coloured Jew from India' called Eleazar Daniel, who was entrusted 'with the superintendence of the transport of our instruments and collections', but was also successfully sent on independent surveys, during which he made valuable observations with European instruments.[96] From the autumn of 1854 until May 1857, he additionally served the brothers as an 'excellent guide' and 'private secretary'. Demonstrating once more the importance of the colonial infrastructure for the recruitment of skilled assistants, Eleazar Daniel was formally 'ordered to join us at the commencement of our journeys'. He had previously worked as 'a guide in the Quarter Master General's Department, of Bombay'.[97]

The fact that several assistants undertook complementary excursions proves that the 'Schlagintweit expedition' was, in reality, not a single, continuous journey. Rather, the enterprise must be understood as a series of various major and minor expeditions, which crisscrossed south and high Asia with a web of itineraries undertaken by different people with different functions. The brothers represented these autonomously undertaken explorations in their published maps with dotted lines (Map 5.1).[98] The activities of the indigenous surveyors and collectors continued at least until March 1858, when the colonial government stopped their payment. The fact that a number of the assistants continued their researches well beyond the departure of the brothers is at odds with Eurocentric readings of the Schlagintweit mission as purportedly a 'German expedition in India' merely from 1854 to 1857. One assistant even continued his work with the Schlagintweits in Europe for over a year, helping the brothers with their philological studies. Sayad Mohammad Said, an erudite munshi from Calcutta, fluent in 'Bengali, Urdu, Persian and Arabic',[99] accompanied the brothers to Berlin, where he was presented to a number of philological and anatomical scholars from central Europe.[100] Paid for

[95] H. Schlagintweit, *Reisen* III, 87; H. Schlagintweit et al., *Results* III, 14.
[96] H. Schlagintweit et al., *Results* I, 37–8. See also 'Bearings with prismat compass by Eleazar, taken on the route from Martoli to Milum in Ghobar' Kumaon (Milum), 3 June 1855, pencil and feather, BSB, SLGA, IV.5.58.
[97] H. Schlagintweit et al., *Results* I, 37–8.
[98] H. Schlagintweit, *Reisen* I, 84.
[99] H. Schlagintweit et al., *Results* I, 69.
[100] On 10 July 1858, the orientalist Theodor Nöldeke wrote to his colleague Goeje about his acquaintance with the munshi during a soirée at the home of the political economist and statistician Dieterici in Berlin; Leiden University Library [LUL], BPL 2389.

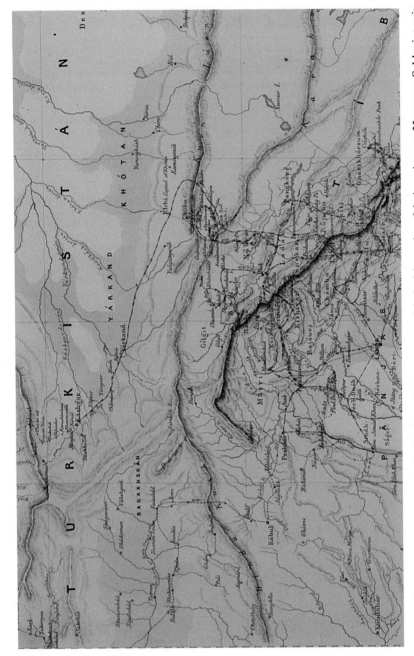

Map 5.1 Extract from the Schlagintweits' route map, with the itineraries of their assistants; Hermann Schlagintweit et al., *Atlas*.
© DAV.

his services by the brothers, Said was primarily involved in compiling the philological sections of the brothers' travels accounts. When he left India, Said 'did not master any European language'. During his stay in Prussia, however, Said learned English, not German, with a view to entering the colonial service in India upon his return.[101] For all these reasons of continued collaboration from near and afar, the brothers perceived of 1858 as the terminal year of the expedition.

Itinerant naturalists – indigenous or European – faced serious health risks during such an expedition, which often led the group through extremely inhospitable terrains. High altitudes, freezing temperatures and malnutrition characterised long stretches of their itineraries in the Asian highlands. The Indian plains, on the other hand, caused problems of a different sort for the travellers who suffered from tropical fevers, heat, humidity and physical exhaustion. While the brothers enjoyed access to British hospitals and sanitary stations within Company territory, such advantages of a colonial infrastructure were not available in the Himalayas and Turkestan. Yet, to have care on hand for those travellers who suffered from the distressing symptoms of altitude sickness, and for those who needed surgeries along the way, the brothers engaged a 'Native Doctor' from the region of Kumaon in the central Himalayas who served the British in India in the rank of a sergeant.[102] Hence, besides the Christian assistant Lieutenant Adams, the Muslim Abdul and the Jew Eleazar, the entourage now also welcomed a Brahmin called Harkishen. Before Harkishen joined the Schlagintweits in April 1855, he was employed at a colonial hospital in Almora. John Russell Colvin, lieutenant-governor of the northwestern provinces, personally recommended Harkishen to them and temporarily released him from his position to join the expedition.[103]

Harkishen's presence was to be of great benefit to the advancing party, as he took over a number of crucial tasks in addition to those for which he was initially hired. He even saved Hermann's life by undertaking a successful yet 'dangerous operation' to cut deep into his back to remove an abscess, which left untreated could have proved lethal.[104] His medical knowledge also qualified him to oversee the various – and potentially poisonous – foodstuffs gathered along the way, and the members of the establishment duly followed *his* 'orders' with regard to all matters

---

[101] Unfortunately, no more information can as yet be found on Said's career; H. Schlagintweit, 'Kru-Neger', 185.
[102] H. Schlagintweit, *Reisen* I, 236.
[103] H. Schlagintweit, *Reisen* II, 359.
[104] H. Schlagintweit, *Reisen* I, 256–7.

nutritional. The brothers furthermore stated that the 'tindal, or patvari (the headmen of the kulis), should be made responsible' for overseeing the food supplies of the porters: 'natives, even those of Tibet, not being disposed to allow a European to inspect and examine their victuals'.[105] This arrangement proves again how cultural and social norms cemented the Schlagintweits' outsider status within their own establishment.[106]

As the 'chief assistant' to Adolph and Robert, Harkishen was temporarily employed 'as superintendent of our plant-collectors'. Since he 'could write, but not speak, English, [he] labelled, in Hindostani and English, our ethnographical collections' – a few thousand objects.[107] The physician was also charged with managing a number of magnetic and meteorological observatories, which the brothers had installed along the way. One of them was erected in the city of Leh in 1856.[108] During his stay there, Harkishen also accomplished other tasks of technical complexity: 'our Native Doctor ... following my instructions, has produced a map of the city with a sextant and a prismatic compass on a scale of 1:1000'.[109] Known and praised by the brothers for 'his zeal and diligence in general, as well as by the accuracy of his observations', Harkishen soon became a key member of the entire operation.[110] He too continued his surveys after the brothers' departure, spending nine months, from June 1857 to March 1858, travelling on a separate route from that of Adolph through Tibet and the Himalayas, where he made a series of observations 'deserving of all praise'.[111] For Harkishen, the expedition was only a lucrative and adventurous intermezzo in his medical career: after the completion of his independent travels, he returned to his former position at Almora. It is an unresolvable contradiction in the Schlagintweits' writings that the Brahmin Harkishen received, on the one hand, open praise, while on the other they starkly condemned the moral corruption of the ruling caste of the 'Hindoo priests'.[112]

The considerable breadth of responsibilities that assistants such as Harkishen, Abdul or Eleazar took on was remarkable and testifies to their curiosity and special aptitude regarding scientific phenomena and practices. Even though the Schlagintweits were keen to highlight their role as leaders and instructors, there was a strong element of independence

---

[105] H. Schlagintweit et al., *Results* III, 16.
[106] H. Schlagintweit, *Reisen* II, 359.
[107] H. Schlagintweit et al., *Results* I, 37.
[108] Ibid.
[109] H. Schlagintweit, *Reisen* III, 278.
[110] H. Schlagintweit et al., *Results* I, 37.
[111] Ibid.
[112] H. and R. Schlagintweit, *Official Reports*, 3. BSB, SLGA, V.2.2.1, 57–8.

and initiative demonstrated by their companions. These were more than 'human instruments', whose pace could be calibrated for the purposes of providing accurate measurements for the GTS and other large-scale colonial surveying projects.[113] The freedom and spontaneity with which the assistants contributed to the various fields of enquiry is among the most striking characteristics of the enterprise. Its execution required all those involved, including the Schlagintweits, to broaden their horizons, learn new techniques and develop a wide range of methods previously unfamiliar to them, as the Schlagintweits' enlarged ethnographic and anthropological programme has shown (Chapter 4).

Throughout, the brothers were keen to tap into the expertise of merchants involved in the central Asian caravan trade. Some of their closest companions made a living as itinerant traders, and their intimate knowledge of the commerce and geography of the trans-Himalayan regions profoundly shaped the most celebrated legs of the enterprise. Jewish and Muslim traders belonged to separate trading communities and together were able to draw on a vast number of local helpers, informants and suppliers to which the Schlagintweits as European outsiders would never have gained access otherwise. Exemplary is a Jewish merchant called Murad from Bokhara (in today's Uzbekistan), whom the brothers first met in Ladakh in 1856 (Figure 5.8). Mutual trust seems to have been established fairly quickly. Murad provided the foreign travellers with 'much good guidance about routes in central Asia', and 'proved to be a very credible and dependable man'.[114] After the first encounter, Murad offered his services for Adolph's final excursion into Turkestan in 1857, and was subsequently enlisted as the 'second caravan leader' for the undertaking. The brothers were acutely aware of the precious topographical knowledge that Murad possessed, since 'he had in his role as caravan leader, and as a fur and silk merchant, already crossed this region [of Chinese Turkestan] a couple of times'.[115] As was clear to all parties involved, the brothers depended on men like this Jewish merchant much more than the other way around.[116]

The brothers profited even more from the expertise of another caravan trader who was mostly responsible for the Schlagintweits' most spectacular journeys in central Asia: the Muslim Mohammad Amin. The brothers' initial mistrust soon gave way to 'a greater liking of our chief

---

[113] Incisive is the treatment of the 'pundits' by Raj, *Relocating Modern Science*, 181–222.
[114] H. Schlagintweit, *Reisen* IV, 222.
[115] Ibid.
[116] H. Schlagintweit et al., *Results* III, 17: leaders of strictly 'secret' expeditions were rightly 'entitled to a high reward; for the personal risks and danger they incur in such expeditions is very great'.

Figure 5.8 Murad, a Jewish caravan trader in Turkestan, drawn by an anonymous member of the establishment, in the Company style, *aquarelle* over pencil sketch.
© Staatliche Graphische Sammlung, Munich, GR 755.

guide' in Chinese-controlled Turkestan.[117] To their relief, '[i]t constantly became more distinctly apparent that he had travelled about in central Asia more and oftener than perhaps any other individual'; he even knew about many abandoned roads 'the existence of which probably only few persons were aware, for this knowledge is handed down from father to son as a family secret'.[118] His acquaintance with trade arteries and alternative paths through Turkestan allowed the party to proceed at first on 'the great and much frequented caravan road between Leh and Yarkand'. Because the group had to avoid contact with the Chinese authorities and their local vassals, they soon, however, left the main road, and 'Amin conducted us along a road running east–south-eastwards,

---

[117] BSB, SLGA, V.2.2.2, 113f.
[118] Ibid.

and which was only known to smugglers'.[119] A 'cordial reception [was] everywhere given' to Amin, which 'plainly showed that amongst his countrymen he was a well known personage, and considered as a man of great respectability and influence'.[120] His smuggling activities, however, which extended up to the Russian borders in the north, and apparently a range of smaller offences he had committed, had incurred the disapproval of the Chinese authorities. During an interview with British colonial officials, the 'Native Doctor' Harkishen even claimed that Amin had been imprisoned when the brothers first took him into their service.[121] According to Harkishen, Amin had even been a criminal, 'a person of questionable antecedents', who had 'acted in the capacity of a gang robber' on the road between Yarkand and Leh.[122] We do not know to what extent Harkishen's account was driven by animosities between the two. Hermann, at least, felt a close affinity to Amin. He later tried to downplay the defamatory statements about the Yarkandi merchant.[123] Whatever his past offences, Amin's future was significantly altered in the summer of 1856, when he had secretly guided two of the brothers into Chinese-controlled terrain.

Some say that the Agents of the Chinese Government in Yarkand having heard of his bringing European travellers across their frontier (which is high treason in their Code) offered a reward of 1000 Rupis for his apprehension ... Gulab Singh ... ordered his arrest and threatened to hang him soon after the Schlagintweits' ... departure.[124]

To escape prosecution, Amin 'fled from Ladák into Kúlu, where Adolph S. found him, at Sultanpur, in April 1857'.[125] Previously, both had come to an agreement in writing. Adolph had secured the services of Amin as his official 'translator, leader, and baggage supervisor' for his pioneering excursion across the Karakoram and Kunlun mountain chains to Yarkand.[126] Amin's importance in this dangerous and secret scheme was reflected in his extraordinary salary, since 'he was to have a

---

[119] Ibid., 99. On the history of smuggling in the highly integrated caravan trade system in central Asia, Blanchard, 'Silk Road'.

[120] H. Schlagintweit et al., *Results*, 1, 39.

[121] H. and R. Schlagintweit, *Official Reports*, 3–4.

[122] Ibid., 3.

[123] H. Schlagintweit, *Reisen* IV, 222.

[124] Harkishen's testimony to Henry Strachey, in H. and R. Schlagintweit, *Official Reports*, 3.

[125] Ibid.

[126] Ibid. Adolph travelled into Turkestan with a range of instruments, among them a barometer (by Pistor), air thermometer (by Greiner), boiling thermometer (by Geißler), earth thermometer, chronometer, sextant, prismatic compass, clinometer and a vertical circle. H. Schlagintweit, *Reisen* IV, 219.

monthly salary of 2000 Rupis whilst traveling with A. S., and a monthly pension of 1000 Rupis after he had brought him back safe to India'.[127]

In the end, this journey proved fatal for Adolph Schlagintweit and some of his companions, as the expedition became entangled in the turmoil of a local rebellion in Kashgar, led by the Muslim warlord Wali Khan in the summer of 1857. Their explorations also coincided with the outbreak of the Great Indian Rebellion in May 1857, which shook the Company rule to its core. At first Adolph had only planned to complement his brothers' observations in central Asia, and to cross the Kunlun again taking a different route. The plan was then to return, as his two brothers had done before him, via steamer from Calcutta to Europe.[128]

Perhaps the news of the violent conflict in northern India had reached Adolph's party, leading to their decision to travel via Kashgar further on in a northern direction into Russian territories, from where they could hope to travel overland to Europe. It is also possible that the independent revolts that took place in Chinese Turkestan at the time were responsible for their change of plans.[129] In any case, the outcome was disastrous. The town of Kashgar was an important nodal point for the trans-regional trade in central Asia on the ancient Silk Roads – a term coined by Ferdinand von Richthofen, a German traveller to China, later president of the Berlin Geographical Society and an influential administrator of international geographical collaboration – which for centuries had linked the Chinese Empire in the east to the Levant in the west.[130] Given Kashgar's geostrategic importance, violent conflicts regularly flared up in the region between Muslim clans with old claims to the city and the ruling Chinese military. When Adolph's party arrived, Kashgar had already been conquered, and Wali Khan had established a short-lived but brutal reign.

Since the German explorer was travelling with official letters of protection issued by the Indian colonial government and a number of instruments and scientific notes in European languages, he was quickly identified as a British agent on his way to the Khan of Kokand; Adolph also presented himself as 'the Honorable East India Company's Envoy'.[131] Without trial, Adolph was subsequently sentenced to death and beheaded in front of the gates to the city. His head crowned a skull

---

[127] H. and R. Schlagintweit, *Official Reports*, 3.
[128] H. and R. Schlagintweit, 'Aperçu sommaire'.
[129] This was usually assumed, Wagener, 'Schlagintweit', 263.
[130] Wood, *Silk Road*.
[131] Mohammad Amin to Colonel Edwardes, 29 July 1858, in H. and R. Schlagintweit, *Official Reports*, 15; also E. Schlagintweit, 'Bericht', 466; Valikhanof and Veniukof, *Russians in Central Asia*, 228.

pyramid erected by Wali Khan. Many of his assistants did not fare much better, even if only Adolph's death is remembered. A Tibetan companion, who was considered Chinese by Khan's followers, was also murdered immediately on a charge of high treason, while three other companions were thrown into dark prison holes.[132]

The fate of the entire establishment was now desperately precarious. A few days after Adolph's murder, 'Murad, the Israelite, converted to Islam, to save his life'. Thereupon he spent the following weeks as Mohammad Amin's cellmate.[133] Adolph's assistant Abdul 'was kept apart, "because he was of Indian origin", and was sold into slavery on the first occasion' for twenty-five rupees to a Yarkandi. Abdul was first forced to travel northwards to Kokand, where he could buy his freedom. Subsequently, he took the long way back by travelling first through Khuchand and Samarkand to Bokhara, then to Balkh, Faizabad (the capital of Badakhstan), and then to Kabul in Afghanistan to Peshawar, which he reached on 15 December 1858. After his return, British officials recorded Abdul's impressions of these little-known regions in a series of interviews.[134]

Back in Kashgar, after more than thirty days in prison, during which two of Amin's servants died, the few survivors of Adolph's late group were released, and managed to escape the city amidst the turmoil during the Chinese reconquest.[135] However, instead of immediately leaving the embattled area for good, Amin hid himself for eight months in different places between Kashgar and Kokand. While doing so, he expressed his motivation and reflections in over twenty letters that he addressed in an increasingly accusatory tone to the British colonial authorities. Amin desperately required 'some written instructions' on the issue of how he should react to the death of his 'master'. Amin angrily noted that 'he has sent twenty-two reports up until now addressed to the honourable government but has not had the honour of having a response to even one of them', and reproached the British officials for their perceived apathy:

I am unsure as to … why the circumstances of the death of the victim [A. Schlagintweit] are being ignored and not being enquired about. Even if I am not trusted any longer or my services are not useful, that man was killed and was a patriot who wholeheartedly sacrificed his life for the good of his government and compatriots … so why are they ignoring him? In return for his sacrifice the least that can be done is to ask about him.[136]

---

[132] H. Schlagintweit, *Reisen* IV, 282–4; subsequent quotations, ibid.
[133] Ibid.
[134] H. Schlagintweit, *Reisen* IV, 282–4.
[135] Hermann refers to '35 days' spent in prison; ibid., 283.
[136] The Persian letter from Amin, 11 September 1860, is reprinted in *JASB*, 29 (1861), 444–6; I thank Ali Khan, Cambridge, for the translation.

Amin had by then even traded his last possessions for information about the whereabouts of Adolph's human remains and notes. Now, completely impoverished, he also had to master his own destiny. As he wrote in 'pain and sadness', Adolph's death represented more than the loss of an intimate friend; he was now also without 'a patron and without any work'.[137] While he had previously traded in goods, Amin's greatest capital was now his valuable knowledge: knowledge about the circumstances of the traveller's assassination, but also his insights into numerous unknown trading and marching routes in central Asia, from Afghanistan up to the Russian border. However precarious Amin's personal lot was, he could nonetheless hope to make effective use of his position in-between different and rival empires in this contested world region.[138]

Amin's decision to lead the Schlagintweits into Turkestan made it impossible for him ever to return to his former life as a caravan trader. After the Indian Rebellion had been crushed by force, Amin therefore enlisted himself as an informant in the service of the British Empire. This move meant more than turning knowledge into income, because he thus also obtained protection from persecution by the Chinese authorities, which still sought to hold him accountable for his betrayal.[139] Amin found support from Lord William Hay, the British Resident in Simla, who later sent out – probably following Amin's advice – a number of search expeditions to investigate the details of Adolph's fate. The concerted efforts by colonial officials ultimately led to the recovery of the traveller's last notes and sketches. From Amin's perspective, his participation in the expedition had thus been both a curse and a blessing: after open threats to his life, he ultimately settled in the Punjab, and received 'the rank of a "station agent"' in the colonial service. 'As such, he had to provide accounts of the transport conditions and the social and political circumstances of the inhabitants in the northwestern Indian provinces and the neighbouring countries' in central Asia.[140]

How valuable Amin's services were for British interests was reflected in a flood of reports and cartographic works based entirely on information he had gathered, which provided useful knowledge on trade and resources available in those regions in which geopolitical rivalries between India, Afghanistan, Kashgaria, China and Russia were played out. As the widely quoted *Punjab Trade Report*, published by the colonial

---

[137] Ibid.
[138] On the opportunities awaiting such Muslim 'outcasts' caught between various great powers in Asia, see the fascinating work by Alavi, 'Imperial Rivalries'.
[139] H. and R. Schlagintweit, *Official Reports*, 11.
[140] H. Schlagintweit, *Reisen* IV, 284.

official R. H. Davies under the supervision of Sir Robert Montgomery, lieutenant-governor of the Punjab, stated: the 'aid rendered in the compilation of the report ... by *Mahomed Amin*, a native of Yarkand, deserves to be prominently acknowledged'.[141] Amin had meticulously described the accessibility of different high mountain passes in central Asia – and this during different seasons (Map 5.2). The Schlagintweits also published a number of route descriptions by Amin, then entirely unknown, such as one 'from Osh to Tashkent, the most northern military frontier post of Kokand, on the Russian frontier'.[142]

The information Amin submitted to his new imperial masters greatly extended the limited experiences gained during his travels with the Schlagintweits: they were the result of a lifetime spent in those areas. This also accounts for the fact that Amin was so intimately acquainted with the political organisation of Chinese rule in Turkestan. He was able to give the precise number of Chinese guards at specific nodal points, and even knew about the troop force in the reconquered Kashgar.[143] His expertise was also expressed in detailed accounts of a broad range of trade products and natural resources and their levels of demand in different regions – ranging from silk to jade, wheat, opium, salt and tea.[144] It is highly likely that the brothers gleaned a significant amount of their colonial knowledge from their interaction with Amin. Yet, they did not acknowledge him as a source for their proposals made in front of the Select Committee on Colonisation and Settlement in India of the British Parliament in 1858 – for which they gained widespread recognition.[145]

To complement his earlier journeys, Amin undertook a number of further excursions into the frontier regions of India, where, according to the Schlagintweits, the inter-imperial rivalries attracted ever more international attention in the second half of the nineteenth century.[146]

---

[141] Davies, *Report*, 'Preface'.

[142] H. Schlagintweit et al., *Results* I, 35.

[143] This and the following account is based on Davies, *Report*, see, e.g., Appendix XX B. I. 'Route from Khokan to Kashgar', cl; Appendix XXXI, 'Caravan Route from Kundu to Yarkand, through Badakshan, the Pamer Steppes, and the Sar-i-Kul', ccclxv.

[144] Ibid.

[145] Numerous British and Indian newspapers and journals took up their suggestions, even if not all were new; e.g., *The Economist*, 24, 14 May 1859, 539; *Journal of the Society of Arts*, 9/427, 25 January 1861, 151; 'The Colonization and Settlement of India', *The Spectator*, 1626, 27 August 1859, 870; Balfour, *Products*, 257.

[146] Robert gave his lectures on central Asia 'at a time when the incessant advance of the Russians towards those parts of Central Asia, which are subject to the Chinese, is justly attracting the universal attention of Europe, and when one hears with astonishment of the success which they obtain there by taking possession of populous towns and important fortresses; at a time too when the names Yarkand, Kashgar, Turkistan, and Samarkand, occur as frequently in the political daily papers, as a few years ago did the name Amur': RS, BSB, SLGA, V.2.2.2, 55–6.

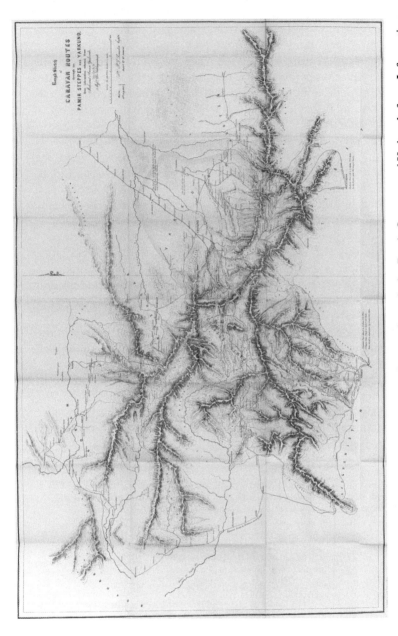

Map 5.2 'Rough Sketch of Caravan Routes through the Pamir Steppes and Yarkund, from Information Collected from Mahomed Ameen Yarkundi; Late Guide to Messrs. De Schlagintweit'; Montgomery, *Maps Accompanying Report on the Trade and Resources*.

© Cambridge University Library, Rare Book OP.3282.382.2, map 4.

It is perhaps of symbolic importance that Mohammad Amin died in the Himalayas 'in the spring of 1870, when he returned from Leh to the Punjab', losing 'his life because of an avalanche, not far away from the milder regions of the foothills'. He had literally taken his final steps exploring the border regions of the British Empire in south Asia.[147]

In view of this multitude of different actors, it would be naïve to assume that all members of the 'Schlagintweit expedition' shared the same motivation. While some joined the scheme to support their relatives back home,[148] others only used it as a paid opportunity to relocate from one place to another. Others, again, sought to satisfy their own intellectual curiosity to travel, measure and explore. Finally, a few assistants used this expeditionary venture as a training ground to acquire the kind of skills and expertise that they could subsequently put to use in making a career within the British imperial establishment. Some former companions, such as Nain Singh, would ultimately outperform the brothers in many respects: in the importance of their geographical discoveries and surveys, in the length of their service for the Empire's cause and the degree of fame they achieved in both the British Raj and in British scientific circles and wider publics.

A Fragile 'Information Order'

As Mohammad Amin's varied career and myriad testimonies demonstrate, knowledge was a key resource in the struggle over markets, natural treasures and political might in central Asia. It was also one of the most capricious resources to handle for statesmen, bureaucrats, colonial officials and travellers whose decisions were based on their access to up-to-date political, social and military intelligence. Indigenous intermediaries could provide Europeans with an opportunity to tap into India's 'autonomous networks of social communicators', who spread news, gossip and rumours from one region to another.[149] Yet, the knowledge gained through brokers was often open to interpretation, making it a vulnerable asset for those who had to rely on it.

For the Schlagintweits, the need to distinguish between reliable and misguided information became imperative to their expeditionary work and personal safety. Moreover, their scientific reputation in British India and in Europe heavily depended on this distinction. Reflecting a higher level of trust, they relied, whenever possible, on the contacts

---

[147] H. Schlagintweit, *Reisen* IV, 284.
[148] H. Schlagintweit, *Reisen* I, 91.
[149] Bayly, *Empire and Information*, 2.

and communications with members of their own establishment for setting the course and pace of exploration. Yet, knowledge of hostilities, the current condition of roads, local trading patterns, the availability of foodstuffs, labour and other issues usually reached them via a chain of intermediaries. Their pooling of such vital information was consciously intended to guide not only future travellers but also colonial schemes and military manoeuvres. At the outset of their published *Route-Book*, the Schlagintweit brothers declared that this compilation 'may be regarded as having a practical bearing upon questions of a more general nature, especially when it is taken into consideration that many provinces of high Asia are of great importance for India, in a commercial, as well as in a military point of view'.[150]

Since British power had begun to spread across south Asia since the 1760s, indigenous rulers, in return, had become increasingly eager to impede and cloud western knowledge of their territories. While they repeatedly limited the freedom to travel for Europeans, south and central Asian rulers also sought to spread false statements on the physical geography of their possessions themselves.[151] This was also one of the brothers' major concerns. It is therefore important to explore the difficult circumstances under which the Schlagintweits tried to acquire new geographical knowledge. The concept most suitable for bringing together their attempts to access, control and use indigenous information as well as the barriers and failures that accompanied this process, is that of a fragile 'information order'. First developed by Manuel Castells, the concept has since been adopted by Christopher Bayly in his seminal study on *Empire and Information*.[152] Bayly showed that colonial state building in south Asia was inextricably linked to the possession of intelligence gained through intimate relations and strategic alliances with influential knowledge brokers such as munshis, messengers, news-writers, spies, postal runners and informal agents including mystics, astrologers, midwives, physicians and barbers.[153] In his reading, early modern India was characterised by a decentralised information order 'consisting of many overlapping groups of knowledge-rich communities'.[154] Bayly showed how the eventual loss of sources of 'information about commercial developments, strategic priorities, and the dispositions of influential

---

[150] H. Schlagintweit et al., *Results* III, 4.
[151] Withers provides some telling examples in 'Margins'.
[152] Castells, *Informational City*.
[153] Bayly, *Empire and Information*, 54; on 'knowledge order', also Ballantyne, *Webs*.
[154] Bayly, *Empire and Information*, 5.

local magnates and moneymen' ultimately led to the catastrophe of the unforeseen Indian Rebellion in 1857.[155]

I seek to develop this framework further by looking not at the information order of British India as such, but more modestly by using it to understand the flow of information within the social configuration of the Schlagintweit establishment. The concept helps to shed new light on the fragile structures and strategies that the brothers used to relate to the unfamiliar world around them. Their changing group of assistants undoubtedly formed the most important frame of reference for the brothers' experiences, scientific results and understanding of the traversed regions, since they mediated the bulk of communication with the local inhabitants. What I call the 'moving colony' of the expedition party formed an – unquestionably incomplete – microcosm of different Asian regions, as a result of the constant influx of a number of local helpers and informants. Crucially, the latter were not only important sources of indigenous knowledge: they gradually turned into objects of study themselves – hence Hermann's description of their entourage as resembling 'an ethnographic museum of living specimens'. The brothers were careful observers of the 'living specimens' in this itinerant 'museum'.[156] Their aim in studying the behaviour, worldviews and social and religious practices of the heterogeneous group of companions was ultimately to gain insights into the highly complex and diverse societies, cultures and 'races' of Asia.

Yet, the validation of information presented a considerable challenge. Given the brothers' enormous itinerary, they were caught in a dilemma: on the one hand, while they were already distrustful of their own establishment, fearing deserters and treason 'from within', they were even more suspicious of the strangers with whom they frequently crossed paths.[157] On the other hand, in order to widen their knowledge base they were obliged to draw on often unverified accounts by mere strangers, whose oral testimony had to be accepted *faute de mieux* as a complementary source of information. The Schlagintweits therefore established strategic, even affective ties with their entourage, as well as persons encountered by chance. 'Strategic intimacy' was thus a vital element of knowledge production in the context of exploration and cultural encounters more generally, but it was also a psychological device for travellers to cope with

---

[155] Ballantyne, 'Strategic intimacies', 5.
[156] H. Schlagintweit et al., *Results* I, 42.
[157] The figure of the treacherous deserter was also a literary trope in nineteenth-century expeditionary accounts, adding further nuance to the cast of characters of a gripping (hi-)story of exploration; H. Schlagintweit, *Reisen* IV, 220–4.

their anxieties by choosing a number of people whom they would trust without reserve.[158] The brothers clearly established such an inner circle of companions.

A telling example of how affective bonds could also be established with outsiders is the 'political agent and representative' of the raja of Sikkim, Chibu Lama, whom Hermann Schlagintweit learnt to appreciate as a trustworthy informant during several stays in the kingdom. Chibu Lama had been instrumental in rescuing the imprisoned Joseph Hooker and Archibald Campbell, who had travelled through Sikkim in 1849 without permission from the raja.[159] Both travellers later recommended Chibu Lama to the brothers (Figure 5.9).[160] Tellingly, the former is still considered a traitor by some Sikkimese today for his service to the British cause, capturing the precarious situation in which some of the brothers' informants found themselves.[161] Hermann clearly profited from the encounter: 'Among the natives of Darjeeling, Chibu Lama emerged as an important contact for me … I must gratefully acknowledge the fact that … he was always willing to engage in conversations about religious and ethnographic matters, but also about mountain designations and topographical details.'[162]

Chibu Lama's elevated social rank, scientific curiosity and learning made him valuable for closer exchanges. His 'permanent stay in the huge mountainous country' had led the political agent to carefully study 'stones, geological layers and mountain shapes', allowing the German travellers to discuss finer scientific 'distinctions with him'. In a passage that sought to convince his European readers of Chibu Lama's trustworthiness, Hermann maintained that by means of 'comparative studies' he was able to establish that his information generally 'proved true'.[163] Indeed, he increasingly came to enjoy 'instructive conversations' with the indigenous diplomat-naturalist, 'the more *I was able* to judge specific issues with any detail myself'.[164] At the same time, perhaps in anticipation of European reactions to his appraisal of indigenous expertise, Hermann was also eager in his writings to ridicule some of Chibu Lama's views of physical phenomena, not least to re-establish his own authority.[165]

---

[158] Ballantyne, 'Strategic Intimacies'.
[159] Hooker called Chibu Lama their 'ally', see Huxley, *Letters of Hooker* I, 310.
[160] H. Schlagintweit et al., *Results* III, 138.
[161] I thank Emma Martin, Manchester, for this information.
[162] H. Schlagintweit, *Reisen* II, 187–8.
[163] Ibid.
[164] Ibid. 189, emphasis mine.
[165] See on the relationship between European identity, memories of explorations and the construction of alterity, Fabian, 'Remembering'.

Figure 5.9 Chibu Lama, a Lepcha, forty-three years old, at Sikkim, 'The political agent of the Rájah of Síkkim at Darjíling', modified photograph.
© BSB, SLGA, IV.2.68.

An important pillar for establishing and maintaining strategic intim-acies was the exchange of material objects.[166] While Chibu Lama was responsible for acquiring a set of otherwise inaccessible religious artefacts and manuscripts from Tibet, he was, in return, presented with a choice of different objects of European provenance.[167] 'Among other things, [Chibu Lama] chose a sun-watch, a compass, and a drawing set, with which I was well equipped in terms of quantity and quality for similar cases.'[168]

While the literature on strategic intimacies is useful for framing the complexity of social relations that emerged in the 'field', the concept sometimes still suffers from the basic assumption that only (itinerant) Europeans were eager to form such ties, and that non-Europeans represented easily manipulated and to some degree powerless actors in the encounter. This was by no means the case. The Schlagintweits were travelling as agents of the British Empire, and were perceived pre-cisely as such by their companions and indigenous rulers. They were thus regarded as potential harbingers of British expansionist designs, and non-European scholars and officials were therefore keen to establish relations with the foreign travellers themselves – but with a view to con-fusing European stores of knowledge about their home countries.

A telling episode occurred with Robert Schlagintweit during his stay in Leh. Amongst the numerous strangers Robert met 'was one distinguished personage' – a high-ranking Chinese Buddhist priest. The odd couple of the European traveller and the lama became closer acquainted in a series of meetings, until the spiritual leader invited Robert into his personal quarters. There, he received him with an exquisite dinner served on the finest porcelain, and entertained him with insightful conversation.[169] Robert, in turn, could not withstand the temptation to enquire about the then almost unknown route from Leh to Peking, the seat of the Chinese government. After the First Opium War (1839–42) had violently opened the Chinese coast and pockets of its hinterland to international trade, the greatest parts of China's interior remained unexplored. After a long discussion of 'the social life and many of the most remarkable manners and customs of the Chinese', the travellers approached more sensitive terrain:

---

[166] See on the multiple functions of the exchange of material objects, not least for establishing relationships in colonial encounters, Thomas, *Entangled Objects*.

[167] H. Schlagintweit, *Reisen* II, 189. On Emil Schlagintweit's acknowledgement of Chibu Lama as an important informant for his work, E. Schlagintweit, *Buddhism*, viii–ix.

[168] H. Schlagintweit, *Reisen* II, 189. On the brothers' gift-giving, see also H. Schlagintweit et al., *Results* III, 20.

[169] On the Leh episode, BSB, SLGA, V.2.2.2, 29ff.

In the course of conversation I could not repress my astonishment that to us Europeans, China Proper herself was kept so closed and made so inaccessible. Then *the Lama offered*, and I accepted his offer with most sincere thankfulness, to give me in writing the whole route from Pekin to Leh; a route, which up to the present time was, as regards its details, as good as unknown, and concerning which, information had decidedly a high value.[170]

Weeks later, Robert received a 'bulky manuscript' written in the lama's own hand. Yet, when the brothers started to decipher the route it described, it became clear to Robert that, in his view at least, 'the Lama, deeply penetrated by the justice of the perfidious policy of his government had intentionally attempted to deceive me'. The entire 'itinerary, from beginning to end, proved to be a fabrication'. Yet, the lama had been so ingenious as to use real names of places from within China to make his attempted deception more convincing. For the same purpose, the first three days of the proposed itinerary were given accurately, as the first legs of the route were much easier to verify by the foreign agents.[171]

Significant also is the scientific traveller's later reflection about the motives behind the deceit. Robert assumed that the lama and the Chinese government were both convinced that 'only with the greatest impossibility will [China] ever be conquered by Europeans, if its geographical and topographical conditions remain unknown to everybody in consequence of entirely misrepresented facts'.[172] In the lama's view, western ignorance perpetuated through deliberate misinformation could be an effective protection in potential conflicts between his own government and expansive western powers in Asia.

This episode leads us to challenge the impression that arises from a number of works on the history of exploration: the impression that indigenous informants were somewhat naïve, easily manipulated sources of information, who unconsciously gave away their knowledge at the cost of their own destruction.[173] Richard Drayton, for instance, makes the observation that '[t]he story of such appropriations' was often tragic, since 'so many cultures were destroyed by the civilizations to which they gave their knowledge'.[174] While this is true in numerous overseas settings across a vast time span, we nonetheless ought to take indigenous initiatives of protecting their knowledge seriously, not least because they demonstrate how fragile the information order of a moving expedition really was. Instead of assuming a pre-established hierarchy

---

[170] Ibid., emphasis mine.
[171] Ibid.
[172] Ibid.
[173] This is an old trope, see Ludden, 'Orientalist Empiricism', 254.
[174] Drayton, 'Knowledge and Empire', 236.

in cross-cultural encounters between supposedly more cunning – and ultimately superior – European actors and an innocent, trusting and in the end hoodwinked indigenous side, the brothers' experiences force us to acknowledge the agency and distinct intentions that indigenous actors also pursued in such contexts.

Robert Schlagintweit's reflections above, as well as other statements made in the wake of the expedition, clearly demonstrate that the brothers were aware of the causal nexus between geographic exploration, acquired intelligence and its potential use in military campaigns. Robert, at least, easily integrated their excursions into a long, colonially motivated history of exploration of high Asia:

100 years ago, no one troubled himself about the height [of the Himalayas]; no measurements were taken ... A few decades later, in one portion of the H[imalayas], namely in Nepal, political events, in which the English were mixed up, suddenly stepped upon the scene; then the necessity at once presented itself of knowing the country not merely superficially but thoroughly; maps, based on scientific data, were indispensable ... [A] long period of peace is to be regarded as one of the most important causes of a slightly extended geographical litera-ture, whilst wars in countries hitherto inaccessible and unknown unquestionably speedily promote geographical knowledge.[175]

What is, however, silenced here is the considerable dependence upon indigenous information that European travellers had to accept – while they were eager to advance and consolidate western knowledge about such frontier regions. Being excluded from local intelligence could, indeed, prove disastrous. Adolph Schlagintweit's demise is a striking case in point. The fatal outcome of his final excursion beyond the contested frontier of British India can be explained by the failure of the expedition's information order. In spring 1857, Adolph, under the guidance of Amin, had opted for an unfrequented route into Chinese Turkestan that avoided another stay at the city of Leh – a crucial hub for the exchange of news from across central Asia.

The decision to travel on unfrequented tracks was the main reason the entourage did not hear about the emerging crisis in Kashgar. Rumours and warnings would soon have reached Leh. It was here that the 'meeting of caravans' brought together 'two to three thousand strangers' at the same time.[176] In later explaining the catastrophe, Hermann concluded: 'News of the rebellion [in Kashgar] may not have been inaccessible in the bazars of Leh in direct communication with people from Yarkand, and [knowledge of it] would then have ruled out any attempt to push forward

[175] BSB, SLGA, V.2.2.1, 103.
[176] BSB, SLGA, V.2.2.2., 27.

into Turkistan.'[177] The Schlagintweits' dependence upon indigenous information, especially for Chinese territories, was thus both boon and bane: while only the guidance of their expedition by experienced caravan leaders had allowed the brothers to explore parts of Turkestan at all, the isolation from indigenous information was ultimately responsible for one brother's death.

Profiting from the kaleidoscopic experience of travelling through south and central Asia, it became the brothers' ambition to formulate theories on the various ethnic groups of the countries they visited and to engage in speculative conjections on human development and biogeography.[178] In doing so, their scientific programme was again enlarged. Having once started as a minor branch of Britain's 'Magnetic Crusade', it became a politico-economic project of the late Company rule that ultimately encompassed macro-physical surveys of natural and social resources and 'racial' variation in territories within and outside British India.

The EIC had long since established complex institutional arrangements to study and engage with subordinate and foreign populations through an elaborate system of agents and different forms of espionage and intelligence gathering – a concern the brothers continued.[179] Yet, their survey work can also be seen as an early example of comparative ethnography, which included the collection of detailed personal and material information about local populations and cultures.[180] In comparing significant amounts of anthropological and ethnographic data about communities across India and the Asian highlands, the brothers sought to develop a 'racial geography' that would account for historical migratory movements of different 'tribes' and 'races' (and their mixing), and the spread of various belief systems and languages across these vast regions.[181] For their studies, the brothers took advice from the British pioneer of colonial ethnography, Brian Hodgson, who had developed his methodology and 'racial' taxonomy during a period when the ethnographic impulse was weaker and less systematic than after the shocking events of 1857–8.[182]

Due to their almost incessant mobility, the brothers adopted a strategy that used observations of sometimes a single 'representative' of a whole

---

[177] H. Schlagintweit, *Reisen* IV, 226.
[178] H. Schlagintweit, 'Ethnographic Summary', *Reisen* II, 25–54.
[179] Cohn, *Colonialism*; the British obsession with the so-called 'hill tribes' is studied by Middleton, *Ethnopolitics*, 62ff.
[180] This is merely hinted at in Brogiato, Fritscher and Wardenga, 'Visualisierungen', 238–42.
[181] H. Schlagintweit, *Reisen* II, 25–54; on the arrival of the 'Aryan tribe' in high and south Asia from the high mountain areas to the west, see ibid., 26; and H. Schlagintweit, *Reisen* I, 487ff.
[182] Middleton, *Ethnopolitics*, 62f., and, somewhat more critical, Allen, *Hodgson*.

ethnic community in order to extrapolate general insights into the 'character' and worldview of entire 'tribes'.[183] To that end, their heterogeneous establishment became a prime and sometimes even the only reference point. Yet, their entourage, while highly varied, never fully reflected Indian or high Asian societies on the whole: it was an entirely male community, hence allowing close encounters with 'representatives' of only one of the sexes. What further undermined the reliability of their ethnographic observations 'on the move' was the fact that while their travelling party did create spaces for close encounters, almost none of the indigenous members of the expedition party could be studied within their original social environment. All assistants were itinerant, and the at times extremely difficult conditions of travel – especially beyond British India – threw all expedition members into unfamiliar constellations that distorted any attempt to study 'traditional' ways of living.

Lastly, sheer ignorance could lead to grossly biased results in the Schlagintweits' ethnographic studies. Their collaboration with Mani's cousin, the learned pundit Nain Singh, provides a striking case through which to understand how the brothers' scientific judgement was at times clouded by personal misconceptions.[184] Nain Singh was recruited into the brothers' establishment in the summer of 1855, fulfilling different functions for the expedition until 1857.[185] Over time, he continuously rose in the brothers' estimation. They described him as 'a very sharp young fellow who has learned with us to read instruments, a little map making and a little English writing'.[186] After he had acquired the necessary skills, Nain Singh was employed to make 'corresponding observations on different stations' along independent routes.[187] He was also placed in charge, for several months, of an observatory established in Leh, where he was said to have 'diligently and carefully' made a series of useful magnetic observations.[188] In return, Nain Singh taught the brothers some Tibetan and was also an important informant about this country. His readiness to collaborate even allowed the brothers to take his facial cast, originally made with plaster and later reproduced in metal (Figures 5.10a, 5.10b). To the extent that such plaster casts can be considered as 'portraits', the

---

[183] The Schlagintweits, like many contemporary ethnographers, used the term 'tribe' without much precision, and it could mean many things, from race to community or group.

[184] The Singh family from Kumaon also provided other personnel, among them Dolpa Singh, initially employed as a low-ranking helper, but who became Adolph's 'interpreter and chief guide for Bálti'; H. Schlagintweit et al., *Results* I, 39.

[185] H. Schlagintweit, 'Bor-Verbindungen', 513.

[186] Schlagintweits quoted from Waller, *Pundits*, 40.

[187] H. Schlagintweit, 'Anlage des Herbariums', 165.

[188] H. Schlagintweit, *Reisen* IV, 201.

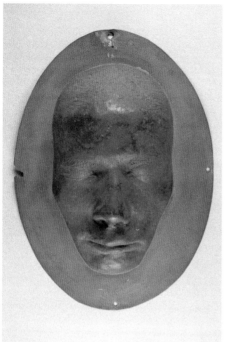

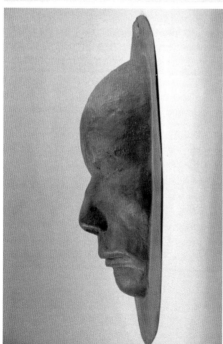

Figures 5.10a and b Nain Singh's facial mask. The inscription reads: 'Nain Singh, Bhot Rajput, Gàrvál Himálaya' (above), and 'Schlagintweit'sche Sammlung Ethnographischer Köpfe' (below). © MNHN-HA-3078.

Figure 5.11 Nain Singh, photogravure. Original title: 'Pandit Nain Singh C. I. E. Survey of India. The first of the pandits of Tibetan Exploration. 1865–75', from *Records of the Survey of India*, vol. VIII, part 1: *Exploration in Tibet and Neighbouring Regions 1865–1879* (Dehra Dun, 1915).
© RGS, S0013206.

mask is – except for a much later photogravure – the only surviving portrait of the young Nain Singh (Figure 5.11).[189]

Yet the brothers' descriptions of Nain Singh were contradictory. His accomplishments during extended travels well beyond his native region, for instance to Ladakh in 1856, were highly praised. Nain Singh 'took a great interest in our operations, and though at first unacquainted with instruments was soon taught their use, as he showed a very great desire to be able to read off the scales and write the reading in English numbers'.[190] His curiosity and eagerness to travel and explore went so far that he 'himself, even though he could only speak Hindostani and

[189] Driver, 'Face to Face'; Brescius, 'Innenleben'.
[190] H. Schlagintweit et al., *Results* I, 39.

Tibetan, for a while reiterated his wish to go with us to Europe'.[191] Since the brothers were dependent upon the philological expertise of a learned assistant for completing their planned travelogues, they proposed to Nain Singh, 'and with apparent acquiescence on his part, to take him with us'.[192] During three years in Europe, Nain Singh should receive 100 rupees per month for his linguistic services, and a further 1,000 rupees in advance for his family in Kumaon.[193]

Further collaboration with their assistant came, however, to an abrupt end. According to the brothers' version, Nain Singh 'unexpectedly went away from us at Raulpindi', leaving behind 'an interminable letter full of apologies and glowing with love of home'.[194] With the letter, Nain Singh had also reimbursed 'a pretty large sum of money' he had formerly received, and 'then fled to the high and lonely mountains, where he hided [sic] himself until after our departure, in order to avoid being made to keep his word'.[195] The Schlagintweits, in turn, used this unexpected turn of events to formulate theories about the distinctions between 'mountain dwellers' and the inhabitants of the Indian plains: 'like *all hill men*', they explained, Nain Singh 'was too much attached to his native mountains to bring himself to leave them'.[196] By contrast, 'it cost us very little trouble to persuade a highly accomplished Mohammedan, living in the hot low-lying plains of Bengal, to accompany us to Europe for half a year' – the munshi Sayad Mohammad Said.[197] The brothers even used this episode to argue for the supposedly unchangeable 'character' of 'all hill men'. Yet, this characterisation stood in an uneasy tension with their previous experiences with Nain Singh, who had been a long-serving, curious and flexible scientific companion during two years of extensive travels.

Many of the depictions the brothers offered of south and high Asians were inextricably linked to their sense of self. It was through the retrospective narration of their experiences that the Schlagintweits could fashion themselves in a specific light in front of western audiences.[198] While the brothers stressed, at times, the supposed lack of courage of their south Asian companions, they could in turn assume themselves to be the dynamic, entrepreneurial side in this cross-cultural encounter, and – despite all contradictory praise – depict their cherished guide Nain

---

[191] H. Schlagintweit, 'Anlage des Herbariums', 165.
[192] H. Schlagintweit et al., *Results* I, 39.
[193] Pathak and Bhatt (eds.), *Life* II, 251.
[194] BSB, SLGA, V.2.2.1, 53.
[195] Ibid.
[196] H. Schlagintweit et al., *Results* I, 39.
[197] BSB, SLGA, V.2.2.1, 53.
[198] Incisive is Fabian, 'Remembering', 50.

Singh as lacking in curiosity and zeal, as an almost pitiable and childlike person bound to his local surroundings by insurmountable fear.

What actually motivated Nain Singh's decision to withdraw was beyond the brothers' imagining. His reasons are explained in a rare document, a collection of his diaries and assorted writings, held today in the National Archives of India (New Delhi). According to his own account, Nain Singh was initially willing to embark upon the projected journey to Europe. He accepted the Schlagintweits' financial offer of three years' service, and informed his cousin Mani Singh. Yet, there existed a personal rivalry and a biting jealousy between Mani and Nain Singh: while (the older) Mani had first been enlisted in a superior position during the Schlagintweit expedition, and had treated Nain Singh almost as his personal servant, the latter had excelled to such a degree that the brothers were keen to take him, and not Mani, as their assistant to Berlin. Mani, feeling slighted, effectively blackmailed his cousin by stating: 'If you leave for England, you shall be as good as dead for us. We shall even conduct your last rites right now.'[199] According to his biographer, Shekhar Pathak, Nain Singh was then still unable to openly challenge Mani, and therefore bowed to family pressure and declined the offer without conveying his real reasons.[200] These family relations and conflicts reflected yet another hierarchy within the establishment that the brothers could not influence, indeed whose existence was entirely unknown to them.

The case of Nain Singh is thus an extreme example of the pitfalls of the racial stereotypes that European explorers could at times develop through cross-cultural encounters. The absurdity of some of the Schlagintweits' ethnographic conclusions is highlighted by the fact that, far from being 'chained' to his hill site, Nain Singh later became the most famous and widely travelled explorer of the trans-Himalayan region, outperforming any western traveller in the geographical exploration of this vast area. Instead of honouring the Schlagintweits, British imperial authorities later lavished medals and praise on Nain Singh for his outstanding achievements – even though the identity and real name of this particular 'imperial hero' had to be kept a secret for fear of his discovery.[201] The secret explorations over thousands of miles of uncharted territories by Nain and Mani Singh, and a host of other pundits under the guidance of Captain Thomas George Montgomerie, demonstrate clearly that we ought not to think of their acquired stores of knowledge merely as 'local'

---

[199] Pathak and Bhatt (eds.), *Life* II, 251.
[200] I am grateful to Shekhar Pathak for this information.
[201] Raj, *Relocating Modern Science*, 220.

knowledge. Here, the 'natives' turned into veritable explorers in their own right.[202]

The expedition's rich cultural and scientific life, the internal dynamics and shifting hierarchies between the Schlagintweits and their key aides provide a starkly different image from Kapil Raj's incisive treatment of the British operations to recruit indigenous pundits for trans-Himalayan surveys in the wake of 1857. Raj suggests that the pundit's training and work passed with 'few signs of controversy', realised under clear chains of command.[203] In his words, it presented 'a textbook account of consensual, cumulative knowledge formation, the success—and indeed the very possibility—of which required the foreclosure of all dispute'. The Schlagintweit enterprise, by contrast, was rife with internal conflicts, mistrust among its members and schemes of deceit. The isolation of the mission's supposed European leaders from their aides and hosts, indeed the Schlagintweits' frequent inability to exert authority thanks to linguistic deficiencies as well as geographical and cultural ignorance, demonstrate their relative powerlessness in the Asian highlands.

In sum, the circumstances under which the Schlagintweits tried to acquire and produce new knowledge, and to rely on the information networks of south Asian informants, reveal the weakness and constant challenge for the explorers to distinguish between reliable testimony, conscious misinformation and mere fantasy. The disorientation, unfamiliarity and inexperience that the Schlagintweits faced during their journey faded, however, once they had returned to Europe. The more time that passed between their actual journey and subsequent occasions to lecture about and publish their results, the more we are confronted with the integration of familial tropes of European supremacy. The final section of this chapter will therefore turn to the curious shifts in perception that later enabled the brothers to again build up a neat hierarchy – between themselves and their indispensable aides.

Reinstating a Hierarchy of Knowledge

'Enlightened science' in the form of empirical and critical observations assumed a critical role in the Schlagintweits' understanding on how to overcome what they saw as the 'frantic' idolatry and 'superstitious' worldviews of many of their companions.[204] '[T]he rich mythology of India' had supposedly led to a state of great obscurity about India's and

---

[202] Driver and Jones, *Hidden Histories*.
[203] Raj, *Relocating Modern Science*, 209.
[204] Cf. Metcalf, *Ideologies* III.4, 135.

central Asia's geography and natural characteristics.[205] In the brothers' published accounts, religiously informed descriptions of lands or natural phenomena often served as anecdotes.[206] This ridicule, however, can be read as an attempt to conceal an uncomfortable truth: the brothers' insurmountable dependence on south Asian informants, many of them Brahmins. This dependence had been a struggle for generations of Company officials. Following military victories, the latter had therefore sought to replace 'secondhand', 'hearsay' and 'traditionary' indigenous accounts with direct and supposedly superior observations and measurements.[207] Given the Schlagintweit expedition's large scale and scope, they had to take a middle ground. While they repeatedly acknowledged the integration of geographical observations and computation from their south and high Asian aides into their accounts, they also frequently juxtaposed their own findings with that of other informants – especially when the latters' trustworthiness could not be guaranteed through personal scientific training and instruction.

At times, the brothers received information about particular routes through the unfamiliar representations of heights and landmarks in Asian cartographic traditions. This is well captured with regard to a Tibetan map that was later integrated into the brothers' *Atlas* of maps and views (Map 5.3). During Hermann's visit to Buthan, one of his assistants, Dávang Dórje, a 'caravan leader' engaged in trade between 'Tibet and the borders of Assam', had drawn a first sketch. Yet, 'in Narigun, we found a rather able Tibetan … and it was he who ultimately brought some order into the different elements of the map, which he also significantly expanded'.[208] In the published *Atlas*, the map was, however, ultimately complemented with 'objective' European data that provided – among other things – the exact geographical position of the mountains and rivers depicted on it. The accompanying data was given in the lower right-hand corner, pointing to fixed stations the brothers had taken themselves. The diagram was entitled 'Sketch of Approximate Geographical Positions', with a scale of '1 to 8.000.000'. The supposedly 'primitive' map was thus transformed in their publication into a 'curious object' of indigenous learning, which had, however, to be made comprehensible to European audiences by inserting the brothers' own geographical co-ordinates.[209] As Hermann furthermore stated: 'My own

[205] BSB, SLGA, V.2.2.1, 99.
[206] H. Schlagintweit, *Reisen* II, 188.
[207] Ludden, 'Orientalist Empiricism', 254.
[208] H. Schlagintweit, *Reisen* II, 102–4.
[209] Bruno Latour discussed at length such a translation process of indigenous geographical information and sketches into western and supposedly universal knowledge, Latour, *Science in Action*, ch. 6; see also Roberts, 'Situating Science', 18.

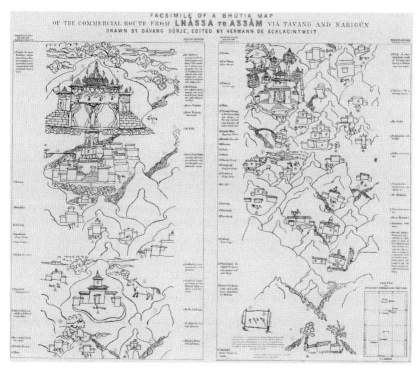

Map 5.3 'Facsimile of a Bhútia Map of the Commercial Route from Lhássa to Assám viâ Távang and Narigún. Drawn by Dávang Dórje; edited by Hermann de Schlagintweit'; Hermann Schlagintweit et al., *Atlas*, Map 3.
© DAV.

travels offered me the opportunity to convince myself of the veracity of the topographical names' given, at least 'for the lower part of the map'.[210] While the brothers thus inserted a hierarchy between their own and indigenous systems of topographical presentation (and different degrees of personal reliability as observers), a wealth of other information about the political and religious landscapes of the depicted regions was, however, lost on them. This information was contained in visual codes across the map, which was decisively more than a tool to depict a commercial route from Tibet to Assam.[211] Given the Schlagintweits' unfamiliarity with indigenous cultural codes, the (as they put it) 'rudely executed' map

[210] H. Schlagintweit, *Reisen* II, 104.
[211] Schwartzberg, 'Tibet', 660–1.

merely seemed to exemplify the superiority of their cartographic modes of representing nature.

Through this account of how the itinerant brothers perceived the indigenous, religiously informed views of Asia's natural worlds, it has become clear what significant role their scientific approach played in their perception of self, especially when contrasted with what they later presented as the 'superstitious' and erroneous 'beliefs' of their assistants.[212] This denigration of non-European systems of ordering the natural world was inextricably linked to the brothers' assertion of authority, especially in front of metropolitan audiences, within and outside the academy. It was precisely in this sense that indigenous ways of seeing contributed to the way in which the German explorers – equipped with European taxonomies and a phalanx of the most advanced instruments and guided by colonial surveying practices – came to develop a belated feeling of mastery over Indian and high Asian nature, as well as over the indigenous inhabitants of those lands.

---

[212] RS, BSB, SLGA, V.2.2.1, 62.

6 Contested Exploration and the Indian Rebellion: The Fateful Year 1857

This chapter argues that the moment of the expedition in the years before and during the First Indian War of Independence needs to be taken seriously as a political conjuncture, as it profoundly shaped the way the Schlagintweits sought to present and legitimise the results of their mission. The Indian Rebellion caused imperial anxieties and nationalist waves among the British public. Those can also significantly account for how the proposed scientific outputs of the Schlagintweits, precisely as non-British travellers in south and central Asia, were received. It is indeed far more illuminating for the 'culture of exploration' and transnational science in this period to ask not what the mounting controversy meant for the brothers personally, but what the Schlagintweits meant for their critics.[1] Such an approach adds considerable nuance to our understanding of the contested meaning of geographical expeditions at the time, which in the Schlagintweits' case cannot be seen as the triumphant expression of European science and global influence. Rather, theirs was as a deeply controversial and fraught undertaking that needs to be placed within the pressing political concerns of mid Victorian Britain.[2] After the summer of 1857, the smouldering controversy over the brothers' employment was to take a decisive public – and vicious – turn. Until then it had largely been a private quarrel. From now on, much more was in question than personal rivalry between competing travellers, the envy of patronage and issues of probity and intellectual property. While such individual spats certainly continued to play a role, the Schlagintweit enterprise ultimately became part of much wider social and imperial debates, even those concerning the very legitimacy of British rule in India itself.

For the Schlagintweit brothers, 1857 became a watershed year. When Hermann and Robert returned to Europe in June, their prospects were wide open as they faced an endless list of questions relating to how

[1] Driver, 'Critics', 137.
[2] Ibid., 142.

they could shape their future careers, make a living and carve out a lasting reputation from their Asian travels. The stakes were especially high since the brothers had given up academic positions to embark on their mission: Hermann had become a lecturer in geography at Berlin University in 1852, and signed his letters 'Professor Schlagintweit';[3] it was only due to his departure in 1854 that the eminent cartographer Heinrich Berghaus inherited his post. Adolph, on the other hand, had failed the *Habilitation* in the Prussian capital, but had passed it at Munich University, where he also briefly taught as a *Privatdozent*. In 1853, Robert still studied in Munich, but received his doctoral degree in July 1854 from the University of Jena. Apparently, he too was eager to pursue a university career, and only weeks before departure decided to join his siblings.

Hence, unlike many other scientific travellers who lacked a secure position, the three Schlagintweits had left Europe not for want of a better alternative. But they had done so, nonetheless, because the Indian and high Asian explorations promised to pave the way to becoming celebrated travellers on a par with the leading Victorian heroes of exploration. The Company commission offered them a faint chance to imitate the remarkable life and international recognition that their mentor Alexander von Humboldt had achieved for his American and central Asian oeuvre. Now, with two of the brothers having returned with the greater part of their collections, 751 sketches and watercolours, a few hundred photographs and over forty 'volumes of manuscripts of observations', and the still fresh memories of their travels, this was their moment to secure a lasting legacy for the enterprise.[4]

However, as foreigners in British service, with multiple loyalties and mixed funding arrangements with multinational sponsors, the way forward was not set in stone. On the contrary, in revealing the motivations and contradictions of transnational science at the time, the brothers had to repeatedly forge their own opportunities, playing once more – if possible – one side against the other. In planning not only their reintegration but also personal advancement within the inner circles of metropolitan science, the Schlagintweits had to find solutions to such vexing issues as: what precise form should their final narrative of exploration take, and how could it be financed? In what language(s) should their accounts appear, and where first? Indeed, could any single publication fulfil the various expectations the Schlagintweits themselves had raised

[3] Hermann to Benjamin Silliman, 18 February 1852, Yale University Library [YUL], Dana Family Papers, MS 164, Box 3, folder 110.
[4] H. Schlagintweit et al., *Results* I, 7–9.

in front of so many different audiences – from Company men to scientific specialists and wider publics? There remained two more pertinent queries: who would claim ownership over their massive, albeit eclectic collections of natural history and ethnography, but also how could the brothers use these valuable artefacts as assets to forge new social and professional standings?

While these issues needed to be urgently decided, other unexpected events intervened – putting their schemes and future careers at risk. As the year 1857 progressed, increasing incertitude arose regarding Adolph's fate during his recent trip into central Asia. What were the consequences, personally, but especially scientifically, if he – arguably the central figure behind the Indian mission – had fallen victim to his zeal and had been killed during his final explorations? Then, in the summer of 1857, news of a severe colonial crisis arrived in Britain: the Indian Rebellion that shook Company rule to its foundations. With the brothers formerly enlisted by the EIC, the outbreak of this uprising also initiated the end of the Company regime in south Asia – and hence the dissolution of the Schlagintweits' employer and supreme financial patron. The Company's impending demise had important repercussions for their careers, but exactly how that played out requires a far more detailed enquiry than has previously been offered.

Negotiating a Reputation

The planning, execution and public consumption of overseas expeditions always entailed the mobilisation of considerable resources, provided by both European and non-European actors and institutions.[5] Finance was essential. During their travels, the brothers' extravagant equipment, huge personal establishments and the acquisition and transport of their collections had required a prodigious expenditure. The need to keep their material patronage intact did not cease after their return. Yet, financial backing alone could not guarantee public appreciation. Non-material resources were equally necessary to secure social and academic recognition by metropolitan scientists, learned societies and broader publics – which could only be acquired through support from a number of advocates. Paying close attention to the communication strategies of the brothers and their committed intercessors yields fresh insights into how the private and public battle for recognition was fought within, and across, national boundaries.

---

[5] Roberts, 'Expeditions', 214; Driver, 'Missionary Travels'.

Since the brothers' appointment in 1854, the nature of their contacts in England had dramatically changed. The much-admired Prussian consul in London, Christian Karl von Bunsen, had been forced to resign during the second year of the Crimean War (1853–6), thanks to an unsanctioned pro-British diplomatic intervention in the conflict. Fervently pro-British in attitude, Bunsen had sought to forge, without official backing, a Prussian–British alliance against Russia. As his initiative failed, and Prussia declared its 'benevolent neutrality', he stepped down in April 1854.[6] His resignation occurred at an untimely moment. During the Crimean War, Prussian–British relations became particularly strained. Prussia's neutrality not only infuriated the British government, but also led 'to fierce anti-Prussian reactions in English newspapers'.[7] While the most potent advocate of the Schlagintweits had now left the island, his successor, Albrecht Graf von Bernstorff (1809–73), was hardly a match for the learned and scientific enthusiast Bunsen, and would never achieve the public acclaim his predecessor had found among Britain's scientific circles. Alexander von Humboldt regarded Bernstorff as 'invisible' and in terms of scientific diplomacy maladroit and insensitive when it came, for instance, to negotiating the future of the Schlagintweits and their collections with British authorities.[8]

To procure recognition for the brothers' exploratory feats in London, their German patrons had to intervene from afar, especially since Humboldt sensed a lack of British appreciation. He even felt that the brothers had fallen victim to a veritable defamation campaign in England. Writing to the RGS president in March 1857, Humboldt stated that the president's recent letters 'have saddened me, they have proven to me that my compatriots have become calumniated in the eyes of Sir Roderick Murchison'.[9] As among the most respected cosmopolitans of his age, Humboldt now sought to overcome those 'poor national rivalries' which he felt had denied his protégés their well-deserved acclaim.[10]

In the mid 1850s, there was hardly a better stage on which to present the achievements of expeditionary science than the anniversary address of the RGS president. The address marked, as Humboldt knew, a yearly 'day of celebration for Geography'. The president's praise of any explorer was especially important, since the address was printed in the society's journal, and thus circulated widely in Britain, its overseas empire and on

[6] Kirchberger, *Aspekte*, 351.
[7] Reichel, 'Responses', 19.
[8] Humboldt to Bunsen, 30 May 1854, in Schwarz (ed.), *Briefe*, 178–83, 181.
[9] Humboldt to Murchison, Berlin, 31 March 1857, EUL, Gen. 523/4; copy at BBAW.
[10] Ibid.

the European and American continents. Indeed, as some of his colleagues reminded Murchison, the address inherently meant walking a tightrope. The botanist Joseph Hooker, for instance, himself an Indian traveller with no small ambitions and self-esteem, once enviously complained about the attention the two Strachey brothers had received in the 1852 speech, which Hooker felt was undeserved. He warned that this would stir up jealousies in India in view of the '*immense* ... *importance*' of Murchison's address, which was 'reprinted in every Indian paper, within a few weeks, & duly canvassed'.[11]

The anniversary address was now precisely the platform Humboldt sought to conquer for the Schlagintweits. In an exchange that shows how affective ties, mutual admiration, but also a sense of obligation overlapped in the Prussian's old relationship with Murchison, the latter sought Humboldt's advice for his upcoming address at the annual meeting, scheduled for 25 May 1857. Six days before the event, Humboldt sent Murchison his commentary on the manuscript. At first, the Prussian geographer assured the president of his undisputed authority, with him being 'right at the sources' of geographical news and undertakings, even leading Humboldt to ask rhetorically 'who could instruct you' on any geographical issues?[12] Humboldt then went on to do precisely that. He kindly requested that Murchison 'grant your praises to the most brilliant success of the travellers, which the brothers Schlagintweit, while defying all dangers, have obtained during their stay at Ladak and in ... Tibet'. But more than that, Murchison was also reminded about the pioneering character of Hermann and Robert's crossing of the Kunlun Shan, which had brought them to Khotan, in today's western China. To underline the pioneering character of this crossing, Humboldt adroitly cited a British Himalayan traveller, who now acted as superintendent of Calcutta's Botanic Gardens: 'A botanist of the highest merit, Dr Thomas Thomson, who has published with my excellent friend Joseph Hooker the *Flora indica* in 1855, writes in the Introductory essay ... "The chain of the Kouenlun ... has not been crossed by any European traveller".' Drawing on this British testimony to the hitherto unaccomplished feat of traversing the central Asian range, Humboldt wrote Murchison that '[t]his statement proves to you the importance of the success by the Schlagintweit brothers'. Humboldt also drew attention to their mountaineering achievements, as they had set a new altitude record in August 1855 when taking observations on the mountain 'Ibi Gamin ... at

---

[11] Quoted from Stafford, *Scientist*, 118, original emphasis.
[12] Humboldt to Murchison, Berlin, 19 May 1856, EUL, Gen. 523/4; copy at BBAW; subsequent quotations, ibid.

the height of 22,260 feet'. 'This is not only a higher point than I reached on the Chimborazo ... in 1802 and Boussingault ... in 1831, but is also higher than the peak of Chimborazo itself.'[13]

Yet, the brothers' mission was, within its imperial context, a scientific venture, and mere physical triumphs were insufficient to prove the success of the scheme. Humboldt thus further enumerated a whole list of other accomplishments, not least the 'very important geological excursions into Tibet', which included pioneering glacial studies on Mount Kamet.[14] The range of activities highlighted to Murchison further included the brothers' geomagnetic researches, especially the study of the effect of great heights on 'the variations of magnetic intensity'. While these were presented as the Schlagintweits' scientific attainments, Humboldt concluded his letter by alluding to the fact that British institutions would also be greatly enriched, since the brothers 'will bring some fine geological collections to England, maybe already this autumn'.[15] Since Humboldt knew about Murchison's passion for geological research, he underlined this branch of their numerous acquired artefacts from Asia.[16]

Friendship and manipulation were not mutually exclusive in Anglo-German scientific networks. What seemed to be merely a polite suggestion to include the brothers in Murchison's address was, in fact, a long-planned and orchestrated campaign. This was especially so since members of the RGS had from early on sought to quash the Schlagintweits' scheme. As early as 1855, Humboldt had written about the hostile reception of the brothers' initial appointment: 'In India prevails a more liberal spirit, nothing of the xenophobia of the [Royal] Geographical Society.'[17] Humboldt had indeed for long been frustrated with an English neglect of what he considered great deeds by his fellow countrymen – including by the African travellers Heinrich Barth, Eduard Vogel and Adolf Overweg.[18] In 1856, Humboldt then lamented to Carl Ritter: 'Is it not disgraceful that, owing to an old hatred of Germans, the great speech by Admiral Beechey [to the RGS in May that year] does not mention with a single word the Schlagintweits and the crossing of the Kuenlun!!'[19] When Humboldt then saw Murchison's manuscript

[13] Ibid.
[14] Ibid.
[15] Ibid.
[16] On Murchison's pursuit of geology as a vocation, without becoming a professional, Porter, 'Gentlemen', 824.
[17] Humboldt to Bunsen, 19 August 1855, in Schwarz (ed.), *Briefe*, 191.
[18] For a useful selection of letters on this issue, Päßler, *Briefwechsel*.
[19] Ibid., letter 155, 196–7.

for the 1857 address, he noted bitterly, again to Ritter: 'Murchison gives his geographical lecture ... asks for advice and does not mention the Schlagintweits – as if they did not exist.'[20] He therefore rushed to get all published accounts of the brothers sent to Murchison, and even suggested making 'an abstract of the content' of 'the most important one', which described the Kunlun episode. As his private communication made clear, the quoting of Thomson in his letter to Murchison was above all 'designed to abash the Geographical Society' by inserting the 'confession of the learned botanist Dr Thomson, who had long been in Ladak and had himself failed in trying to go across the Karakorum pass' – a feat the brothers had achieved.[21] While Humboldt adhered to the conventions of gentlemanly conduct in his correspondence with Murchison, the intentions behind his communication thus markedly differed from their outward appearance.

Though Murchison, in his 1857 address, actually gave a long elaboration of the brothers' activities and successes, citing entire pages from Humboldt's letter, we cannot take such praise at face value.[22] Rather, it was the outcome of a long process of negotiation, in which subtle pressure was exerted to secure an acknowledgement of the Schlagintweits in front of London's scientific circles. In fact, Humboldt had already written to Murchison some weeks earlier on the brothers' success in having crossed the Kunlun and other feats, which Murchison had first chosen to ignore in his manuscript.[23] When Humboldt, some weeks later, feigned innocence and thanked Murchison 'in the name of the King for your active and benevolent interest for the laborious Schlagintweits', we see how some of the few positive voices that were heard in England about the brothers were actually manufactured by transnational negotiations.

While such pressured acknowledgement by the RGS president was important, the prime means to establish one's scientific achievements was the printed word. The two brothers who returned lost no time in taking the first steps towards publication. Once again they relied on one of their intermediaries with British circles of power and patronage, and it was Bunsen who addressed Sykes, the EIC director, in July 1857, informing him that the 'Messrs. Schlagintweit have sketched out a prospectus of the work which is destined to make known to the world the results of their expedition. They want to have your advice before working it out & will take no steps without your sanction.'[24] This was mere rhetoric to flatter Sykes.

[20] Humboldt to Ritter, 18 May 1857, ibid., 199–201.
[21] Originally, 'zur Beschämung der Geographischen Gesellschaft'.
[22] Murchison, 'Address', 1857, cl–clviii. Cf. Finkelstein, 'Mission', 198.
[23] Humboldt to Murchison, 31 March 1857, EUL, Gen. 523/4.
[24] Bunsen to Sykes, Heidelberg, 19 July 1857, GStAPK, FA Bunsen, folder 5.

The brothers had already decided on the shape of their publications in India: The work was to be divided into nine volumes, each dedicated to a different field of knowledge, and accompanied with a number of maps, charts and lithographic prints.[25] Yet the question of material patronage was still unresolved. However, during an audience in June 1857, they had convinced the Prussian king to support the work – albeit only in the sum of 3,000 thalers. Sykes replied to Bunsen that he acknowledged the brothers' 'well-deserved success', and agreed that '[w]e must now try to get the results of their Scientific labours published'.[26] Sykes further promised to use all his political weight to realise the projected oeuvre, adding: 'I trust I may not meet with difficulties in the Court of Directors on the subject and particularly with Mr Vernon Smith President of the Board of Control on the ground of expence [sic].'[27] The signs were thus favourable in the summer of 1857 that the Schlagintweits could secure further, long-term employment on British grants. Yet, unexpected events in India, and their repercussions in the metropole, brought further negotiations to an abrupt halt.

The Great Indian Uprising

The Schlagintweit brothers' plea for renewed employment by the East India Company coincided with one of the worst crises for the British Empire. Sykes, the Company's loyal director, therefore cautioned against holding any high expectations, stating in July 1857 that as a result of 'the present confusion in our provinces owing to the unhappy military mutiny … objections may be raised in the Court and at the Board to engage at present in a scientific work the expense of which cannot be calculated'.[28] The Court indeed faced considerably more urgent concerns than securing another treatise on India when the very existence of this vital colony was under threat following the outbreak of the Great Rebellion in its northern parts on 10 May 1857.[29] On that day, three Sepoy regimens rebelled in the city of Meerut near Delhi, killed their European officers and started to march on Delhi. The news of their actions spread quickly and aroused anti-British violence across northern India, where British rule in large parts quickly collapsed. While much nineteenth- and early twentieth-century scholarship on the Indian Rebellion has underplayed

---

[25] H. Schlagintweit, *Reisen* I, vii.
[26] Sykes to Bunsen, 28 July 1857, GStAPK, FA Bunsen, folder 5.
[27] Ibid.
[28] Ibid.
[29] On the causes for its inception, Osterhammel, *Transformation*, 552–3.

its significance and anti-colonial character, the literature produced after Indian independence and then after the end of the British Empire as a whole has provided different interpretations of the cataclysmic events of 1857–8. The spirit of revolt was not limited to the ranks of the British Indian army, but rather found wider approval 'in significant sections of the civilian population', while this support was to be found not only 'in and around the most active sites of the "mutinying" sepoys in northern India but also in other more distant parts'.[30]

When details of the scale of the imperial crisis gradually became known in Britain, East India House in London was thrown into an existential crisis. Indeed, the Company's authority as the purported legitimate ruler of the Indian Empire was severely, and ultimately irreparably, damaged. In such troubling times, the fact that Sykes was nonetheless willing to give the Schlagintweits' scheme his personal protection speaks volumes about the sense of obligation he felt towards the brothers, but also towards his old acquaintance, Alexander von Humboldt. Humboldt, in turn, showed such a degree of gratitude towards Sykes for supporting the Schlagintweits that he sought to secure him a civil knighthood from the Prussian king.[31]

While the Indian Rebellion complicated the sealing of a renewed contract, the imperial crisis of 1857–8 affected the Schlagintweits' career in another, more fundamental regard. While ever-more details on British military defeats, such as the fall of Delhi, and news of the war atrocities committed on both sides slowly reached the European publics, the revolt soon assumed significant meaning for British and continental contemporaries. Being more than a military affair, the rebellion prompted general discussions on British rule in south Asia, including its failures and achievements. In Britain, regardless of certain critical, anti-colonial voices, public opinion saw a wave of patriotic sentiment sweeping across newspapers and journals. Countless articles glorified, as *The Times* had it, 'the unconquerable fortitude of our isolated countrymen', who defended the precious colony against the vicious, 'mutinous spirit' of the 'natives'.[32]

Yet, when it became clear that British hopes for a quick suppression of the rebellion were unrealistic, a number of continental newspapers and writers began to question the very legitimacy of British rule in India. Even the question of Britain's future power and influence in Europe

---

[30] Mazumdar, 'Introduction', 1. Bayly, *Birth*, 151–4.

[31] 'Rote Adler-Orden 2. Klasse', Sykes to Humboldt, 27 June 1857, copy at BBAW; see also GStAPK, I. HA Rep. 89, Geh. Zivilkabinett, AKlC, 77f. Bernstorff to Illaire, 3 September 1857.

[32] *The Times*, 26 October 1857, quoted in Mazumdar, 'Introduction', 3–4.

became an issue of discussion – much to the disdain of the British press. 'Such an affair as the Indian mutiny was not likely to pass without some malicious comments from our ill-wishers abroad.'[33] Foreign criticisms appeared especially spiteful, as again *The Times* asserted: 'Our opponents cannot openly and straightforwardly pray that we shall be beaten in India, because this would be simply siding with barbarism against civilisation.'[34] As a response, the Indian Rebellion marked a moment when patriotism and xenophobic tendencies towards 'ill-meaning' foreigners became more pronounced in Britain.

This outburst of patriotic sentiment did not leave the Schlagintweits untouched, but significantly contributed to their reception. After the return of Hermann and Robert to Europe, they had seemingly kept a low profile, and had only given papers at a meeting of the British Association for the Advancement of Science in Dublin, in August of 1857. There, the brothers' reports on their Indian travels had been well received, and even occasioned the conferral on Hermann of an honorary degree from Trinity College, Dublin.[35] However, the initial appreciation took a critical turn. In October 1857, *The Athenaeum* published a lengthy, caustic commentary on the brothers' enterprise, and on the Schlagintweits' recent initiative to secure further patronage for their publication. The article started out by acerbically stating that: 'These German gentlemen were sent ... on a mission which, as no Englishman could understand any reason for it, was mysteriously – and, we have no doubt, very erroneously – referred to an occult influence.'[36] By this was implied that their generous employment in 1854 had been arranged behind closed doors, through the silent efficiency of patronage networks that it was argued 'deprived' British subjects of 'those lucrative and honourable employments'.[37] It was above all the brothers' claim to scientific discovery that drew biting mockery:

Well, the Messrs. Schlagintweit have come back, and have told the world their secret. They have been, it seems, on a voyage of discovery ... they claim to have found a range of mountains in Upper India called the Himalaya, and to have crossed the country between Bombay and Madras. Their travels in well-worn roads are styled 'a careful exploration of Asia' ... The Prussian gentlemen, we find, have opened up Thibet, and are about to make India known to Europe. We in England fancied that we knew a little about India, and that we had done something towards laying open its physical and geographical features ... But we were labouring, it would now appear, under strange illusions. Doubtless the two

---

[33] *The Times*, 4 July 1857, quoted ibid., 3.
[34] *The Times*, 10 October 1857; quoted ibid.
[35] 'Gossip of the Week', *Literary Gazette*, 5 September 1857, 857–9.
[36] 'The Latest Indian Mission', *The Athenaeum*, 1566, 31 October 1857, 1358–9.
[37] Ibid.

Gerards, Vigne, Moorcroft, Thomson, the two Cunninghams, Hooker, and the two Stracheys – all the men that *we* fancy opened up Thibet – were all myths![38]

The Athenaeum's critique was prompted by a 'summary report' the brothers had delivered of their explorations, routes and 'discoveries' to the Académie des sciences in Paris only two weeks earlier and soon printed in newspapers and the Académie's journal.[39] The report mentioned not a single British traveller. The brothers' account indeed gave the unjustified impression that almost every aspect of their researches and itineraries was pioneering. The facts that the Schlagintweits had received substantial amounts of corresponding data and observations from British officers and stations, not to mention significant personal support from other British travellers, the Surveyor-General's Office, from political Residents and even the Governor-General himself throughout their expedition was entirely silenced. In the 'report', nearly all traces of earlier British scientific engagement with the natural history and ethnography of the subcontinent and the Himalayas were effaced.[40]

Suddenly shifting the gear from biting sarcasm to scornful review, *The Athenaeum* continued that 'we have no hesitation in saying that the facts claimed as discoveries by [the Schlagintweits] were *all* known to English scientific men', with their mission having thus 'terminated in pretensions which are ridiculous and disgraceful'.[41] While such a critique seemed immature, given that the brothers had not even published their official account, it is nonetheless worth considering why the British paper displayed such ferocious aggressiveness when rejecting all the claimed achievements of these 'foreigners'.

The critique is only understandable in the context of contemporary British fears about the current and future state of the Indian Empire. As *The Athenaeum* indeed continued, '[o]ur scientific corps in India consists of men unequalled in their own studies and their own work ... Their Trigonometrical Survey is one of the noblest scientific labours of our generation.' Why, then, would the British imperial authorities adopt 'the policy of engaging foreigners to do what they could have done so well? Is this the way in which Leadenhall Street hopes to gain affection for the service [in India]?'[42] Yet, the appointment of non-British subjects seemed even more flawed, indeed dangerous, as the article asked: 'Is this the way

---

[38] Ibid., 1358.
[39] H. and R. Schlagintweit, 'Aperçu sommaire'. Finkelstein, 'Mission', 198.
[40] First, French newspapers printed their 'report' (e.g., Galignani's messenger), which in turn was reissued in British papers; *Literary Gazette*, 24 October 1857, 1023.
[41] 'The Latest Indian Mission', 1359.
[42] Ibid. The EIC's headquarter, East India House, stood in Leadenhall Street, London.

to impress the native mind with the superiority of *English intellect* and with the justice of *English rule?*'[43]

With the present insecurity over British rule in India, the nature of the Schlagintweit controversy thus underwent a fundamental shift. The *Athenaeum* article still captured the patriotic sense of entitlement that proclaimed an unwritten duty of the EIC to enlist primarily national subjects. Its author still condemned the 'cruel neglect of unfriended genius' among British men of science. But science now seemed to matter not only as a career opportunity that should above all be reserved to fellow citizens. Rather, in the current crisis, the pursuit of science in the Indian Empire was portrayed as a way of establishing, and maintaining, British superiority over indigenous peoples. Science, in this reading, was more than just a 'tool of empire' to better control indigenous populations, implement European health regimes for the sake of Anglo-Indians and maximise the Company's profit by improved exploitation of the land and its resources.[44] For contemporaries, while 'British science' in India certainly had this utilitarian dimension, it had also assumed an equally important symbolic meaning. Scientific activity in the colony was portrayed as a form of rule, as a continuous demonstration to 'the native mind' of how British dominion over the Empire's 'primitive' Indian subjects was indeed of a justified nature.

British contributions to European knowledge of India thus became an important pillar for British imperial identity. 'By the early nineteenth century, following a period in which they showed themselves relatively receptive to Indian ideas and practices, the British saw science, technology, and medicine as exemplary attributes of their "civilising mission", clear evidence of their own superiority over, and imperial responsibility for, a land they identified as superstitious and backward.'[45] By alluding to the high-precision results of the GTS, which also measured the highest peaks in the world, *The Athenaeum* sought to give ample proof of how the British imperial project on the subcontinent was indeed an enlightened undertaking. While it is a familiar trope that wars often reinforce nationalistic sentiments, much less scholarly attention has been devoted to the importance of science as a rhetorical battlefield, on which the virtues of a nation's imperial enterprise had to be defended against the imposturous claims of 'outsiders'. The supposed neglect of British accomplishments by the German brothers was thus portrayed not only as a break with the conventions of gentlemanly science; much worse, it now appeared also as

---

[43] Ibid., emphasis mine.
[44] Headrick, *Tools.*
[45] Arnold, *Science*, 15.

'a gross insult to the labours, merits and memories of the scientific men of India, living and dead'.[46]

Not only the actual colony, but also the historical place of British science in India seemed suddenly challenged. In response, *The Athenaeum* urged that '[i]n days like these Englishmen should hold together'.[47] They reassured their readers that: 'The civil servants of the East India company, labouring at their work seven thousand miles from London, may be deprived by occult influences of some of those lucrative and honourable employments ... but they may rest assured that a watchful press and a generous public will not suffer them to be defrauded of their well-won reputation at home.'[48] Securing British scientific achievements was now seen as an act of piety towards those serving, fighting and already deceased Britons in India. To commemorate their triumphs against the pretentious claims of foreigners thus became a self-assumed duty at the home front during a time when the Empire was at war.

Knowledge Gaps

The Indian Rebellion impacted on the brothers' lives in another fundamental way. Whilst Hermann and Robert completed their travels shortly after the outbreak of violence, the third brother, Adolph, remained in northern India, the western Himalayas and the Kunlun range to complete their observations. In April and May 1857, he was crossing the Punjab, reaching Lahore and Sultanpur. In response to the turmoil in the northern provinces, Adolph was suddenly forced to abandon his plan to return to Europe via Bombay.[49] His diary entries in the summer of 1857 indicate an increased awareness of his fragile position. He was travelling with a small establishment of servants through Chinese Turkestan, spending sleepless nights with a loaded gun in his hand as he feared assaults by the peoples of 'dubious character' he frequently encountered.[50] His plan was, it seemed, to continue on a northern route at this time of intensifying conflicts in trans-Himalaya and Turkestan, to complete the expedition's measurements and eventually reach the Russian territories in central Asia. From there, the land route to Europe was again considered safe.[51]

---

[46] 'The Latest Indian Mission', 1359.
[47] Ibid.
[48] Ibid.
[49] H. and R. Schlagintweit, 'Aperçu sommaire', 1.
[50] 'Zeitungs-Nachrichten', *Bonplandia*, 9 (1862), 160.
[51] Wagener, 'Schlagintweit', 263.

Yet, even after almost a year from the other Schlagintweit brothers' return to Europe, no further sign of life had been received from Adolph – his last letter dated from late June 1857.[52] That letter had reached Germany through a communication by German missionaries in Lahaul in the Himalayas. In it, Adolph reported that he had spent the last few weeks recrossing the Kunlun Shan in different directions and now planned to visit Tibet. Yet, he also alluded to the geopolitical conflicts in this area: 'My itinerary nonetheless depends ... very much on the events in Yarkand; there, a war has been waged for three months, and a large part of the country has been temporarily taken away from the Chinese.'[53]

The initial excuse for Adolph's long silence was that he was exploring regions that the communication lines of the imperial government did not easily reach. As newspapers throughout the British Empire noted on his fate: 'It is a pity our arms are not long enough to touch those distant barbarians. We can readily pounce upon Sikkim, but Kokan[d] is beyond our reach.'[54] International concern soon set in and, encouraged by the brothers in London, led to the preparation of a first search expedition by the British authorities under the direction of Sir John Lawrence and Lord William Hay. A party was sent off from the hill-station of Simla in mid July 1858.[55] The great interest of Leadenhall Street in the whereabouts of the German traveller was, however, hardly shared by the popular press in Britain. As the widely-read newspaper, *Allen's Indian Mail*, sourly noted: 'The British Government have shown more interest in the scientific German than in their own officers, Connolly, Stoddart, and Wyband' – all of whom had fallen victim to their excursions into central Asia.[56]

Adolph's disappearance into the 'unknown' was also alluded to in the most successful literary work that ever appeared on the Indian Rebellion in Germany, *Nena Sahib, or the Uprising in India: Historical-Political Novel of the Present Times*, by the German writer Hermann Goedsche.[57] It immediately became 'a runaway best-seller'.[58] In the novel, a fictitious Prussian doctor encounters a Muslim 'warrior', Fattih Murad Khan, in the 'deserts of [northern] India'.[59] In the ensuing dialogue, the

---

[52] Ritter, 'Sitzung', 87–88.
[53] 'Mannigfaltiges', *Erheiterungen, Beiblatt zur Aschaffenburger Zeitung*, 43, 19 February 1858, 171–2.
[54] *The Moreton Bay Courier* (Brisbane), 30 March 1861, 5.
[55] 'Deutschland', *Laibacher Zeitung*, 194, 25 August 1858, 777.
[56] *Allen's Indian Mail*, 28 May 1859, 456.
[57] *Die Empörung in Indien*, published under the pseudonym Sir John Retcliffe.
[58] Mazumdar, 'Figurations', 110.
[59] Retcliffe, *Die Empörung in Indien* II, 31; subsequent quotations, ibid.

Prussian, a doctor who had been pressed into the British Navy, tells the Muslim prince: 'I have heard that three of my countrymen are presently engaged in a scientific expedition through northern India and Thibet.' When the Khan enquires about the name of these travellers, the Prussian doctor responds: 'The itinerant savants are the brothers Schlagintweit.' Thereupon, the Muslim warlord 'recognises' their name, and replies: 'Two of the three brothers have returned to Calcutta. I guided the third one into the mountains of Thibet.' Here, the narrative on Adolph abruptly ends, capturing the German concern for the disappearance of the explorer. The literary reference to the brothers makes clear that the Schlagintweits had by then become household names in Germany, and how their fate gripped the public imagination.

In the spring of 1858, news of Adolph's potential murder and some details about the last leg of his journey finally reached Europe. It fell to Hermann to compile an account of the events that had led to Adolph's capture and eventual killing in Chinese Turkestan, based on the scattered information that came to light. According to his report, Adolph 'was recognised as a European after having passed the Karakorúm and Küenlüen, in disguise, where before us no European had ever travelled; he had taken a route more westerly than ours, and had succeeded in penetrating far into Central Asia'. Hermann further declared that 'the political condition of these countries', and the fact that Adolph had been identified 'as *an officer of the Indian Government* ... essentially contributed to his tragic end', as he was executed as a British agent.[60] This identification of Adolph as a British agent by his own brother is significant, as it refutes the often repeated but proofless claim that the explorer was murdered on suspicion of being a 'Chinese spy'.[61] After all, all three brothers were travelling with European instrumentation, making notes and observations in European languages and were always equipped with official letters of introduction from the Government of India.[62] Further evidence removes any doubts of Adolph being identified and murdered as an agent of the Indian government: he also saw himself as such.[63]

Given the importance of Adolph's pioneering excursions, British officials invested considerable resources in recovering his last notes and sketches. The experienced Himalayan traveller Henry Strachey was

---

[60] H. and R. Schlagintweit, *Official Reports*, 1, my emphasis.
[61] Cf. Polter 'Nadelschau'; Sarkar, 'Science', 12; Mayr, 'Schlagintweit, Emil'.
[62] The governor of Bombay, Lord Elphinstone, provided the brothers with an official letter to enter Nepal, see HS to Governor-General Dalhousie, 31 January 1855, BSB, SLGA, II.1.43, 37.
[63] 'Letter from Mohammad Amin of Yarkand, to Colonel Edwardes, 29 July 1858', quoted in H. and R. Schlagintweit, *Official Reports*, 15.

appointed to gather information and help to rescue Adolph's collections, records and instruments. British officers also interviewed a number of members in the German traveller's last establishment of translators, porters and guides with a view to obtaining further intelligence on his murder and his surviving observations.[64] The RGS president, Roderick Murchison, prominently declared in his 1859 anniversary address that 'it is most distressing to have to record that he of the three brothers who pushed his adventure the farthest should have been cut off at a time when his note-books and observations must have been of the highest value'.[65]

The death of the German explorer at the geographical extremes of empire was a cruel twist of fate. It meant not only a great personal deprivation, but it also endangered the scientific legacy of the Schlagintweit expedition. Humboldt himself confirmed the seriousness of the loss when he wrote in 1858 that 'Adolph was, also in the eyes of Leopold von Buch, the most distinguished of the three brothers.'[66] His demise left a void – a knowledge gap – that the two surviving brothers could not fill. From the start of the Indian mission, Adolph had been its leading spirit. He had also undertaken the longest and most thorough preparations. It was Adolph who had already settled in London in February 1854 to be trained in the use of instruments and, more importantly, to peruse British scholarship on the geographical features of the subcontinent in the East India House library. This institution gave him access to British literature on India that was often 'exceedingly difficult' to obtain for continental geographers.[67]

During the period that Adolph immersed himself in scholarship about India, his brothers Hermann and Robert were, by contrast, still finishing the second Alpine treatise. This task was only completed in June 1854 – just three months before the mission's start.[68] Hermann only left Berlin for London to prepare for their travels in mid July.[69] Robert, too, had very little time for preparation.[70] Thus only Adolph was sufficiently

[64] The transcript of such an interrogation is kept in the Punjab Archives, Lahore [PAL], Political Department 1859, Proceedings for the week ending 14 May 1859, nos. 31–3, esp. the 'Statement of Abdullah' to colonial authorities. I thank Irmtraud Stellrecht for generously providing me with these sources; see also Strachey's report on Adolph's last journey, *JASB*, 27/4 (1859), 375–86.

[65] Murchison, 'Address', 1859, 260. The Russian imperial government also commanded its officers in central Asia to enquire into Adolph's fate, 'Die Gebrüder Schlagintweit', *Gartenlaube*, 1858 (unknown date), in ÖAV, CC, 1848–65, 29.

[66] Humboldt to Illaire, 15 April 1858, GStAPK, AKlC, 126.

[67] See Petermann to Sykes, 31 March 1860, and Petermann's endless book requests to Company men in London, in FBG, SPA, ARCH PGM 353/2, 56.

[68] Published as A. Schlagintweit and H. Schlagintweit, *Neue Untersuchungen*, ix.

[69] Humboldt to Joseph Hooker, 16 July 1854, UGL, Ms 2153, box 11, folder 14.

[70] Humboldt to Illaire, 11 June 1854, GStAPK, AKlC, 24.

familiar with the substantial amount of British knowledge on Indian natural history and central Asia's history of exploration. The thick layer of previously accumulated expertise led Adolph to state, for instance, that 'my feeble exertions to add to the scientific knowledge of so interesting a country as India have met in Madras with much more attention than they deserve'.[71] By contrast, Hermann and Robert went to India merely as experts on the European Alps. Their almost constant movement during the Indian and Himalayan expedition hardly allowed them to develop an intimate knowledge of the rich and highly specialised British scholarship that had been accumulated, over the course of more than a century, on south and high Asia.

This stood in a marked contrast to many British Company men resident in India as surgeons, officers or political agents for much longer periods of time. This gave them the chance to develop a specialised knowledge of particular territories. Brian Hodgson, the government agent in Kathmandu, was the epitome of the Company servant turned scholar-naturalist with considerable local expertise. He spent half a lifetime in Darjeeling, turning himself into an acknowledged expert in such diverse fields as Buddhism, Tibeto-Burman languages and Himalayan natural history.[72] His advice to the itinerant Joseph Hooker, whom he had hosted during the latter's travels, was taken to heart: 'Hodgson dwells strongly on the simple fact that it is better to explore one district well than to wander.'[73] Hooker, like the two surviving Schlagintweit brothers, had left Britain without a profound knowledge of Indian natural history or ethnography. However, his scientific authority was protected from damage precisely because he *limited* his fields of engagement, and refrained from advancing, for instance, speculative theories on Himalayan ethnography grounded only on superficial impressions.[74] By contrast, the Schlagintweit brothers covered a dozen fields of enquiry and, driven by their ambition to scrutinise large parts of India and trans-Himalaya, they frequently only hastened through vast regions of the subcontinent. Adolph and Robert's geological findings, for instance, were later regularly critiqued as superficial and erroneous surveys, as being 'too brief, to permit of his [Adolph] making any detailed examination of the country'.[75]

To a degree, the Schlagintweit expedition was indeed a voyage of discovery, but, owing to their insufficient knowledge in many of the fields

---

[71] AS to Sykes, Pondicherry, 26 February 1856, FBG, SPA, ARCH PGM 353/1, 18.
[72] Hodgson has been rediscovered as a scientific polymath in recent years, Waterhouse (ed.), *The Origins of Himalayan Studies*.
[73] Hooker quoted from Arnold, 'Frontier', 194.
[74] Ibid.
[75] Blanford, 'Geological Structure', 217.

of knowledge they engaged in, it appeared to be one of self-discovery, as was noted time and again by metropolitan scientists and journals in Britain. Even satirical magazines picked up this recurrent critique. Especially in the *Kladderadatsch*, the Berlin counterpart to the British *Punch*, the brothers' earlier ridicule in *The Athenaeum* served as the template for its own derision. While giving it a further twist by alluding to the great expenses the brothers' 'discoveries' had by now incurred, it noted:

It is indeed outrageous to read how the three poor Schlagintweit brothers are now torn to pieces by British papers. Ten thousand pounds sterling for three poor travellers is certainly not too much if one considers the thoroughness with which the learned savants have embarked upon their duty, and have always been anxious to include only those findings in their reports, whose reliability had already been proven by the most trustworthy testimonies of other, more important, scientific authorities.[76]

Following this first skit, the brothers made something of a second career appearing in the satirical paper, which now changed and ridiculed their name to 'Schnabelweit' – a euphemistic variation of what in German would be understood as 'plapperhaftes Großmaul', a big mouth.[77] Making several appearances in the *Kladderadatsch*, the magazine's most witty ridicule of the brothers' grand claims was a 'fictional paper' that the Schlagintweits were said to have given at the (imaginary) 'Academy of Science at Disteldingen'.[78]

In the fictional account of the satirical journal, the brothers 'stated' that their academic lecture took place after 'our great travels to the Himalayas, which, as you will know from different newspapers and advertisements, we have undertaken as much for the sake of science as for our very own interest'.[79] In the ensuing farce, almost every single aspect of their research – as described by the reports the brothers had sent home – was sneered at. The biting mockery touched on the Schlagintweits' habit of continually taking measurements of the most apparently insignificant details during their passage to India, and using elaborate instruments to detect such banal circumstances as the fact that it was 'raining'. It also parodied their wide interests in 'racial varieties' – which, as the article had it, had already begun in Brandenburg in the one-horse town

---

[76] *Kladderadatsch. Das deutsche Magazin für Unpolitische*, 51, 8 November 1857 (Berlin), 206.

[77] The made-up name wittily played on two German sayings at the same time: 'nicht sein Schnabel halten können', talking incessantly, and not necessarily with great substance, and 'sein Mund/Maul zu weit aufreißen', to talk boastfully.

[78] 'Bericht der berühmten Reisenden, Gebrüder Schnabelweit, über ihre berühmte wissenschaftliche Reise nach dem Himalaya, erstattet in der Akademie der Wissenschaften zu Disteldingen', *Kladderadatsch*, 37, 9 August 1857, 146ff. Subsequent quotations, ibid.

[79] Ibid.

of 'Erkner' just outside Berlin. In the mock lecture, the Schlagintweits also applauded the neutral stance of the 'Austrian Government, one of the few which was not at all complicit in the considerable increase of costs for our scientific expedition' – hence, a jibe at the Prussian, British and Indian governments for having been fooled by the brothers into accepting such spiralling expenses. Yet, the most piercing passage came at the very end of 'their' account:

We thus arrived in Asia and went on the shortest way to the top of the Himalayas. There, during an extended stay of several years, we found all the information given in Brockhaus's and Pierer's Encyclopaedias so thoroughly confirmed that we avoid any useless repetitions ... and confine ourselves to refer you to the respective articles in both [works] as regards the details of our travels and our scientific results. Dixi et salvavi animam![80]

The Latin saying stands for a symbolic act of catharsis, meaning: 'I have spoken; and by so doing have delivered my soul from all responsibility, which I might have incurred by silence.' The joke was that the brothers did exactly the opposite: in the eyes of the satirists, the Schlagintweits seemed to claim as scientific discoveries what others considered to be received wisdom.

That such a poignant satire would appear in Berlin adds further nuance to the Schlagintweit controversy. It shows that, at least in the beginning, there was no black-or-white reception of their works, with the fault lines running along the Anglo-German boundary. German papers, even if it was only a few, also initially criticised the lavish patronage of their many sponsors, and highlighted how the Indian mission was, not least, intended to fill their own purse and nourish their personal vanity. What is more, the Prussian satirists' reference to the body of knowledge that did exist in German learned works and encyclopaedias proved that India and the Himalayas were, at least for specialist audiences, not the blind spots that the brothers at times portrayed them to be for their German audiences.

The *Kladderadatsch* article furthermore demonstrates that German newspapers and satirical magazines were closely following the reputation and press coverage the Schlagintweits received in Britain. This instance of mutual observation of the British and German press underlines a central argument of this book, namely, that the scientific authority of scientific travellers was never universal; on the contrary, it was forged within a landscape of multiple public spheres, while the boundaries between

[80] Ibid.

those spheres were porous and open to outside influences.[81] During the most heated points of the controversy, the majority of magazines and newspapers in the German states were in some way responding to the hostile reactions to the brothers in the British Isles. Apart from their recurring ridicule in Prussia, the German scientific community and popular press generally sought to vehemently defend the brothers' merits. The following analysis of how the brothers faced formidable hurdles to secure their publications, and how they became defamed in Britain, thus sets the background against which their later glorification in the German states must be understood.

Securing a Written Monument

On 6 February 1858, another 'scandalous discovery' about the Schlagintweits was published in the London *Athenaeum* that gave rise to a new wave of dispute over the value of the mission in general and the brothers' ungentlemanly behaviour in particular. A document written by Hermann Schlagintweit fell into their hands. The paper was entitled 'Practical Objects connected with the Researches of the MM. Schlagintweit, under the Orders of the Hon. Court of Directors'. Hermann had personally submitted the document to East India House in September 1857. The intention was to secure a new contract with the Company for the publication of the brothers' results. They wanted to secure the necessary funds in London, and then disappear to Berlin for several years to get the work done without there being any public record of the financial and publication arrangements.

However, this time the Schlagintweits' negotiation and communication strategies failed dismally, since an anonymous source leaked their confidential submission to the press.[82] The brothers knew that a faction within East India House opposed their schemes, including the plan to take the bulk of their collections to Berlin. Hence, they informed the Prussian minister of culture that while they had succeeded in shipping most of the artefacts to the Prussian capital, this was done 'even though many members of the Court of Directors … were against this', and were said to have induced 'some public organs' to dismiss their plans.[83] This instance

---

[81] An important theoretical basis for such an analysis is Muhs et al. (eds.), *Aneignung und Abwehr*. An insightful analysis of Anglo-German 'newspaper wars', in the late nineteenth and early twentieth centuries, is Geppert, *Pressekriege*.

[82] 'Messrs. Schlagintweits' Indian Mission', *The Athenaeum*, 1580, 8 February 1858, 178–9.

[83] H. and R. Schlagintweit, report to von Raumer, Berlin, 19 February 1858, in GStAPK, 'Acta Commissionis', supplement 'Wissenschaftliche Reise' [WR], 5.

showed that the brothers' opponents also tried to use the press to build up support for their position in negotiations with the Schlagintweits over a new contract. Now, with the contents of their 'memorandum' made known, the brothers were not able to present the already agitated British public with a *fait accompli* of renewed employment. Rather, delicate details about how they ingratiated themselves with the Company directors, and specifics about their extensive financial requests, suddenly became public. For *The Athenaeum*, the leaked proposal was grist to their mill. The editors therefore reprinted the entire original 'petition these gentlemen prefer to the India Company for more money, patronage, and power'.[84]

The petition to the Court of Directors demonstrates that the Schlagintweits themselves believed that their mission and publication schemes were indeed dependent on colonial and imperial interests. The assumption that their researches ought first and foremost to meet the material aims of the tumbling Company, if further patronage was to be won, also shows how adaptable the definition of the expedition's aims and scope really was. The petition proves that the Schlagintweits showed a remarkable interpretative flexibility over what the key objectives behind their scheme were, while the implied audience often dominated the question of the accuracy of their accounts. Whereas they stressed in their Paris lecture to scientific peers in 1857 (and on many other occasions) Humboldtian global physics and climatological physiognomy as their principal concerns, the leaked petition set their researches in a completely different light. It seemed to suggest that the driving forces behind their work were, above all else, questions of scientific utility.

Since so much of the scholarship on the Schlagintweit expedition has ignored or denied any relation between their enterprise and concerns of scientific imperialism, it is salient to take a closer look at the petition.[85] It is, first of all, remarkable to what extent the two surviving brothers therein downplayed any 'philosophical' aspects of their mission. On the contrary, the proposal stressed, almost exclusively, the potentially useful implications of their researches, as if their exploration was merely driven by colonial interests, and not also by Humboldtian models of enquiry in physical geography. With the confidential petition, intended to secure their income for years to come, the Schlagintweits therefore did not appeal to the Company as an 'enlightened patron of the sciences' – a role

---

[84] 'Messrs. Schlagintweits' Indian Mission'. The original petition is held at the BL, IOR, MSS EUR F 195/5.
[85] Cf. Finkelstein, 'Mission', esp. his conclusion.

in which the EIC sometimes sought to portray itself, 'if only to advance its commercial interests and protect its privileged political position'.[86]

The brothers' eagerness to emphasise the imperial value of their data, observations and collections suggested – at least on paper – another major reinvention of their scheme. The first had taken place when the brothers submitted their 'proposed plan of operations' to East India House in 1854. This had successfully transformed their initial employment as leaders of the more narrow Indian geomagnetic survey into an interdisciplinary study of Indian and trans-Himalayan natural history. Whereas this enlarged scientific programme had still been committed to the idea of writing a natural philosophical treatise about their eastern travels (which was later, in the field, complemented by extensive ethnographic and anthropological enquiries), their submitted plan for publication now implied another marked change of direction.

Hermann and Robert dedicated each paragraph of the petition to one scientific discipline (or a few related subjects) that they proposed to cover, while attempting, sometimes awkwardly, to gear each field of enquiry towards colonial science, material development and Company profits. Under the first heading, 'Magnetism, Meteorology, Physical Geography', they sketched their findings and research objectives in these diverse fields as such:

Besides the more accurate determination of the magnetic elements in general, their relation to the magnetic laws in Europe, and the declination of the needle, especially, the well-defined characterization of the climate of India, in general, may be named as the most important practical result. From the difference of climate depends so much the selection of [military and settlement] stations and the cultivation of certain crops. Cotton had been particularly kept in view in reference to places inhabitable by European colonists.[87]

This was followed by their assurance that the 'Fixing of hill stations, of easy access, in a most favourable climate, and in proximity to the rich treasures of mineral and hot waters in the Himalayas' would be a key concern in their work.[88]

This opening was surprising, since one of their prime subjects of investigation should have been the completion of the geomagnetic survey of the eastern Empire. Yet, this initially central aspect was immediately sidelined by their much larger emphasis on medical topography and

---

[86] David Arnold, quoted from his keynote lecture at the conference 'Colonial Careers: Transnational Scholarship Overseas in the 19th and 20th Centuries', EUI, May 2012.

[87] H. Schlagintweit, 'Practical Objects', BL, IOR, MSS EUR F 195/5.

[88] Ibid.

questions of European health and settlement opportunities in conjunction with projects of agricultural expansion. To be sure, this was a moment in British India that saw a 'trend towards an expansion of the cultivated area of land'.[89] The brothers had become well informed about such contemporary concerns during their travels, and now used their knowledge about British interests and needs as assets to support their case.[90]

For the purpose of agricultural expansion, the Schlagintweits had even brought hundreds of water samples and over 1,200 types of soils with them to Europe for analysis, with a view to extending and improving the agrarian output of British India. The soil samples were later thoroughly analysed for the International Colonial Exhibition in London in 1862 by the 'Reporter for the Products of India', John Forbes Watson. While the results secured Forbes Watson a gold medal, it is unclear if and to what extent the insights gained from the Schlagintweits' samples improved the cultivation of cash crops such as tea, indigo and flax in India.[91] Yet, whatever the ultimate application of their results, the brothers, at least, were convinced that an 'Examination of soils is … very important for all questions of cultivation and manuring' in the British colony.[92]

The next set of disciplines addressed in the Schlagintweit petition was 'Topography, Hypsometrical Observations, Maps'. This part of their work was to 'contain a detailed account of heights, – on which the choosing of roads, as well as agriculture, equally depend'.[93] The Schlagintweits proposed to the Court that they would compile a 'Route Book (Military and Commercial)' with detailed descriptions of existing passes, bridges and roads. This they believed to be of considerable strategic significance, since 'We can complete it [the Route Book] for parts where nothing similar has been tried, over Central India, the Himalayas, and the important country adjoining India to the north-east and north-west, viz., Burmah, Assam, Tibet, Kashmir, and the Turkistan Provinces, in Central Asia.'[94]

Military journals, which later reviewed the published *Route-Book*, noted that the brothers' observations 'on the physical conditions of the paths, the nature of the passes, the availability or absence of foodstuffs and the like' was both 'of importance for war operations, and also for

---

[89] Pouchepadass, 'Agrarian Economy', 308.
[90] For the example of tea cultivation, H. Schlagintweit, *Reisen* I, 444–5; II, 177–8.
[91] Forbes Watson, *Classified and Descriptive Catalogue*, Class I: India, Subdivisions VI. Soils and Mineral Manures, 23–6.
[92] H. Schlagintweit, 'Practical Objects'.
[93] Ibid.
[94] Ibid.

the critical assessment of military events in these regions'.[95] While it is unclear how much use was ultimately made of their extensive topographical descriptions, such an evaluation of the book's strategic value is consistent with the work of the historian James Hevia. He argued with regard to imperial intelligence gathering: 'The documentary products of the security regime – the intelligence genres of route books, military reports, handbooks and precision maps – organised East, Central and South Asia as legible space, making Asia visible to the eyes of military planners and strategists concerned with the protection of British interests in the region.'[96] Since hardly any comparable works existed at the time, the Schlagintweits' *Route-Book* added a wealth of new information on Indian and Himalayan topography and route accessibilities to the archive of the Indian government.

Complementing their detailed description of innumerable roads within and beyond the British spheres of influence, the brothers also offered to the Court that they would produce two large-scale maps – of 'India proper' and 'high Asia' – that would include 'the politically most important neighbouring provinces as well as the territories chiefly adapted for European colonization'. In addition, they proposed compiling 'a most detailed account of the discharge of rivers, their motive powers, and the questions of their navigability and irrigation properties'.[97] The brothers were convinced that knowledge of the characteristics of rivers was important for trade. Indeed, the possibility to transport commercial goods via rivers (and the barely exploited timber resources in the Himalayas and Dehra Dun in particular) was one of the Schlagintweits' recurring concerns.[98] Likewise, a better understanding of the problem of the 'discharge' and sudden floods of rivers was considered by the brothers (and contemporary officials) to be important for decisions over new settlements, irrigation works and agricultural schemes.[99]

---

[95] For example, review of *Route-Book*, in *Österreichische militärische Zeitschrift*, 5/3 (1864), 101.

[96] Hevia, *Imperial Security State*, 152; also ibid., 74–8.

[97] H. Schlagintweit, 'Practical Objects'.

[98] H. Schlagintweit, *Reisen* I, 304–9; R. Schlagintweit, 'Geographische Schilderung', 9. Overlapping with the Schlagintweit travels, there were several river surveys conducted in British India looking into using the discharge of rivers for perennial canal systems; Dickens, *Canals of Irrigation and Navigation*. On how British engineers launched the construction of large-scale systems of dams and canals across India from the 1850s onwards, see Teisch, *Engineering Nature*, 29–31.

[99] On large-scale British irrigation works and artificial water reservoirs in India, see H. Schlagintweit, *Reisen*, 1, 172–3; and 250–84; also Kreutzmann, 'Habitat Conditions and Settlement'.

The rest of the memorandum was equally marked by a striking contrast between those more 'philosophical' concerns of metropolitan men
of science who were eager to detect the general laws of nature, and the
brothers' mantra-like allusion to specific material gains that could be
derived from their investigations. When they had transformed their
modest Company appointment into a great scientific enterprise in
1854, these more speculative, natural philosophical concerns had held
centre stage. On geology, for instance, the brothers had noted: 'We shall
endeavour as much as possible to collect fossils, for the accurate determination of the comparative age of the different sedimentary strata, and
to ascertain their order of superposition.'[100] This natural philosophical
interest was now scaled down to mere issues of agricultural 'improvement', colonial infrastructure and profit:

The general practical results, everywhere indispensable, from geological
researches, allow particularly brilliant hopes for India, where the riches of ores
in the Himalayas, long expected, could be confirmed in numerous instances ...
Besides ores and coals ... the determination of the best materials for roads and
buildings may be mentioned.[101]

This strategy of overemphasising the practical implication of their
researches tells us as much about the Schlagintweits' own colonial imagination than about the real needs of their imperial benefactors. Indeed, as
a later appearance of the two brothers in the House of Commons will
further demonstrate, they excelled at suggesting far-reaching projects of
white settlement and British trade expansion across south and high Asia
whose novelty or feasibility were, however, doubtful at times.[102]
    Since the Munich-born travellers knew that the Company gave only
irregular support to more abstract works of science, their constant
appeals to the Court's commercial considerations were influenced by
yet another motive. They wanted to ensure that their vast collections
and field notes would not go to waste and deteriorate, but would be
thoroughly analysed – which guaranteed to be a costly undertaking. As
The Athenaeum remarked in their critique of the leaked petition, 'the
greater part' of collections made by British travellers 'are deposited
in the library and vaults of the India House and elsewhere', a rotting
symbol of the neglect by the Company of its many deserving servants.[103]
As the editors scornfully added, 'many' of the Company servants, for
want of patronage, had to publish 'their observations and collections

---

[100] NA, Records of the Meteorological Office, 1849–54 (Indian Subcontinent), BJ 3/53.
[101] H. Schlagintweit, 'Practical Objects'.
[102] House of Commons, *Fourth Report*, 1–10. Cf. Finkelstein, 'Mission', 200.
[103] 'Messrs. Schlagintweits' Indian Mission'.

at their own expense'. The brothers ended their petition with a section
on 'Botanical Geography'. Here, they drew particular attention to 'our
complete collections of woods', whose value they hoped would persuade
the Directors to sponsor the laborious preparation of their gathered
specimens – which ultimately amounted to prepared wood samples in
boxes weighing '3 tons'.[104]

In rather boastfully considering 'the value of our own labours',
Hermann Schlagintweit concluded the petition that '[t]his large work,
when completed in the manner proposed, will not fail to be most
important, by attracting general scientific and practical attention to
India'.[105] This was the brothers' honest conviction: their work was to
be scientific by any means, but the experience of travel had also infused
them with a sense of duty to highlight and 'make use' of those natural resources and trade potentials they had identified. However, several of their 'improvement' schemes were not so new, after all. On the
contrary, the brothers often adopted ideas about south Asian agriculture and high Asian commerce that were already practised, or at least
discussed, in colonial circles. This included their reference to the vast but
underexploited timber resources of the Himalayas. Their later nemesis,
Joseph Hooker, had already noted in 1848 that he had 'memorialized'
the Governor-General 'on the advantages that would accrue from an
investigation of the Timber-trees & capabilities of the Lower Himalaya,
& of this district [Assam] where there are thousands of woods, in danger
of being ruined'.[106] In the entire petition, however, not a single British
naturalist or administrator was named. As in the case of their controversial 'report' to the Paris Académie, this petition too projected an image
of British India's northern borderlands as a supposedly unknown territory. The brothers alone seemed to have unveiled their almost infinite
scientific and utilitarian possibilities for European enterprise. Together,
the Schlagintweits' 'practical objects' amounted to a far-reaching vision
of how Indian and high Asian territories could become the object of
further colonial development and increased exploitation of their natural
resources.

The Schlagintweits' own petition to the Company presented, however, a heavily distorted and one-dimensional picture of their expedition
and its varied scientific concerns. The brothers' attempt to associate

[104] J. S. Gamble Esquire, Asst. Conservator of Forests, Bengal, to the Secretary to the
Government of Bengal, Revenue Department, 'Forests', 5 January 1882, NAI, Revenue
& Agriculture, Branch: Forests; 1882-01, File no. 87-92, Part B, 'The Schlagintweit
Collection of Indian Timbers', no. 133.

[105] H. Schlagintweit, 'Practical Objects'.

[106] Joseph Hooker to William Hooker, Assam, July 1848, RBG, JDH/1/10, 61.

every single branch of enquiry, all activities and observations, with the strategic and profit motives of the Court meant that important facets of their mission were silenced. Several aspects of their meteorological, botanical, geological and other disciplinary researches undoubtedly combined science with considerations of imperial benefit. Despite that, their multifarious activities in these diverse fields never exhausted themselves with concerns over colonial science and material advancement. Their submitted petition also left out entire fields of enquiry, in particular the considerable ethnographic and anthropological observations. In addition to hundreds of pages filled with notes on ethnographic encounters, their anthropological programme had included the production of a widely acclaimed series of plaster casts of 275 individuals from different Asian 'races'.[107] These ranked among the most precious parts of their collections. What is more, while the petition highlighted the brothers' cartographic projects, it did not even mention their series of photographs and over 750 sketched views and watercolours across south and high Asia. These 'views' were of great aesthetic quality and equalled in beauty and execution any comparable work by contemporary British artists. Even if their views did depict a series of hitherto unknown landscapes, and thus also provided – in conjunction with explanatory sheets – useful topographical information to the Company, they clearly exceeded mere strategic value.[108] The Schlagintweits' highly selective description of their miscellaneous activities turned the memorandum into a strategic travesty of their mission. And yet, except for the unintended leaking of the proposal, their plan to thus persuade the Company to another act of largesse struck a chord with the imperial directorate.

In contrast, *The Athenaeum* did not accept a single word of the petition. To expose the 'effrontery' of the application, the article went on:

We need scarcely say that the Messrs. Schlagintweit are Germans, living in Germany ... that their expedition has cost England as many thousand pounds as it has cost His Majesty of Prussia hundreds, – and that the whole of their collections, said to be contained in 300 cases, have been transferred to Germany. His Majesty of Prussia, we are told, no longer allows them one shilling. Messrs. Schlagintweit are therefore craving the further patronage of the Government which has treated them with such mistaken liberality.[109]

---

[107] H. and R. Schlagintweit, *Prospectus of Messrs. de Schlagintweits' Collection of Ethnographical Heads*.
[108] Kleidt, 'Kunstwerk'.
[109] 'Messrs. Schlagintweits' Indian Mission'.

The fact that the brothers were now 'craving' further patronage seemed to prove again that their expedition on British pay was a 'job'.[110] Since the nineteenth century, the word 'job' has undergone a remarkable semantic, moral and lexical shift. At that time, a 'job' was defined as 'a piece of work ... whether of more or less importance; a lucrative business, an undertaking with a view to profit', commonly regarded as 'a low, mean ... affair'.[111] Through this careful wording, their whole scheme was thus publicly portrayed as above all motivated by personal gain – the antithesis of respectable, gentlemanly science.

Some of *The Athenaeum*'s claims and accusations were, however, false libels. First, the Prussian monarch had by no means abandoned the brothers (whom he affectionally called the 'Indo-Bavarian Trimurti'), but had already promised his renewed financial backing for their planned publication: 3,000 thalers, regardless of whether their work should only appear in Britain.[112] Second, as Sykes informed the Prussian administration, it had been in the Court's very own interest that the brothers took most of their collections to Prussia. The reason was that 'the costs of renting [and] of the analyses, etc. etc., which are necessary in order to bring the results of the magnetic observations and physical researches ordered by the Directory into a shape worthy of publication, would be four times higher in London than in Berlin'.[113] Lastly, a good number of duplicates and unique objects from the Schlagintweit collections had actually remained in Britain, and were prominently displayed in the Company's India Museum in Leadenhall Street.

Ultimately, the expedition's perceived programmatic overreach drew the sharpest mockery from *The Athenaeum*:

That their three years' scamper from Tibet to Assam, and from Madras to Nepaul, should enable them to construct a map of all India and Central Asia, – determine new lines of roads, military, commercial, and political, – fix new hillsites, sanitary and colonial, – open up new agricultural districts and resources, temperate and tropical, – develop new mines, lodes and seams of coal, iron, and other ores, – and, finally, illustrate the botanical geography of India in 'all its details,' is the

---

[110] Already in *The Athenaeum*, 1566, 31 October 1857, 'The Latest Indian Mission', 1358–9, 1359; again, *The Athenaeum*, 1764, 17 August 1861, 215; see also the valuable analysis of this point in Finkelstein, 'Mission', 199f.

[111] Wright, *Royal Dictionary-Cyclopædia*, 331.

[112] The king to Humboldt, Marienbad, 24 June 1857, GStAPK, AKlC, after fol. 120; and fol. 55, order of the king, 18 July 1857. This also disproves the repeated claim in the literature that the brothers were supposedly unable to make any positive impression when giving the Prussian monarch a personal report after their return.

[113] Sykes to Sabine, 1 March 1858, GStAPK, AKlC, appendix IV, German copy of English letter, my translation.

grossest imposture that has ever been laid before any Board or dignified with the name of science.[114]

The editors showed little discretion in their revelations. They rather drew specific attention to the delicate financial arrangements 'the two Messrs. Schlagintweit have the effrontery to ask' for. These included a 'salary for each, for an indefinite period; the amount to be left to the munificence of the Court of Directors'. Yet, a thorough analysis of their collections of over 40,000 objects also required the mobilisation of broader networks of expertise. Therefore, they had applied for '[a]n establishment, *at the rate of* 150*l. a month*, for the first year – (with a promise that the sum shall be reduced in future years – for how many is not stated!)'.[115] The brothers also wanted the 'money to be paid in advance, and accounted for afterwards!'. This would give them considerable freedom in preparing their publication – an arrangement that echoed the exceptional financial liberties they had already secured during the expedition. To secure large readerships, the Schlagintweits also asked for the 'assurance that the East India Company shall take a large number of copies' – they settled ultimately on sixty copies of a projected nine-volume work. If the Court were to accept all the terms, this arrangement would secure the Schlagintweits' income for years; conversely, from *The Athenaeum*'s perspective, a significant amount of money would, again, be lost to many British men of science.

This perceived threat led *The Athenaeum* to go one step further and to openly demand political action. The editors – as the self-appointed voice of the British people – started to put considerable pressure on both the British government and the scientific community to intervene in this scandal. 'We most sincerely hope, – and indeed cannot doubt – that the Messrs. Schlagintweit will be forthwith ordered to return their collections to the country to which they belong.'[116] Following this call for immediate repatriation of their artefacts, it was also expected 'that no steps whatever will be taken by our Government towards publishing them or the maps until they have been inspected and reported on by competent scientific Englishmen'.[117]

The Athenaeum was not a solitary voice in the wilderness. While it is difficult to determine exactly the public impact of the paper's campaign against the foreign travellers, some of the consequences can be established. First, the paper's criticism lastingly tarnished the brothers'

---

[114] 'Messrs. Schlagintweits' Indian Mission', 179.
[115] Ibid., 178–9, original emphasis.
[116] Ibid., 179.
[117] Ibid.

reputation among several British men of science. This was even the case with those who had been perfectly unacquainted, personally and scientifically, with the Schlagintweits and their work. After the critical piece, William Henry Harvey (1811–66), professor of botany at Trinity College, Dublin, informed a colleague about his plans to write a new botanical treatise. But he already cautioned that he could not match the magnificent *Flora Indica* by Joseph Hooker and Thomas Thomson, on whose oeuvre he commented: 'What a pity that it is stopped – for want of support – while the Slangtenthweil [sic] humbugs get £*10,000* of John Companys money – & are now *asking* for as much more "to enable them to publish their discoveries"! – See a few words in this week's Athenaeum.'[118]

Moreover, other newspapers picked up *The Athenaeum*'s patriotic mission on the home front – which was that journalists 'cannot admit that English merit and English service ought to be forgotten'.[119] The *Gardeners' Chronicle*, in February 1858, published a review of Mary Somerville's *Physical Geography*, stating that her work 'now takes its place as the most complete compendium of Physical Geography in any language. We recommend it to the perusal of the Messrs. Schlagintweit, who seem to have returned from India in a state of the most happy ignorance of the labours of former travellers.'[120] It testifies to *The Athenaeum*'s influence that the journal's pressure on the EIC was, to a degree, successful. After the latest scandalous revelation, the Court saw itself forced to consult British scientific experts and consider their judgement on the value of the Schlagintweits' proposed work. One expert report has survived. In February 1858, Colonel Sykes received an answer from the Himalayan traveller Joseph Hooker, who was 'really glad to have the opportunity of stating my own views'.[121] Hooker believed his opinions to be 'quite in accordance with those of the most distinguished scientific Englishmen regarding the labours of the Messr. Schlagintweit in India, & the unbounded liberality' of the Court in supporting 'their journies & their experimental researches'.[122]

At the outset of his evaluation, Hooker gave the rather questionable impression that he was not prejudiced against the brothers in any way: 'I have the pleasure of a personal acquaintance with 2 of the brothers Schlagentweit.' He even conceded that '[n]o one, who has been in their society, can fail to be struck with their scientific ardour & intelligence'.

[118] Quoted from Ducker (ed.), *Letters*, letter 95, 3 November 1857, 303–4; original emphasis.
[119] As stated in the midst of the Indian Rebellion, *The Athenaeum*, 30 October 1857, 1358.
[120] 'Notices of Books', *Gardeners' Chronicle*, 27 February 1858, 155.
[121] JH to Sykes, February 1858, RBG, JDH 2/9, 76.
[122] Ibid. Subsequent quotations, ibid.

Yet, in the remaining parts of his report, Hooker used his botanical expertise to confine himself 'mainly to the Plant Department', and there found considerable fault with the brothers – and the Company's system of patronage. Hooker candidly criticised the fact that 'there are persons, equally deserving of such encouragement as the Messr. Schlagentweit have received, & who have been overlooked on behalf of them' – which above all included his ally Thomas Thomson and himself. While the entire 'Scientific World has, without a dissenting voice, acknowledged the great value of' Hooker and Thomson's joint work on Indian flora, he scornfully noted that they have 'met with no favour or encouragement from the India Company'. Understandably, Hooker's chance to evaluate his competitors' collections and results gave him the opportunity to vent years of personal frustration and barely concealed anger.

Yet, Joseph Hooker was directly informed that more than the Schlagintweits' own reputation was now at stake. In the wake of the *Athenaeum* revelations, William Henry Sykes, with whom Hooker shared both personal and scientific ties, was suddenly afraid of becoming the object of public scandal himself. With the EIC already subjected to intense public scrutiny because of the current war in India, the last thing Sykes needed was 'being disgraced in the eyes of Europe, by permitting the matter [the Schlagintweit publication] to drop through, after such an outlay'.[123] Hooker thus had to walk a tightrope between his personal interest to secure Company grants for British colleagues, and his attempt to maintain good relations with such influential patrons as Sykes. Hooker therefore suggested letting them 'send their Plants, Woods etc. to this country, for inspection & to be reported upon: This is only reasonable; when it is considered that public money has been spent upon them.'[124] This was above all a precautionary measure since a 'public Body may disgrace itself … by spending money without a competent adequate object'.[125] Hooker made it clear, however, that he did not expect the great expenses to justify the Schlagintweits' meagre botanical results, as the brothers' 'Herbarium' seemed 'neither very extensive, nor well preserved'. He thus portrayed the brothers not as skilled and respectable colleagues, but more as well-meaning but ultimately useless amateurs.

In view of such attacks and ridicule among the British public, it is salient to explore how the Schlagintweits reacted to these waves of criticism. Fighting a battle it seems they had already lost, the brothers tried to de-escalate the public debate about their supposed ineptitude and

---

[123] Sykes is quoted in Hooker's reply, RBG, JDH 2/9, 77.
[124] RBG, JDH 2/9, 76.
[125] Ibid.

ungentlemanly conduct largely by remaining silent. While other German travellers and scientific experts publicly defended themselves against similar defamations in the British press, as the example of Dietrich Brandis has shown, the brothers refrained from publicly commenting on the accusations. By contrast, they once again relied on private channels of communication to position themselves and issued an 'excuse' to influential British scientific administrators – above all to Murchison, RGS president. This seemed adequate, as Murchison had come under considerable pressure from his peers, following his generous praise of the Schlagintweits in his 1857 anniversary address – praise encouraged by a particular Prussian friend. In his 1858 address, the president was thus forced to qualify his earlier statements on their achievements:

> In alluding ... last year to other labours of these gentlemen, I much regret to have unwittingly attributed to them geographical results ... which it is well known were mainly accomplished, more than thirty years ago, by the very able British officers of the Trigonometrical Survey of India ... Nothing could be farther from my thoughts than not to sustain the hard-won laurels of the many British subjects who have earned great scientific reputation in the Trans-Himalayan regions.[126]

While it thus seemed that Humboldt's intervention had thus partly misled the RGS president into giving credit where it was not due, the Schlagintweits' petition on *Practical Objects* had added even more fuel to the fire. Murchison, in the same address, bowed to public pressure and stated that 'I have been the more called upon to correct this erratum' about their findings, 'and to register the antecedent labours' of several British 'geographers and engineers, in consequence of a document presented by the MM. Schlagintweit (in September last) to the East India Company, in which they specify all their intended publications'.[127] However, the Schlagintweits had done so 'without referring to the labours of their numerous predecessors in the regions through which they travelled', thus giving 'umbrage to those who thought that numerous observations of our countrymen were slighted'.[128]

Murchison, with his affectionate ties to German geographers, now tried to occupy the middle ground and exculpate the brothers from charges of fraud. He declared that '[i]n justice, however, to MM. Schlagintweit, I must state that they have assured me of their having always intended to enumerate the labours of their predecessors ... and they claim to be not judged by a mere [manuscript] announcement of their own researches',

---

[126] Murchison, 'Address', 1858, 302.
[127] Ibid.
[128] Ibid.

which was 'not intended for publication'.[129] It is no small irony that their private excuse was – against their own intentions – thus again made public, which caused in turn further private and public criticism of their behaviour.

After this act of public mediation, Joseph Hooker wrote a scornful letter to Murchison and openly attacked the 'private vs. public' games the brothers constantly played:

The British-Govt. has spent (I am told) now fully £18000 on these men; they have been 5 years in our service, and what are the results? but a manifesto [the petition] ... that is disgraceful to themselves & dishonoring to our countrymen. I can well understand what you imply – that it was not intended for the public eye – and this is its worst feature. It was however intended to procure public money ... – was presented to every member of the India House & Board of Control – was distributed where its pretensions were not expected to be called in question.[130]

Hooker's critique was justified. The brothers had precisely sought to circumvent any public or expert evaluation of their petition, hoping that the patrons at East India House would be impressed with their numerous claims to scientific discoveries and schemes of colonial settlement and resource extraction.

Unsurprisingly, the exposure of their petition and the negative reports that followed left their mark also within the Court of Directors. As the new Prussian consul in London, Graf von Bernstorff, informed his home government, the president of the Board of Control, Robert Vernon Smith, had openly refused to sanction any further expenditure on the brothers as a result of 'insinuations that the Schlagintweit were charlatans'.[131] Commenting on the brothers' pending situation, the consul further noted that it was due to multiple resignations that, at the moment, 'the Directory of the EIC is not ... capable of acting'.[132] The Indian Rebellion entailed a major reshuffling of posts and imperial responsibilities that caused an administrative nightmare for the British, one that also created unexpected openings for the Schlagintweits.

The current Governor-General of India was forced to resign and the installation of his successor, Lord Stanley, in early June 1858, opened a brief window of opportunity for the brothers to conclude a formal agreement with the dying Company (Figure 6.1). Through renewed

---

[129] Ibid.
[130] JH to Murchison, 19 July 1859, RBG, DC 96, 406. This sum would be roughly equivalent to £2,150,000 today.
[131] 'Bericht des Gesandten Bernstorff an Illaire', 7 July 1858, GStAPK, AKIC, 131.
[132] Ibid.

Figure 6.1 'Execution of "John Company;" Or, The Blowing up (there ought to be) in Leadenhall Street'; *Punch*, 15 August 1857.
© DAV.

submissions of proposals, several visits to East India House and William Henry Sykes' unwavering and, as we have seen, self-interested support, the Schlagintweits finally managed to secure a lavish offer on 8 July 1858.[133]

The timing of the East India Company's decision might well have been in connection to an interview that the two surviving brothers gave in the House of Commons only two days earlier. Robert and Hermann were invited on 6 July 1858 to answer the questions of the Select Committee on Colonisation and Settlement in India, which comprised an illustrious group of British politicians, Anglo-Indian judges and leading British manufacturers. Their testimony was published thereafter and contained some of the detailed information they had promised in their earlier petition to the Court.[134] The specifics of their intelligence on military, commercial and political questions now occupying British minds proved to the MPs and company directors that they had indeed gathered useful knowledge – even if far from all of the Schlagintweits' schemes were feasible. The Schlagintweits' elaborations on suitable regions for white settlement, trade opportunities in high and central Asia (especially, as they proclaimed, to counter Russian advances in these regions) and natural resources to be exploited in the future matched the interests of MPs now preparing themselves to play a more active part in the rule of the Indian Empire.

Despite all public indignation, the brothers' proposed terms were largely met in the formal letter of appointment. As J. D. Dickinson, secretary of East India House, informed them, the Court sanctioned their request that 'the work should be printed in English, and should consist of eight or nine Volumes which you estimate can be completed in three years and that you should be allowed to publish it at Berlin'.[135] As Adolph's death had by then been confirmed, the Court granted each of the surviving brothers 'for a time not exceeding three years a personal allowance of £25 per month', together with an 'outlay not exceeding £150 a month' for hiring an 'Establishment' of specialists to assist in the production of the work. In addition to the sum of £800 that was 'authorised for engraving & lithographing the Maps and Plates', the Court 'subscribe[d] for 60 Copies of it'.

---

[133] Copy of communication by J. D. Dickinson, secretary of East India House, to the Schlagintweit brothers, 8 July 1858, GStAPK, I. HA Rep. 89, Geh. Zivilkabinett, AKlC, 177.

[134] House of Commons, *Fourth Report*.

[135] Dickinson to Schlagintweits, 8 July 1858, GStAPK, I. HA Rep. 89, Geh. Zivilkabinett, AKlC, 177f. Subsequent quotations, ibid.

After signing the generous agreement, the Schlagintweits immediately informed Humboldt that 'notwithstanding all political storms ... our arrangements with India House could be completed just in time in a fortunate manner'.[136] The Government of India Act, which came into effect only a few months later, maintained that not only 'the properties' and material collections of the former EIC had to be transferred into the hands of the British Crown. Rather, '[a]ll treaties made by the said Company shall be binding on Her Majesty', too, 'and all contracts covenants, liabilities, and engagements of the said Company made, incurred, or entered into before the commencement of this Act may be enforced'.[137] After the abolition of the EIC in November 1858, a newly created department, the India Office, from now on oversaw the Schlagintweit publications. The legal arrangements attending the Company's suppression thus meant that while the brothers outlived their former imperial patron, they were not cut off from British means.[138]

The Schlagintweits took other monumental works as the inspiration for their projected nine-volume enterprise, the *Results of a Scientific Mission to India and High Asia*. As they noted in relation to the high 'price of 162 Thalers for the whole oeuvre ... with an atlas of 100 lithographs', this considerable sum might be excused when compared with 'similar works'.[139] These included nothing less than the 'Description[s] de l'Egypte' – the massive series of publications that not just two savants, but a small army of French scholars and naturalists had produced in the wake of Napoleon's ill-fated Egyptian campaign (1798–1801). The Schlagintweits' other model was the 'great work' by Karl Richard Lepsius on Egyptian culture and architectural monuments, a masterpiece in twelve tomes.[140]

In 1860, the brothers published a *Prospectus* that announced their forthcoming oeuvre to the world.[141] A pompous preface stated:

The scientific mission to India and High Asia, with which from the year 1854 to 1858, Messrs Hermann, Adolph, and Robert de Schlagintweit were charged, has been universally acknowledged to take a prominent rank amongst those

---

[136] HS to Humboldt, 21 July 1858, London, Deutsches Literaturarchiv Marbach [DLAM], Humboldt letters, 62.2276.

[137] Quoted after Muir, *British India*, 389.

[138] Ibid., 339. Yet it took until 1874 for the EIC to be entirely wound up.

[139] HS and RS to Humboldt, 20 March 1859, copy at Humboldt Research Centre, BBAW.

[140] Lepsius, *Denkmäler*, published between 1849 and 1859.

[141] The brothers had also planned a German version, *Resultat einer wissenschaftlichen Sendung nach Indien und Hochasien, 1854–58*, and had been in negotiations about a bilingual publication with F. A. Brockhaus of Leipzig, but the projected costs of '21,720 Thalers' seemed to have frustrated their plans. HS and RS to Humboldt, 20 March 1859 and 24 March 1859, BBAW.

exploring expeditions which have added, within the last twenty or thirty years, so essentially to our knowledge of the distant parts of the globe.[142]

Suggesting both a delusional and unabashed ignorance of the waves of British criticism over several years, this opening claimed exceptional scientific authority and achievement. The Schlagintweits' 'exploring expedition' was indeed held to be on a par with other major overseas travels executed during the previous decades to foreign continents, and thus aligned their travels with those of Barth, Franklin, Livingstone and others. The Schlagintweits' *Prospectus* was also remarkable in other ways. After acknowledging crucial support from British imperial institutions and the Prussian and Bavarian governments, the brothers immediately added that their travels had been accomplished 'at the earnest recommendation of Colonel Sykes on the part of the East India Company, and of General Sabine and Sir Roderick Murchison on the part of the Royal Society'.[143] By rhetorically connecting their travels with such illustrious names as Sabine (and with him the Royal Society), Humboldt and Murchison from the RGS, the brothers subtly appropriated the prestige of these leading scientific men and institutions. They further claimed that 'the arrangements for this expedition were made on a scale well worthy of the East India Company, that illustrious body, which has, for all ages, enhanced and graced its political importance by furthering the great ends of science and connecting its name with many works of a high artistic character'.[144] The Schlagintweits thus portrayed their *Results* and the accompanying *Atlas* as an artistic and scientific milestone in the history of European science in India. Even in the wake of Adolph Schlagintweit's murder and the vital loss of his experience and precious insights, there was no scaling down of the scope and ambition of their written monument. As a result of Adolph's death, and the ferocious public controversy in Britain that had left their reputation in tatters, the completion of the work indeed assumed a higher, existential meaning for the surviving duo, who themselves came to 'regard the publication of our travel accounts as one of the most essential parts of our life's work'.[145]

What gives the Schlagintweit enterprise particular significance is that it was the last great scientific commission by the imperial Company – tellingly given to recruited *foreign* specialists, the likes of whom had

---

[142] H., A. and R. Schlagintweit, *Prospectus: Results of a Scientific Mission*, 3; a copy has survived in the archives of the APS.

[143] Ibid., 4.

[144] Ibid.

[145] H. and R. Schlagintweit to the Minister of Culture, von Mühler, 28 April 1862, GStAPK, Akt Kultusministerium, WR ['einen der wesentlichsten Theile unserer Lebensaufgabe'].

accompanied the EIC regime in south Asia almost from the outset. With the Schlagintweit commission, the whole ambivalence of the East India Company's relation to science – both as an instrumental tool of enrichment and rule and as an enlightened embellishment to its oppressive regime – was neatly captured. The brothers' enlarged enterprise, with its many mentors and sponsors (and their not always identical interests), encompassed both dimensions of European scientific activity on the Indian subcontinent and beyond its northern frontier. Gathering trade intelligence, detecting exploitable resources and mapping new routes of commercial and military potential on the one hand, and the pursuit of more abstract and philosophical enquiries on the other, were, to be sure, not mutually exclusive. Indeed, there existed a long tradition of Company-backed science and surveys in south and high Asia that saw 'the intermingling of the functional and the ornamental'.[146] What makes the Schlagintweit case particularly salient for an understanding of the constraints and opportunities of transnational science in India is, additionally, the way the brothers chose to convey, over time, quite different private versus public representations of their mission's objectives and significance. The thorough analysis of their repeated negotiations with the dying Company has clearly shown that the brothers themselves presumed their expeditionary project should also advance the utilitarian ends of the EIC, not least to justify the Court's prolonged magnanimity during a moment of great imperial crisis. At the same time, this allegiance to the assumed interests of their imperial patrons was often minimised or denied in public accounts of the enterprise.

[146] Arnold, *Science*, 23.

7 The Schlagintweit Collections, India Museums and the Tensions of German Museology

The legacy of an overseas exploration could take many forms. While the Victorian culture of exploration necessitated the publication of a travel account, the Schlagintweits tried to go one step further and pursued the plan to found an India museum in Berlin. This would be entirely devoted to their Asian expeditions and be filled with their heterogeneous collections. The two surviving Schlagintweits would also act as directors of the new institution. While there existed widespread interest among German savants in philhellenism throughout the nineteenth century, the brothers knew that India, too, attracted considerable attention from German oriental scholars, philologists, archaeologists, historians, medical practitioners, missionaries, scientists, merchants and investors.[1] Their engagement with south Asia was driven by different agendas – but above all by curiosity, creed and commerce. The projected Schlagintweit museum precisely sought to exploit this widespread attraction of India and its rich natural and cultural worlds for different visitors, and to advance German engagements with the subcontinent in the future. The potential of such an institution was considerable, since such an explicitly East India museum, which combined humanistic collections and science with considerations of commercial opportunities abroad would have been unique. The Schlagintweit museum ultimately demonstrates that their German fatherland did not need formal overseas possessions to bring home and display the material cultures of European imperialism.

To add this new establishment to Prussia's cultural landscape, the brothers tried once again to reap the benefits of their bi-national constellation of benefactors. These had not directly communicated with each other in the past, and had thus allowed the brothers to pursue a series of double games and manipulative manoeuvres. Yet, the following analysis will also explore what transpired once their communication monopoly with their British and German sponsors broke down. This loss, in turn,

---

[1] A masterful treatment of German graecophilia is provided by Marchand, *Philhellenism*.

led to further movements of their vast collections across the German lands, and ultimately to their dispersal into very different institutional arrangements.

After their return from India in June 1857, Hermann and Robert Schlagintweit immediately began trying to alter their social and professional status. The projected India museum would not only celebrate their scientific achievements, but also secure a paid position for decades to come. In portraying the brothers as scientific entrepreneurs, this chapter shows that they were not the passive victims of public defamation campaigns. Indeed, they never ceased to pursue underhand plans to maximise their personal advancement, especially once their name became lastingly tarnished in Britain. At the same time, their museum project in the Prussian capital, and later in Bavaria, demonstrates how British imperial institutions impacted on the brothers' imagination and career strategies. The Schlagintweits were undoubtedly influenced by the East India Company's museum in London, which they had frequented during their preparation for the eastern journey in the summer of 1854. The brothers' repeated encounter with colonial collections in Britain, and their successful public display at EIC headquarters, were to shape their own collecting practices in the field.

Following a description of the Company museum as the Schlagintweits' prototype, the analysis moves on to show that while scholars have previously assumed that their large-scale accumulation of data, samples and collectables in the field had 'exceeded any rational measure', an in-depth examination rather suggests that their wide-ranging collecting efforts were part of a long-term plan.[2] It is thus crucial to regard the brothers' initial preparation and collection diplomacy, their acquisition strategies during the expedition and the later attempts to harvest the results from these earlier efforts as much more intimately connected than has previously been acknowledged. At the same time, the ultimate failure of the brothers' far-reaching museological ambitions sheds light on the growing power of state administrations to shape cultural and scientific institution building, which implied a move away from the effectiveness of earlier and often unaccounted schemes of royal patronage. The story of the exceptional Schlagintweit collections ultimately reveals a number of little-known connections between transnational exploratory schemes, the culture of expeditionary science in India and German museology in the mid nineteenth century.

---

[2] Finkelstein, 'Mission', 183; Uhlig claimed the brothers were 'almost inebriated by an eagerness to collect', 'Das Neue Schloß', 94–5.

Conflicts of Collecting

Prior to departure, Adolph, and later also Hermann, spent time in London to prepare for their upcoming voyage, using the full breadth of specialised libraries and collections in Britain's imperial capital. Besides trips to the Royal Botanic Gardens in Kew (*the* hub of global networks of economic botany at the time), the Schlagintweits regularly worked at East India House, the Company's headquarters on Leadenhall Street. It not only contained a vast library, but also housed the EIC's India Museum, which surpassed all other contemporary south Asian collections in Britain and continental Europe.[3] Naturalists, travellers and scholars preparing for an overseas voyage to the East could consult its extensive oriental collections, which included manuscripts, books, maps, war memorabilia, fine art and many objects of Indian craftsmanship as well as natural history specimens. While the museum was open one day a week for the general public, it could be visited 'on other days by the possessors of tickets, obtainable from members of the Court, or other authorities'.[4] The selected view (Figure 7.1) shows the shared space of the library and the display of the collected artefacts. Spectacular showpieces such as 'Tipu's Tiger', a life-sized mechanical carved tiger that imitated the sounds of a European soldier being devoured by the creature, which the Company had pillaged as a war trophy from the sultan of Mysore after his defeat in the Fourth Anglo-Mysore War (1798–9), were displayed in close proximity to the Company's collection of journals, maps and books.[5]

In addition to such individual pieces that glorified imperial expansion and successful military campaigns against the Company's archenemies, the museum also encompassed a bewildering array of natural historical and cultural objects from India. These were complemented by artefacts from other countries or regions of British overseas trade and influence, including the Middle East, Afghanistan, Burma, Tibet, but also the Dutch East Indies and China.[6] While this testifies to the, at least in part, eclectic composition of the museum's holdings, it was 'in essence a living memorial to the British Raj and to those officials of the East India Company and the Indian Civil Service who were attracted by the subcontinent, its history, culture and natural resources'.[7]

[3] Desmond, *Museum*, 1; Ratcliff, 'Museum'.
[4] 'A Visit to the East India Museum', *Leisure Hour*, 29 July 1858, 469–73, 469.
[5] The most comprehensive account of the war, and the acquisition of numerous war trophies, is Stronge, *Tipu's Tigers*.
[6] Desmond, *Museum*, vi.
[7] Ibid.

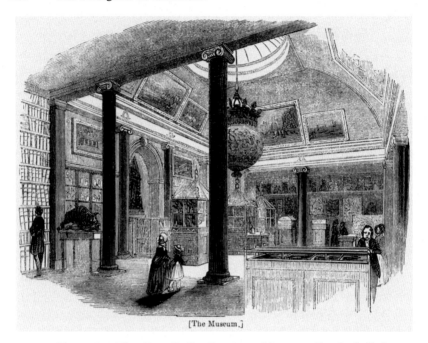

[The Museum.]

Figure 7.1 The East India Company Museum, Leadenhall Street, London, 1843.
Knight, *London* V, 67.

The very nature of London's India Museum, and its role for British commercial expansion overseas, had significantly changed over time. Founded in 1801, it was initially intended to house and display Asian manuscripts and any 'articles of curiosity' gathered by Company servants, hitherto scattered among East India House's different departments and the Company's warehouses.[8] Sir Charles Wilkins, the museum's first director, foresaw its scientific and industrial uses, yet his advice to establish 'a new System for ingrafting the knowledge of India on the commercial pursuits of the Company' was only fully realised later on.[9] The gradual shift towards the museum as a prime site for promoting British trade with the East was completed in the 1850s, precisely the decade when the Schlagintweits frequented its assorted collections.

---

[8] Foster, *India House*, 149. Armitage, 'Schlagintweit Collections', 67; Desmond, *Museum*, 4–9.
[9] Desmond, *Museum*, 8.

A watershed event for this transition was the Great Exhibition of 1851, which heralded the era of modern world exhibitions. The Great Exhibition's massive Indian Court, covering some 30,000 square feet, was organised by the EIC and displayed a wide range of oriental treasures. The event 'was highly significant in popularising Indian design for the British consumer market'.[10] The wider significance of the 1851 exhibition was indeed that it 'provided a benchmark in changing popular attitudes to Britain's colonial possessions, and its organisers emphasised the commercial importance' of overseas territories, in particular their raw products and manufactures.[11]

The increasing trade orientation of the East India House museum was further captured in the appointment of a formal 'Reporter on the Products of India' in the newly established India Office in 1858. The first 'Reporter' was John Forbes Watson (1817–92), a former surgeon in India, with whom the Schlagintweits collaborated for a number of schemes related to making their collections useful to British imperial concerns.[12] Under Forbes Watson's leadership, the India Museum ceased to be, in the words of the under-secretary of state for India, M. E. Grant Duff, 'a mere museum of curiosity, [or] even primarily a museum intended for the advancement of science'.[13] Rather, it truly became 'the reservoir, so to speak, that supplies the power to a machinery created for the purpose of developing the resources of India, and promoting trade between the Eastern and Western empires of her Majesty, to the great advantage of both'.[14] When the Schlagintweits worked in the Company's museum in 1854, it housed both *ethnographica* and *naturalia* – including precious stones and timber sections, plants, soil samples, silks, cotton and woollen fabrics, carpets and flags, oriental arms, paintings and many other objects. As the brothers knew from firsthand experience, the museum enjoyed enormous public interest, as thousands of visitors came to see the displays each month.[15]

The Company museum had a lasting impact on the Schlagintweits, whose collecting efforts in Asia went considerably beyond whatever Prussian officials and the EIC directorate had originally envisaged. Alexander von Humboldt had encouraged the Schlagintweits from the

---

[10] Barringer, 'South Kensington Museum', 12–13.
[11] Ibid., 12.
[12] H. Schlagintweit et al., *Results* I, 9; 'Indische Producte auf der Londoner Ausstellung des Jahres 1862', *Das Ausland*, 1 April 1865, 13, 299–303; 'Ostindiens natürliche Reichthümer und deren Verwendung', *Aus der Natur*, 30, 1866, 465–69; ibid., 31, 481–5.
[13] Quoted from Driver and Ashmore, 'Mobile Museum', 361.
[14] Ibid.
[15] Stronge, *Tipu's Tigers*, 66.

start to always collect artefacts in duplicates so as to be able to augment existing Berlin and London collections.[16] Humboldt drew on his own experiences during the American travels, stating 'I have simultaneously collected rocks for Berlin, Madrid and Paris, even two boxes for Sir Joseph Banks, who provided them to the Brit[ish] Museum.'[17] To that purpose, Humboldt also made a plea for considerable Prussian grants to be put aside, reminding the Prussian Consul in London, Bunsen, 'how indelicate it would be to leave all costs to the English Government ... since it would later become anyway public that the travellers did not only enrich the British Museum, but also the collections in Berlin'.[18] Humboldt expected the brothers to amass geological and botanical specimens in elevated regions in particular, 'in order to complete our mountain collections, which are only rich for America and Northern Asia'.[19] Yet, what the brothers ultimately gathered on the spot, and how they anticipated their collectables would be used in Berlin, differed markedly from Humboldt's modest vision.

During their south and high Asian travels, the Schlagintweits collected artefacts in bulk. Impressed with the holistic approach of the London collections, the brothers and their dozens of hired hands amassed articles that reflected the diversity of Asia's flora and fauna, human cultures and religious practices, as well as numerous scientific, agricultural and commercial items. Stephanie Kleidt, a great expert on the Schlagintweits, assumes that the collections amounted to over 40,000 objects.[20] Since the Court of Directors claimed ownership over the artefacts, the brothers followed Humboldt's advice.

[O]ur collections ... as regards geology, geographic botany and zoology, but also ethnography are, I think, pretty complete. During March and April 1856, we have sent 210 large boxes to the India House, and recently 109, containing all the collections of this year. Of all [items] always exist duplicates, and we hope that we will manage to receive from the Court of Directors a great portion for Prussia.[21]

The acquisition and transport of several tonnes of artefacts were all realised on British credit and generated costs that were far beyond the sums the British authorities had initially agreed to cover.[22]

[16] Humboldt to Bunsen, Berlin, 20 February 1854, in Schwarz (ed.), *Briefe*, 170–8. Kirchberger, *Aspekte*, 390.
[17] Humboldt to Bunsen, 20 February 1854, in Schwarz (ed.), *Briefe*, 174.
[18] Humboldt to Bunsen, 30 May 1854, ibid., 181. Bunsen, too, was always eager to acquire 'oriental' objects for the collections of the Berlin Museum.
[19] Humboldt to Bunsen, 20 February 1854, ibid., 175.
[20] Kleidt, 'Sammlungen', 117.
[21] HS to Humboldt, 11 December 1856, BSB, SLGA, II.1.43, 369.
[22] Frederick William IV to Humboldt, 29 March 1857, GStAPK, Zivilkabinett, AKlC, 121.

It is difficult to determine when precisely the idea for their own India museum in Berlin emerged. However, just as the three Schlagintweits had used their last reunion in November of 1856 to determine the future shape of their publication, it is reasonable to assume that by then, having already despatched hundreds of boxes with duplicates they felt entitled to, and which they knew covered numerous branches of the sciences, the brothers had also forged definitive plans for these collections. Once back in Europe, Hermann and Robert immediately set about making arrangements to keep their collections in close proximity to their manuscripts.[23] Humboldt rejoiced in a letter to the Prussian monarch that their collections 'currently stored in the Indian House [in London], but half of which shall be brought hither, will most brilliantly enrich our museums'.[24] However, the Schlagintweits did not want their artefacts to disappear anonymously into already existing institutions. In late June, they therefore met with Frederick William IV. The king promised the Schlagintweits preliminary rooms in two representative buildings for setting up and analysing their collections: Schloss Monbijou and the Berlin Bourse, right at the heart of Berlin. With this asset in hand, the brothers mobilised their German benefactors for the upcoming negotiations with the EIC.

In July, the former Prussian consul in London, Bunsen, intervened from afar. On behalf of the travellers, he proposed to the EIC director, William Henry Sykes, that the brothers should be enabled to compile their publication and study the *entire* collections in Berlin, 'where they have help & assistance more easily at hand and where the King has generously offered them the Palace of Monbijou for putting up provisionally their collections till the work is finished'.[25] However, once this was accomplished, Bunsen assured the director that the brothers would then be 'enabled to present the whole to the Court, to whom their collections of course belong, save such <u>doublettes</u>, as may be spared for the Museum Royal'.[26]

At this point, the chronology of the Schlagintweits' intended India museum accelerates. On 13 August 1857, they requested from the Court of Directors and were granted 'free permission to receive the Manuscripts, and Collections sent by them from India', to be taken to Berlin.[27] Their

---

[23] The travel manuscripts and 751 landscape views, which were not officially part of the 'Schlagintweit collections', were kept in a special room in Berlin's Neues Museum, under the supervision of Ignaz von Olfers.

[24] Humboldt to Frederick William IV, 20 June 1857, GStAPK, Zivilkabinett, AKlC, 51f.

[25] Bunsen to Sykes, Heidelberg, 19 July 1857, GStAPK, FA Bunsen, folder 5, 272.

[26] Ibid.

[27] BL, IOR, B/234, Court Minutes, 1460 and 1628a.

request was 'read in Court' on 19 August, and immediately approved. The same day, the brothers started an orchestrated campaign for their museum, informing many leading Prussian minds about their successful negotiations – including Alexander von Humboldt, the Prussian privy councillor Illaire and the Prussian prime minister von Manteuffel.

It was in their despatches from London that the brothers, for the first time, openly mentioned the project of an independent India museum. Hermann boasted that they had finally managed to secure 'half of the collections' to stay in Berlin on a permanent basis. To this news, he added that '[t]he only *condition* the India House made was that the [Prussian half of the] collections will not be broken up, but will rather remain as one whole', 'to be displayed in a distinct Museum'.[28] Hence, the foundation of a new museum was presented as an absolute requirement. Yet, the British demands on this point were more ambiguous. At first, however, the Company's supposed insistence on a newly founded institution was not doubted or commented on by the Prussian king or his government. Yet, the brothers' risky communication strategies – to present half-truths as unalterable conditions – could, and indeed would, eventually backfire, to the point that the brothers ultimately lost their integrity within Prussian scientific and government circles.

The transport of the collections to Berlin took place in early October 1857 and comprised 510 boxes in total. Those objects that could not be stored in-house were given to other Berlin institutions such as the zoo, including a living pair of Tibetan ghorkars (or wild asses), and a pair of horses and camels from Turkestan, which had all survived the shipment via Bremen. Presenting themselves as grateful employees, the brothers gifted the horses to the Prussian king.[29] The shipment of all objects was paid with Prussian credit.[30] Taking up Frederick William IV's earlier offer, the two brothers began to unpack their boxes in Schloss Monbijou and the Berlin Bourse. They assembled the 40,000 Asian natural specimens and ethnographical objects not only in preparation for writing their *Results*, but also turned the 'Tanzsaal' of the Palace into the core of their own, albeit only temporary, museum (Figure 7.2).

---

[28] Hermann Schlagintweit to Illaire, GStAPK, Zivilkabinett, AKIC, fols. 61f.; Schlagintweit to Prussian prime minister Otto Theodor von Manteuffel, London, 19 August 1857, ibid., 63–5: 'zur Bearbeitung herüberzunehmen und die Hälfte derselben zur Aufstellung in einem eigenen Museum in Berlin zu behalten'.

[29] Description of their collections by RS, arguably intended as an aide mémoire for Humboldt, ibid., 53f.; also: 'Ein Besuch im zoologischen Garten zu Berlin', *Gartenlaube*, 48 (1858), 687.

[30] 'Legationskasse: Transportkosten für die Sammlungen', 29 December 1857, GStAPK, MdA, III, 18929, HEBS.

Figure 7.2 Schloss Monbijou, 'Tanz- und Festsaal', photograph by 'Preuß. Messbildanstalt' between 1900 and 1940.
© Fotoarchiv Marburg, no. 1.251.966.

For this purpose, the brothers commissioned display cabinets (for 800 thalers) for the objects, and had some Indian paintings including three large portraits of Indian rulers framed and hung on the walls.[31] They also ordered the Berlin foundry to produce the first series of their 'ethnographical heads in copper' (resulting in the considerable personal expense of 2,000 thalers), the first forty of which would be displayed in Monbijou in early 1858.[32] In addition, the brothers engaged the services of a taxidermist for their numerous specimens of quadrupeds, including a giant Bengal tiger, which were, alongside those preserved in spirits, eventually exhibited. The animal collection was complemented by a natural-historical department, as some '600 tree sections', as well as '11,000 plants' amassed during their travels, were prepared to be analysed and

[31] HS and RS to Prince William of Prussia, 12 March 1858, GStAPK, Zivilkabinett, AKlC, 121f. The brothers possessed thirty-one Indian miniatures, and three large portraits of Indian rulers.
[32] H. and R. Schlagintweit, 'Menschenracen', 250.

partly displayed.[33] The tree sections, illustrating the natural treasures of south and high Asia, had one side 'left in its natural state when sawn', whereas the other was pleasingly 'polished, to show the grain and colour of the timber'.[34] Within only a few months, the brothers had spent over 6,500 thalers (over £1,000) on analysing and bringing growing numbers of their artefacts into presentable shape.

While the display of their travel treasures in Berlin was well received by visiting naturalists and German royals, the brothers' goal was to turn this fleeting arrangement into a permanent legacy.[35] However, in a twist of fate, before any formal written arrangements had been made, their longstanding and committed patron Frederick William IV suffered a stroke on 23 October 1857. The Schlagintweits had hoped to use the royal connection to overcome any administrative hurdles and realise their museum in direct communication with His Majesty. As any hopes for Frederick's recovery faded, his brother, William, prince of Prussia, became the *de facto* ruler and eventually king in October 1858. With Frederick's unexpected departure from government affairs, their projected permanent museum – so close to its realisation – was suddenly left in limbo. Contrary to their initial plan, the Schlagintweits now had to take on the Prussian administration. On 10 December 1857, they submitted an 'Immediateingabe' to 'His Majesty the King'.[36] Informing the cognitively impaired Prussian king that parts of the collection were already on display at Monbijou, they now asked for royal 'funds' to retrospectively cover the costs incurred for both the transport from London to Berlin, and the preparation and arrangement of their artefacts. The brothers had run into considerable financial difficulties, precisely because they had been the driving force behind the project, and had advanced large sums of money. In the submission, they repeated the claim that 'the Court expressed the wish and expectation' that the Prussian half of their collections 'should form an independent geographical museum'.[37] The reason was that 'the individual objects … as regards their completeness over such vast, hardly known countries, [being so] distinct from collections of a related character, would form a very good whole'.[38] To

---

[33] Together some 2,300 thalers, see Schlagintweit to Prince William, 12 March 1858.
[34] 'The New Museum at the India House', *The Observer*, 23 May 1858, no page numbers, in ÖAV, CC.
[35] For instance, Humboldt noted that the grand duke of Weimar 'was surprised about the richness of their collections in Monbijou', and bestowed a medal on both brothers. Humboldt to RS, 15 March 1858, GStAPK, Zivilkabinett, AKIC, 120.
[36] 'Immediateingabe' by the Schlagintweits to His Majesty the King, 10 December 1857, ibid., 149f.
[37] Ibid.
[38] Ibid.

further their cause and impress the king and Prussian administrators, they submitted a preliminary inventory of all the branches of their artefacts, amounting to many thousands of objects.[39]

In their official communication, the brothers emphasised the 'practical and scientific interest' that their projected India museum would arouse.[40] They also argued that 'the ethnographical objects, because of their completeness and extension over a territory of 200 million peoples' would give the new museum a distinct 'accessible character' to capture broader, including mercantile, audiences.[41] Underlining this point, they soon published detailed descriptions of their cultural artefacts that made unmistakably clear that the brothers sought to use their collectables to encourage further European commercial expansion into and beyond the trans-Himalayan regions. Exemplary is their collection of 281 samples of Eastern 'woven manufactures' – in cotton, leather, wool and silk. These samples of 'native cloth' were bound in nine luxurious volumes entitled *Technical Objects from India and High Asia*. Each volume was dedicated to a particular region – e.g. 'Cashmere' or 'Tibet' – and was complemented by a description of what kind of garment was produced out of the different textile samples (Figure 7.3a).[42] The types presented ranged from 'shawls' to 'festival-coats' of 'Lamas' dresses', and valuable Alpaca silks 'used by Europeans and rich Natives'.[43] Forbes Watson considered them 'a valuable addition to the industrial museum'. They later served him as a model for his own series of bound textile samples, 'illustrating the Textile Manufactures of India, and promoting trade operations between the East and West'.[44]

Echoing the model of the imperial India Museum in London, the travellers claimed that their industrial collection 'not only shows what the different nations do make', but could further demonstrate to European manufacturers what various indigenous societies 'consider best suited for their climate and taste in reference to strength, texture and colour, and in what forms a cheaper and in consequence more generally used dress might be offered to the little cultivated tribes surrounding the Indian empire'.[45] Indeed, the south and high Asian textiles – and other technical

---

[39] Ibid. 'Zusammenstellung der Schlagintweit'schen Sammlungen aus Indien und Hochasien', draft by Illaire to von Raumer regarding the brothers' Immediateingabe, 16 December 1857, ibid., 110f.

[40] Ibid.

[41] Ibid.

[42] This hitherto neglected prospectus, printed in Berlin, April 1859, survives as a copy in Innsbruck, ÖAV, CC, 283–4.

[43] Ibid.

[44] Quoted after Driver and Ashmore, 'Mobile Museum', 365–6; 'Ostindiens natürliche Reichthümer und deren Verwendung', *Aus der Natur*, 30 (1866), 466–9.

[45] Prospectus to *Technical Objects*, ÖAV, CC.

objects such as different types of paper made of 'animal and vegetable materials' (Figure 7.3b) – were considered to be 'not without practical value for extending international trade' into hitherto unexploited markets in the regions the Schlagintweits had explored. While their Berlin museum would, unlike the Company headquarters in London, not display the material cultures of peoples ruled by a German overseas empire, it was nonetheless intended to exhibit natural and cultural artefacts from the East 'in a luxurious manner' for the furtherance of European – including German – trade with this world region.

On the other side of the Channel and only a few months before its suppression, the East India Company employed, from May 1858 onwards, three different curators – 'Drs. Horsfield, Watson, and Wilson' – to arrange several key components of the Schlagintweit collections at its London headquarters.[46] Over time, the display at the India Museum included the entire series of 275 ethnographic heads and several large religious masks – 'most tastefully arranged in the sculpture room'.[47] Moreover, several hundreds of polished tree sections, plants, flowers, religious objects, samples of Indian marbles and the brothers' commercial items from India and high Asia, among them the entire textile series, went on display by 1859. The Company trade in India's woven manufactures had, of course, a long tradition. But the Schlagintweit series of textiles added to Western commercial knowledge because it provided new insights into the tastes and fashions, and hence demands, of communities *beyond* the Indian Empire, and into markets hitherto only targeted by the Chinese and Russian Empires. As a British paper noted: 'This collection of manufactures [from Thibet and central Asia] may prove perhaps soon very important for our trade, which is daily more increasing into these regions. Till now, Russia only supplied and knew the wants of the people of those regions.'[48]

Despite the acclaim from visitors for their collections in London and Berlin, the Schlagintweits' ambition to found their own museum in Prussia proved difficult to realise.[49] The reason was that the Schlagintweits' application to the king caught the attention of some of the brothers' old

[46] *Illustrated News of the World*, 2 July 1859, copy in ÖAV, CC.
[47] *Illustrated News of the World*, 23 July 1859, copy in ÖAV, CC. Nowadays ethnographic heads from the Schlagintweit collection can be found at the Bhau Daji Lad Museum, Mumbai; in the Peter the Great Museum of Anthropology and Ethnography (Kunstkamera), St Petersburg; and especially at the Muséum national d'Histoire naturelle, Paris, and the National Museum of Ireland, Dublin.
[48] 'New Contributions to the India House Museum', *Star*, June 1859, ÖAV, CC, 103.
[49] See the report by von Reizenstein to Baron Malortie, Berlin, 10 October 1859, Königliches Hausarchiv, Hanover [KHA], Depot 103, XX, no. 320.

Coll. Schlagintweit:

Technical objects from India and High Asia.

Woven Manufactures.

Vol. II.

Assám and Tribes of the Northeast Frontier.

Cotton worn like Shawls over the shoulder
(by men and women).

Figure 7.3a 'Technical Objects from India and High Asia', *Woven Manufactures*, vol. II, no. 11, 'Assám and Tribes of the Northeast Frontier'. © BL, IOL, X 366.

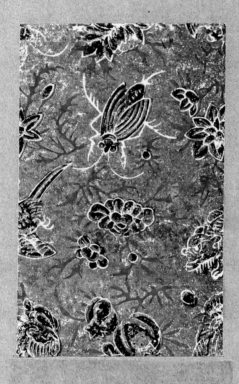

Technical objects from India and High Asia.

GROUP I.

Samples of Paper.

Déhli, Hindostán (N. W. Prov.)

Figure 7.3b 'Technical Objects from India and High Asia', Group I, Samples of Paper, no. 26, 'Déhli, Hindostán (n. w.-prov.)'.
© MFK, II/5,2,2.

enemies in the Prussian administration. The culture minister Karl Otto von Raumer, with whom Humboldt shared a deep mutual contempt,[50] and who had earlier declined any state-funding for the Schlagintweits' proposed Prussian Himalayan expedition in 1852, also got his hands on the proposal. It testifies to the specific communication strategies of the brothers, who had sought to involve only their closest allies Bunsen, Humboldt and the king in their scheme, that von Raumer had to make formal enquiries about the museum project, as 'all further negotiations, which have taken place with the brothers Schlagintweit concerning their travel and collections, have until now remained entirely unknown to me'.[51] In reply, even Privy Councillor Illaire could not provide satisfying information about the brothers' plans, proving how far the museum was a project pursued behind the back of even the ministry of culture.[52] The brothers had pursued a death-or-glory approach: if they could have reached an official agreement with the king via private communication, the creation of the India museum would have been a stroke of genius. By contrast, once this shortcut had failed, and influential administrators found out that the brothers had left them in the dark for months, the Schlagintweits' prospects of success dramatically faltered.

In early 1858, things became increasingly uncomfortable for the ingenious projectors. For one thing, the brothers were under considerable pressure for failing to defray the considerable costs for the transport of the collections from London, which had temporarily been covered by the 'Königliche Legationskasse' up to 2,425 thalers.[53] The negotiation about settling the debts with the 'Speditionshause Phaland & Dietrich' became a small affair of state, entailing numerous private and public interventions from various actors, with the brothers continuously applying for further government grants to clear the debt. (The affair continued for several years, well into the premiership of Count Otto von Bismarck, who in the mid 1860s still had to deal with the brothers' stubborn unwillingness to refund the Königliche Legationskasse. The conflict only ended when the new Prussian king refused to launch a legal case against the Schlagintweits and paid off the debts out of his *Privatschatulle*.)[54]

---

[50] See, e.g., Humboldt to Bunsen, 27 March 1852, in Schwarz (ed.), *Briefe*, 148; 19 August 1855, ibid., 193.

[51] Raumer to Illaire, 11 January 1858, GStAPK, Zivilkabinett, AKlC, 116.

[52] Illaire to Raumer, 17 January 1858, ibid., 117.

[53] 'Legationskasse: Transportkosten für die Sammlungen'.

[54] See enquiry by von Bismarck, finance minister von Bodelschwingh and culture minister von Mühler to the Prussian king, 1 March 1865; and order of King William, 20 March 1865, to liquidate the brothers' debts; both in GStAPK, Kultusministerium, WR, no page numbers.

In February 1858, the brothers even tried once again to play one side against the other, informing the culture minister that '[if] the bills for the setting and working up [of the artefacts] ... were presented to the Court of Directors', this would imply 'that the collections would have to be sent back to London in their entirety', thus putting pressure on the Prussian officials to sanction their demands in order to avoid a scandal.[55]

Amid their struggles over debts and the Prussian administration's increasing distrust of the brothers, it was again Alexander von Humboldt who intervened. He rightly sensed that the whole museum project, and the securing of their exceptional artefacts for Prussian collections, had reached a critical point. First, Humboldt invited influential naturalists and Prussian government officials to visit the Schlagintweit collections in Monbijou.[56] His reasoning was to build up a network of supporters now that the planned museum had become a bone of contention. The invited dignitaries included the general director of the Prussian museums, Ignaz von Olfers (1793–1871), as well as the acclaimed travelling naturalist, geologist and anatomist Christian Gottfried Ehrenberg (1795–1876).[57] Humboldt also directly addressed his archenemy von Raumer. In his letter, he warned that public disgrace would be brought upon Prussia if the museum project failed, and if the entire enterprise, 'after the reluctant Company had finally agreed that the greatest natural historical and ethnographical collection which ever reached Europe from Inner Asia could here be divided, may now come to an end in a widely perceived, disgraceful manner'.[58] Humboldt's belief in the great value of the Schlagintweit collections was genuine: 'you have achieved in very distant countries more than ... Professor Ritter and I, judging by your two great works on the western and eastern Alps, could have expected from you'.[59] He specified that 'the glory [das Rühmliche] of your great and dangerous expedition is based on the plentiful collections you have brought home'. Convinced of their (otherwise often contested) achievements, Humboldt further urged the Schlagintweits to seek an audience with William, the

---

[55] Declaration by the Schlagintweits to von Raumer, 19 February 1858, GStAPK, Kultusministerium, WR, no page number.

[56] German royalty also visited their collections in Monbijou, e.g., the Grand Duke of Mecklenburg during a two-hour visit; RS to Humboldt, Berlin, 12 February 1858, copy of letter at BBAW.

[57] Humboldt to Illaire (?), Berlin, 4 February 1858, GStAPK, Zivilkabinett, AKIC, 118.

[58] Humboldt to Raumer, 8 February 1858, ibid., 119: 'auf eine schmachvolle, weit ertönende Weise ihr Ende nehme'. Months later, Humboldt still praised the museum idea, letter to Illaire, 28 August 1858, ibid., 167.

[59] Humboldt to Robert (?) Schlagintweit, 15 March 1858, ibid., after 120.

prince regent, whose goodwill and material patronage now ought to be secured.[60]

The brothers now went on the offensive. In direct negotiations with von Raumer, they suggested on 19 February 1858 that 'a commission consisting of scientific men [shall] examine the already exhibited objects of our collections … in order to prove [their] full value'.[61] The Schlagintweits also enlisted their favoured members of the committee: 'die Herren von Humboldt, von Olfers, von Ledebur, und die Herren Professoren Ehrenberg, Rose, Poggendorf, Klotsch und Dove'.[62] This handpicked selection of scientific experts would cover most of the disciplines touched by their artefacts. Crucially, many of these candidates were close allies of Humboldt's. Gustav Rose, a mineralogist, had accompanied the Prussian naturalist on his Siberian expedition in 1828–9. During this expedition, Humboldt had also collaborated with Dove and Poggendorf, who had assisted him in undertaking geomagnetic observations in Saxony 'during his absence' in Russia.[63] Ignaz von Olfers, now general director of Prussia's museums, was also personally committed to Humboldt, who had openly supported his appointment to his current post in the late 1830s.[64] Following their bad experience with unknown advisory experts in 1852, the brothers now tried to play the game of bureaucratic manoeuvring that decisively shaped Prussia's museum policy at the time.[65] However, even well-disposed allies of Humboldt's may not act as desired, since hired experts often 'delighted in demonstrating their skills', using such commission appointments as a stage for professional self-representation.[66] What was more, there was no way around Raumer. In consequence of the cultural ministry having acquired greater powers in the 1830s and 40s, high officials such as the acting *Kultusminister* were confident of being in a good position to try to shape the kingdom's museum policies according to their own visions.[67] Raumer was also little impressed by the brothers' claim that the Prussian state had paid only '1/40th' of the costs carried by East India House for

[60] Ibid.
[61] Copy of letter by HS and RS to von Raumer, Berlin, 19 February 1858, GStAPK, Kultusministerium, 'Beiheft', Acta Commissionis [AC], prod. 5.
[62] Ibid.
[63] Bruhns, *Life*, 147.
[64] Sheehan, *Museums*, 106.
[65] Ibid.
[66] Ibid., 91.
[67] Königliche Museen, *Museen*, 50–5.

the collections, which suggested that this really was a golden opportunity to permanently found an Indian museum in Prussia.[68]

Raumer announced the founding of an expert commission of four professors to evaluate the Schlagintweits' collections and proposal on 4 March 1858. However, he and Councillor Lehnert organised the committee on very different terms than the brothers had hoped for. Crucially, while the zoologist Wilhelm Peters, the botanist Alexander Braun, the meteorologist Heinrich Wilhelm Dove and the mineralogist Gustav Rose were formally enlisted, the well-meaning von Olfers and Alexander von Humboldt were excluded from the body.[69] Further weakening the Schlagintweits' position, no report was to be written on their important ethnographic and anthropological collections, undoubtedly one of their strongest assets, since both were strongly collections-based disciplines.

In a scenario which speaks volumes about the supposedly objective stance and evaluations of such expert commissions, Raumer provided its four members with a set of comments and questions. In these, the commission was briefed with the opinions and judgements Raumer had already formed, which made the commission's involvement somewhat resemble a show trial over the brothers' application. While Raumer raised the question of whether it was indeed 'advisable to take over the collections' once their real value had been ascertained, he also immediately noted that 'the Company's purported condition to form a distinct museum out of the entire collections ... appears inadmissible and ought therefore to be declined'.[70] Having thus already rejected the independent museum, another question touched on the issue of what expenses were to be expected each year 'after the integration of the [Schlagintweit] collection into the already existing institutions here'.[71] In short, the minister's founding document for setting up the commission contained no subtle hints about his own position on the issue. His official document, in a sense, effectively barred any chance that such an *Indisches Museum* was to be established.

Yet, the commission members added insult to injury. Following his appointment, the meteorologist Dove broke the brothers' privileged communication position with their British benefactors. Dove wrote to

---

[68] Schlagintweits to the 'Geh. Räte Lehnert und Knerk', 25 March 1858, GStAPK, I. HA Rep. 76, Kultusministerium, 'Beiheft', AC, appendix V.

[69] Lehnert, letter to the professors of the commission, 6 March 1858, ibid., prod. 7.

[70] Assignment of von Raumer to Councillors Lehnert and Knerk for the establishment of the advisory commission, 4 March 1858, ibid., prod. 1: 'erscheint unstatthaft und wird daher abzulehnen seyn'. Note the subtle challenge to the brothers' trustworthiness.

[71] Ibid.

the Company director Sykes through his friend and colleague Edward Sabine of the Royal Society, and stated that the Court's supposed insistence on a new museum would 'considerably decrease the value' of the offered collections.[72] The reason was that Dove believed that 'in England, as in the British Museum, heterogeneous objects are united in a common, great whole'. This was contrasted with the Prussian museum policy, where, echoing Raumer's position,

the zoological, mineralogical ... botanical and ethnographical collections were completely separated, which are again distinct from art museums. This facilitates the scientific engagement with the materials and, therefore, all acquired collections by itinerant naturalists have always been incorporated into ... these museums.[73]

The museological specialism of German sciences and models of disciplinary order were thus advanced to block the Schlagintweits' proposal. As the Prussian scientist made explicit, the great 'running costs' for the existing Prussian scientific collections would further hinder 'the creation of a new museum'.[74] Yet, the most harmful effect of Dove's letter was not his complaint about the unfavourable museum demands; it was the breaking of the Schlagintweits' communication monopoly with their British employers, which had never been threatened by Humboldt or Bunsen as their well-meaning intercessors.

By effectively undermining the brothers' privileged negotiating position, Dove's direct intervention with the British side revealed that the brothers were involved in a swindle about the British terms as regards the collections' ownership and the precise 'conditions' for the Prussian half. William Henry Sykes and Edward Sabine informed Dove, and through him the entire commission, that things were perceived differently in the British Isles. Sabine saw the supposed insistence on a new museum as a 'misunderstanding', stating that the Prussian king could receive the duplicates as a gift 'without any tedious conditions', to be used for Prussian museums 'at their own discretion'.[75]

Sykes, by contrast, stated that Dove's letter 'has occasioned me no small surprise', as he was 'not aware that the directory had gifted the collections, or any part of it, to the Prussian king'.[76] Rather, following the considerable costs for the Court, the Company director claimed that 'the collections always were, and still are, property' of the EIC. In

---

[72] Dove to General Edward Sabine, 12 March 1858, ibid., appendix I.
[73] Ibid.
[74] Ibid.
[75] Ibid., appendix IV, translation of the letters from Sabine and Sykes, 26 March 1858.
[76] Sykes to Sabine, part of appendix IV, 26 March 1858, subsequent quotations, ibid.

Sykes's understanding, the Schlagintweits had only been allowed to take all the collections to Prussia for greater convenience. Following this liberal arrangement by the Company, the 'results of their labours could [then] be published exclusively in the name of the East India Company'. Sykes would, however, soon change his position, as he later confirmed that a full set of Schlagintweit duplicates were to be gifted to the Prussian monarch.[77] He further stressed that while it was not an absolute condition that an independent India museum was to be erected in Berlin, he made clear that such a memorial to the liberality of the Court was nonetheless 'desirable', indeed it was the Court's 'decided wish'. Yet, whatever concessions William Henry Sykes was soon willing to make, the damage to the brothers' credibility was already done. Dove's first communication in March 1858 had exposed the Schlagintweits, in the eyes of Prussian scientists, as fraudulent schemers.

Once the brothers learned that their half-truths had been revealed, as Sykes had immediately informed them of Dove's request, they themselves felt maliciously betrayed. Consequently, instead of opting to claim there had been a misunderstanding, they sent a terse note to the Prussian government officials in charge of the advisory committee. They noted with indignation that they, and Humboldt too, were greatly surprised at Dove's unsanctioned enquiry. Exposing the extent to which they may not have been aware of the internal workings of the Prussian administration, the brothers now claimed that they had intended to directly negotiate the whole museum affair only with the king, not with the culture ministry.[78]

Frustrated that their communication tactics had so bitterly failed, the brothers then issued an ultimatum. Therein, regardless of the British positions, they first demanded that the creation of a new India museum in Berlin was to be confirmed. Second, if this was accepted, they requested precise information about 'what medals, salaries and official positions we can expect'.[79] As they rightly stressed, it was only due to their own schemes and fighting, not least against British protest, that 'such a [valuable] collection' could have been taken to Berlin, thus revealing a moral sense of entitlement over the artefacts. In case their bold claims were not met, the brothers openly threatened to send all the duplicates back to Britain, leaving Prussian institutions not only emptyhanded, but also publicly ridiculed. In fact, the Prussian government's fear that the brothers would portray them in a compromising manner in

---

[77] Report by Bernstorff, from London, to Illaire, 7 July 1858, GStAPK, Zivilkabinett, AKlC, 133.
[78] Schlagintweits to the 'Geh. Räte Lehnert und Knerk'.
[79] All quotations, ibid.

front of the Company led Prince William to request the Prussian Consul von Bernstorff in London to supervise and, if necessary, correct the Schlagintweits' version of events in future communication with the EIC. However, with this ultimatum, the brothers had overplayed their hands. Only four days later, on 29 March 1858, the Prussian councillor Lehnert compiled a dismissive twenty-four-page report for the culture minister, signed off by all the advisory experts.[80] This final report contained a scathing review of the brothers' personal conduct as men of science; it also questioned the value of several parts of their collections, and rejected the idea of a separate museum as a whole. Reflecting von Raumer's personal agenda, the report stated that such '[a]n Indian museum in Berlin would be a mere oddity'. Such an institution, for which even 20,000 thalers 'would hardly suffice to cover the expenses', would only 'excite the curiosity of the prying masses'.[81] While expressing an elitist stance regarding the gathering of non-European collections and their displays, the Berlin professors and officials openly rejected the usefulness of the intended 'practical' applications the brothers sought for their Indian collections. According to the report, all that mattered for keeping up the state collections was the pursuit of 'pure science' ['wissenschaftliche Zwecke'] – hence not commercial encouragement or the spurring of popular interest.

In further condemning the Schlagintweits' scientific reputation, the commission's final report stated that even if an 'Indian museum' was to be created, its supervision should be placed in the hand of 'excellent scholars [Gelehrte], proven scientists, in particular thorough specialists of zoology, botany, meteorology, and geography'.[82] By contrast, the Schlagintweit brothers, on the basis of 'the results of our intercourse with them', were not considered by the experts to possess the necessary qualities for such a 'directorship'. Indeed, it was not only their 'presumptuous behaviour' towards 'several members of the undersigned' that led the commission to plea for the abandoning of any further negotiations with them. Rather, the advisory committee also concluded that 'the collectors have considerably overestimated the scientific value' of the objects. The brothers' demand to receive a new building to house their India museum that would offer 'three times as much space as they currently occupy in Monbijou' was also declared to hinder any chance of success.[83] Even

[80] Report of the commission by Lehnert, 29 March 1858, GStAPK, Kultusministerium, 'Beiheft', AC, Prod. 10.
[81] Ibid.: 'die Schaulust der neugierigen Menge zu reizen'.
[82] Ibid. Subsequent quotations, ibid.
[83] Ibid.

Humboldt's urgent support for a museum filled with their 'East Indian collection', expressed in numerous government submissions and private letters, this time failed.[84]

Faced with this spectacular fallout, how can we make sense of the Schlagintweits' far-reaching, uncompromising claims, and their ultimate Icarian fall? The brothers' achievement in having traversed such vast and often dangerous regions in south and high Asia, and having succeeded in getting their collected treasures to Berlin, had simply gone to their heads. This explains their haughty confidence about the value of their collectables, which was also shared by Humboldt. Their sense of great achievement was also reflected in their condescending treatment of their critics in Prussia, which ultimately contributed to their downfall. However, the desired creation of the Schlagintweits' own museum also mirrored their historical consciousness – their willingness to create a lasting legacy to their contested feats. As James Sheehan has argued, '[m]useums promised permanence and preservation; they also provided immediate recognition and material rewards' for those who could stage their own works or scientific collections in such 'holy halls'.[85] The reason was that museums – as the public guardians of precious objects – 'carried the prestige of official, even royal patronage, but at the same time they were instruments of public culture, accessible to everyone'.[86] A museum dedicated to their adventures and accessible to broader publics remained an ambition the Schlagintweits sought to realise throughout their lives.

With hindsight, one could argue that the brothers' pursuit of an interdisciplinary East India museum in Berlin turned them, in a sense, into unrecognised pioneers. While the Prussian expert commission had entirely ignored their rich anthropological and ethnographic materials, Berlin would get a discrete museum of 'Ethnologie, Anthropologie und Urgeschichte' (Museum für Völkerkunde) in 1874. This was partly filled with Schlagintweit objects later acquired from the brothers through the initiative of Rudolf Virchow and the museum's founding director, Adolf Bastian.[87] What is more, it was around the same time that Berlin University started to plan a new, representative building to host its diverse natural

---

[84] Humboldt even suggested the brothers should be rewarded with Prussian honours for their service, Humboldt to Illaire, Berlin, 18 August 1858, GStAPK, Zivilkabinett, AKlC, 164f.

[85] Sheehan, *Museums*, esp. ch. 3: 'The Museum Age, 1830–1880', 95.

[86] Ibid.

[87] While Hermann demanded in November 1876 no less than 226,710 marks for their ethnographic collections, including sketches, maps and Tibetan manuscripts, the Berlin museum ultimately acquired some 335 objects for 8,000 marks in November 1881; see Hermann's report to the Berlin museums, 7 November 1876, Berlin, Museum für Völkerkunde [MVK], vol. 1, doc. 24; see also Kleidt, 'Sammlungen', 132.

history collections.[88] The *Museum für Naturkunde*, inaugurated in 1889 as the largest building in Berlin at the time, would unify the collections of three formerly distinct museums: the Anatomical-Zootomical Museum, the Mineralogical Museum and the Zoological Museum of Berlin University, which together echoed the diverse composition of the Schlagintweits' temporary museum in Schloss Monbijou.

The Flight Forward: A Private Schlagintweit Museum in Bavaria

The Prussian museum episode, with its many discreditable incidents and revelations, was never made public. Unlike in Britain, where Company officials and men of science had leaked discrediting material to the papers, in Prussia the brothers were mostly protected from such critical scrutiny by the press. However, their status had clearly suffered, as government officials noted that, especially after Humboldt's demise in 1859, interest in the Schlagintweits and their travels had 'considerably cooled down'.[89] The conflict over the unsuccessful museum project shaped their trajectories in other unforeseen ways. Having become *personae non gratae* among the Prussian administrators and the prince regent, the earlier offer of free exhibition space to the Schlagintweits in Schloss Monbijou was withdrawn in 1860, and was not renewed for another location. Now, instead of surrendering their treasures to existing collections and thus allowing their anonymous fragmentation, the brothers aspired to find a home for their artefacts elsewhere, evidently feeling they had a moral alibi to proceed as they pleased with the Asiatic booty.

While the Schlagintweits had claimed to be financially unable to hire exhibition rooms in Berlin, Hermann and Robert settled on the purchase of a castle in southern Germany that was large enough for the brothers and their entire collections.[90] They moved out of Schloss Monbijou and back to Bavaria in the autumn of 1860, leaving some surplus collectables behind, while taking the most valuable parts of the Prussian 'half' with them to Castle Jägersburg.[91] Apparently, Prussian officials were only informed about the relocation of the majority of the collections to Franconia almost a year later. While still refusing to settle their debts with the Prussian administration and offering to grant the abandoned collectables in Monbijou as a repayment, the Schlagintweits

---

[88] Kretschmann, *Museen*, 37–8.
[89] Report by Reizenstein to Baron Malortie, 10 October 1859, KHA, Depot 103, XX, 320.
[90] HS and RS to Illaire, Jägersburg, 14 October 1861, GStAPK, Zivilkabinett, AKIC, 204f.
[91] Report by the new culture minister, Bethmann-Hollweg, 19 March 1861, ibid., 191–3.

plainly informed Berlin officials in 1861 that, as a result of monetary concerns, they had to further pursue the preparation of 'our book in the countryside, where we have a small property'.[92]

Through this move, the Schlagintweits effectively 'privatised' the collections, which could not be placed so easily in any university or royal museum in Prussia. Rather, the artefacts were stored on private grounds and used to decorate their castle. Soon, the Jägersburg became a semi-public museum space open by appointment to collectors, scientists and scholars, and to the wider public from May to October each year. As different visitors made the tour through the first two floors, the exhibition of their Indian landscape views, oriental arms, models of Indian mausoleums, and with a number of Indian textiles and printed papers decorating entire halls, the objects on display left a lasting impression on the visitors' minds.[93] Although the brothers had by no means personally collected all of the Asiatic objects, they were all silently appropriated under the deceiving label of the 'Schlagintweit collections'. In that sense, the travellers' privately funded museum became an architectural statement that was to represent their achievements in expeditionary science.

The fact that neither Prussia nor the British side formally recalled the objects that the Schlagintweits had taken with them to Bavaria was pure luck. After the brothers had published their scientific *Results* between 1861 and 1866, it would have been feasible for the official owners to insist on the return of their respective shares of the collections. Yet, only the most valuable and rare objects found their way back to the India Museum in London. The bulk of the artefacts remained with the brothers, who treated the collections as their personal property, not least by selling increasing numbers of artefacts – originals and replicas – for their own profit.

While it remains unclear when and why the Prussian king abandoned any direct claims to ownership, there is some evidence that helps to explain why the East India Company, and later the India Office, dropped their legitimate demands to receive half of all the duplicates and *all* unique pieces. The reason was that the Company's India Museum was already overcrowded with objects, and became even less able to house all its collectables after its move to a new site in Fife House, Whitehall Yard, in 1861.[94] This lack of space played into the hands of the brothers.

[92] HS and RS to Illaire, 14 October 1861.
[93] E. Schlagintweit, 'Besuch'.
[94] Digy Wyatt, memorandum to Lord Ellenborough, 4 March 1858, BL, IOR, PRO/30/12/22.

Fife House seemed absurdly unsuited to be used as a museum, even though some objects enjoyed pride of place. For instance, the 'large collection of ethnological specimens' by the Schlagintweits was 'arranged in the entrance hall', where their entire set of heads was prominently displayed.[95] The Schlagintweit samples of 'Indian marbles' were also permanently 'deposited' there, furnishing 'proof of the varieties to be met with' for British exploitation'.[96] Despite the unsuitable building, the imperial museum still attracted a flood of spectators; some 175,000 people visited within the first two years at the new site.[97] The collections of London's India Museum moved again in 1867 to the India Office, at Whitehall. However, there likewise existed no appropriate display space, so that only parts of the former Company's 'treasury of rarities' could be displayed.[98] The former Company's collections were ultimately divided up between the South Kensington Museum, built in 1857, and since 1899 the Victoria & Albert Museum, the British Museum and the Royal Botanical Gardens at Kew between 1879 and 1880, when a great number of the Schlagintweits's items were often anonymously subsumed under their holdings.[99]

Some objects even went full circle back to Asia, like the enormous 'Collection of Indian timbers (comprising 557 cross-sections) made by the brothers Schlagintweit during their journeys'.[100] First deposited at the museum at Kew, they were considered 'of little or no value owing to the want of a catalogue'. Yet given the Indian government's interest in exploitable resources, the brothers' wood samples were, with heavy costs, shipped back, in nine packages, and sent 'to the Dehra Forest School, where the specimens can be identified by Mr. Smythies with the help of the Dehra type collection'. They were ultimately 'incorporated in the School collection', and could thus be used for pedagogical purposes in scientific and commercial forestry.[101]

While British authorities, over time, lost interest in the more mundane objects of the Schlagintweit collections, the Prussian king could

[95]   *The Times*, 22 July 1861, copy in ÖAV, CC, 280.
[96]   Forbes Watson, *Classified and Descriptive Catalogue*, Class I, 16.
[97]   Altick, *Shows*, 509.
[98]   *The Nation*, 27 July 1865, copy in ÖAV, CC.
[99]   See, e.g., *Ms. Catalogues and Memoranda of the Zoological Collections in the India Museum Transferred to the British Museum in 1879*, 2 vols., Zoology Library, Natural History Museum, London.
[100] NAI, Revenue & Agriculture, Branch: Forests; 1882-01, File no. 87–92, Part B, no. 1069.
[101] J. S. Gamble, assistant conservator of forests, Bengal, and author of the widely consulted *Manual of Indian Timber*, also inspected their woods samples; Gamble to the secretary to the government of Bengal, Revenue Department, 5 January 1882, ibid., 133.

have insisted that the artefacts belonged to Berlin museums, since the Schlagintweits had no right to dispose of any parts of the objects, either as gifts or sales.[102] However, while the brothers had no legal ownership, they simply gambled and used delaying tactics to reach their goals nonetheless.[103] They strategically used the artefacts' multiple relocations, and simply the passage of time, to sit it all out and to slowly turn informal arrangements into accepted realities.

Between Memory, Science and Commodification: The Schlagintweit Collections at the Jägersburg and Beyond

While the Schlagintweit objects, their marbles, manufactures and mammals, added to the imperial inventory of the East in London, the practical commercial implications of the brothers' collections were not lost on German audiences, either. German paper and textile producers, for example, used the brothers' collections, including the seventy rare indigenous paper samples. This part of their collection had been inspired by the researches of Joseph Hooker and Thomas Thomson into usable plants for paper manufacture in India.[104] In a report containing a technical analysis of their samples, a German magazine noted that the famous paper maker Dr Alwin Rubel possessed the entire set.[105] Rudel was not only among the two owners of the 'Papierfabrik Königstein' in Saxony, but also acted as editor and publisher of the *Central-Blatt für die deutsche Papierindustrie*. The published analysis on the Schlagintweit paper samples from Asia concluded that one 'ought to know that they are made out of different plants, whose study ... is of the utmost importance to European manufacturing', on account of the 'different character' of the papers they helped to produce. The same applied to their samples of indigenous textiles and shawls, lauded as 'being certainly not without practical value for the growth of trade between peoples' from western Europe and Asia.[106]

German promoters of industry and trade indeed undertook similar efforts to those made by Forbes Watson, the Reporter for the Products of India at the India Office in London, to use the Schlagintweit samples

---

[102] Report by Bernstorff, from London, to Illaire, 7 July 1858, GStAPK, Zivilkabinett, AKlC, 131f.
[103] Second report by Bernstorff to Illaire, 30 July 1858, ibid., 168–72.
[104] Behrnauer, 'Schlagintweit'schen Sammlungen', 374; Kew then housed a great collection of plant fibres for papermaking.
[105] Ibid.
[106] Ibid., 376.

of Indian papers and textiles to improve domestic manufacturing.[107] A German manufacturing journal noted: 'the importance of this collection is also not insignificant for the Arts and Crafts, whose different aspects are sometimes even richly represented. We mention in particular the textiles and embroideries [from India], but partly belonging also to China and Japan. It is compelling to compare these fabrics with samples of Indian textiles that are part of the Bavarian Museum of Arts and Crafts' in Nuremberg.[108] The value of the artefacts proved so significant that in 1887, 'the sample collection' [Mustersammlung] of the same museum, founded in 1872, formally 'received a collection of Indian papers, which the brothers Schlagintweit had once collected. These papers present an attractive addition to the Oriental papers it already held.'[109] According to its founding members, the Bavarian museum's specific goal was, 'through periodical travelling exhibitions to all parts of the country, through the foundation of a technical information bureau and through the establishment of experimental stations, through comprehensive Arts and Crafts collections and through the support of technical education among the tradesmen und workers of the country, to improve and diffuse technical skills and artistic taste'.[110] The Schlagintweit objects were thus used to demonstrate the skills and properties of south Asian manufacturers to German industrialists and designers. This goal was achieved by integrating them into prominent sites of industrial research and display.

At the Jägersburg, the Schlagintweits also hosted visits from several 'improvement societies' such as the powerful Agricultural Society of Bavaria, whose representatives were furnished with 'several seed samples and economically useful plants' from the Himalayas.[111] The brothers were also proud to receive further royal and scientific visitors from afar, whose presence added new lustre to their damaged reputation. It was widely reported that even King Otto (1815–67), the Greek monarch of Bavarian origin, together with Queen Amalie, visited the castle and

---

[107] On the Schlagintweits' plant-based papers in Britain, Ramasheshan, 'Paper in India', 114.

[108] 'Schlagintweit Sakünlinski'sche Sammlung', Kunst und Gewerbe, 12 (1878), 133.

[109] 'Bayerisches Gewerbemuseum, Nürnberg', Kunst und Gewerbe, 21 (1887), 78. The museum bought fifty-nine paper samples on 28 January 1887 for 30 marks; letter from Hans Kellner, the 'kgl.-bayr. Inventarsekretär', to an unknown recipient (likely Emil Schlagintweit), Germanisches Nationalmuseum [GNM], inventory no. LGA 7095.

[110] Verwaltungsausschuss des Bayer. Gewerbemuseums, 'Bayerisches Gewerbemuseum in Nürnberg', Kemptner Zeitung, 16 September 1869, 907.

[111] Anon., 'Jahresbericht pro 1859', Zeitschrift des LandwirthschaftlichenVereins in Bayern, 50 (1860), 9–36, 12. In 1853, the society had 13,140 members, ranging from government officials, priests and schoolteachers, to an increasing number of farmers and agricultural experts. Regensburger Zeitung, 323, 23 November 1853, 1123.

perused the collection.[112] While such visits bestowed an august impression of their museum, the brothers also portrayed it as a site for international scientific collaboration. On the occasion of a scientific gathering of the 'Deutsche Naturforscher-Versammlung' in Giessen in 1864, they received a number of scientists from other German states, but also England and Holland.[113] The Schlagintweits also opened their doors to oriental scholars from all over Europe, and even India itself, who came to study the Schlagintweit collections and manuscripts and engage in prolonged scientific conversations.

Yet, it was the youngest Schlagintweit brother, Emil, who arguably made the most thorough use of their ethnographic and linguistic collections.[114] Emil studied law, but his final grades in 1859 did not allow him immediate entrance into state service. Encouraged by Alexander von Humboldt to learn eastern languages, Emil therefore started to help analyse his brothers' eastern artefacts. Tibetology as a university-anchored discipline was then still in its infancy, its first chair having only been established in Paris in 1842.[115] What was more, collections from the barely accessible country of Tibet were exceedingly sparse in Europe at the time. It was therefore fortunate for Emil that the Schlagintweit expedition had provided a wider framework to collect larger amounts of Tibetan ritual, religious and artistic objects, together with dozens of precious texts – both printed and handwritten, and often acquired with the help of intermediaries. The brothers ultimately listed 207 works from Tibet, among them Buddhist sutra and sagas, manuals on burnt offerings and a set of Arabic manuscripts.[116]

Emil used these rare materials to turn himself into an internationally acclaimed Tibetan scholar and expert on Buddhism in both its ancient and modern forms. From 1860 onwards, he learned Tibetan with the help of the St Petersburg-based philologist Franz Anton Schiefner. In the following years, his analysis of the materials stored at the Jägersburg led him to write his own book under the working title 'Objects of Buddhist Worship to Illustrate the Buddhism of Tibet'.[117] By 1863 Emil's interests had considerably widened, so that his oeuvre appeared as *Buddhism in*

[112] 'Vaterländisches', *Ingolstädter Tagblatt*, 129, 31 May 1864, 513; 'Lokal- und Provinzial-Chronik', *Regensburger Anzeiger*, 148, 31 May 1864.

[113] E. Schlagintweit, 'Besuch', 1115.

[114] Emil's career as a significant orientalist scholar still calls for a wider treatment; an important introduction is Neuhaus, 'Tibet-Forschung'.

[115] Held by Philippe-Édouard Foucaux: Neuhaus, 'Tibet-Forschung', 220.

[116] See their enumeration at BSB, SLGA, VI.1.23. These manuscripts were purchased in 1885 by the Bodleian Library, Oxford; Driver, *Descriptive Catalogue*.

[117] Hermann to the editor Cotta, 10 May 1862, DLAM, Cotta private papers, Schlagintweit letters.

Tibet.[118] To describe the historical development of Buddhism in that country, Emil made use not only of the existing Tibetan scholarship in Europe, but also offered many new insights gained from the analysis of the Schlagintweit artefacts, especially from several Tibetan inscriptions, statues and religious treatises. He combined these materials with oral evidence given to him by the two surviving brothers – a rare case of a famous sedentary oriental scholar profiting directly from the observations and collections of itinerant naturalists, collectors and ethnographers coming from the same family and generation.

The Asian expedition of his older siblings thus became a turning point for Emil Schlagintweit's life as well. In seeking to turn his rare expertise into salaried employment, he applied, in 1864 and again in 1866, for a university chair in 'Sanskrit and Oriental Languages' at Würzburg, yet twice unsuccessfully.[119] Despite this failure, and the many official duties Emil later assumed in the Bavarian administrative service, he remained deeply interested in south and high Asian societies, religions, military events, trade and politics. Besides learned treatises, Emil later also compiled a more popular work, *India in Words and Images*, which again drew on the material objects and firsthand experiences of his brothers.[120] Emil was the only Schlagintweit brother to become a member of the respectable Deutsche Morgenländische Gesellschaft founded in 1845, in whose organs he published. He was a prolific and significant oriental scholar, but clearly one with more than just biblical or philological interests.[121]

While the Schlagintweit collections were used as raw materials for different disciplinary engagements, science and commerce clearly colluded in them. Especially after the removal to the Jägersburg, Hermann and Robert Schlagintweit continuously explored commercial avenues for their collections. Their castle functioned, in a sense, as a display and distribution centre, where reproductions of many of the artefacts could be purchased. Amongst the objects for sale were the 275 ethnographic heads that were replicated in plaster or, more exclusively, in copper: 'the original moulds have been reproduced by making strong metallic casts of zinc the basis, and coating them with a galvanoplastic deposit of copper varied in tint according to the different degrees of colour of the native Tribes'.[122] The commodification of their collections also included replicas

---

[118] E. Schlagintweit, *Buddhism* (1863); translated as E. Schlagintweit, *Bouddhisme* (1881).

[119] See for Emil's multiple efforts, ARS-Akte 1589, Universitätsarchiv Würzburg [UAW].

[120] E. Schlagintweit, *Indien*; he later also compiled a description of the life of Padmasambhava, the legendary founder of Buddhism in Tibet, *Lebensbeschreibung*. Another significant work is a printed lecture, *Gottesurtheile*.

[121] None of the Schlagintweits is considered in Marchand, *German Orientalism*.

[122] 'Monthly General Meeting', *JASB*, 28 (1859), 266f.

of religious masks, 'facsimiles' of playing cards, prints from wood-blocks (all collected in Tibet), twenty different human skeletons and skulls in 'papier mâché', as well as curious objects such as the 'brain of [an] elephant skull in plaster'.[123] Their collections provided a source of income for the brothers, even if their financial ambitions were only rarely realised. To promote the sales, the brothers placed numerous advertisements in newspapers and magazines. Through their publisher, Johann Ambrosius Barth at Leipzig, they further printed catalogues, introducing the different artefacts. One of these catalogues, a *Prospectus* on their series of 'heads', went through several editions and provided ethnographic background information on every plaster cast and the sitter.[124] Such commercial prospects were also printed for the paper samples, the woven fabrics, the brothers' zoological and botanical collections, as well as the human body parts they had acquired (Figure 7.4).[125] The Schlagintweits' entrepreneurship undoubtedly 'thrived in an age in which the worldwide quest for rare and beautiful plants [and ethnographic objects] was burgeoning and where science and capitalism constantly colluded'.[126]

Sales catalogues were only one way to attract potential consumers, as the Schlagintweits also used their scientific network to connect with potential buyers internationally. Through personal communications with Italian, French, American and Anglo-Indian museum directors and private collectors, they marketed their objects abroad. The attraction of their collections was enhanced because buyers were convinced of the originality of the objects bought by them from the brothers themselves. This source seemed more trustworthy than the judgement of dealers and intermediaries in an increasingly crowded market for ethnographic objects and natural history specimens.[127] The American scientist Theodore Lyman, for instance, introduced the Schlagintweit collections to a potential buyer, the Smithsonian in Washington, thus: 'the fact that [the specimens] were collected by the Schlagintweits in person gives them a full guarantee'.[128] Even in the case of reproductions, it was generally claimed that the brothers had guaranteed that those could still be

---

[123] Letter from HS, 7 November 1862, Smithsonian Institution, *Annual Report* (1863), 84–5; also, 'Zehnter Jahresbericht der naturhistorischen Gesellschaft zu Hannover', *Bonplandia*, 21 (1860), 333–41, 340.

[124] The second edition of the *Prospectus of Messrs. de Schlagintweits' Collection of Ethnographical Heads* was printed in November 1859 (Leipzig), BSB, SLGA, VI.5.6.1.

[125] H. and R. Schlagintweit, *Naturgeschichtlicher Catalog*.

[126] Arnold, 'Plant Capitalism', 918.

[127] See the authoritative study by Penny, *Objects of Culture*.

[128] Theodore Lyman to the Smithsonian Institution, Paris, 24 November 1862, in Smithsonian Institution, *Annual Report* (1863), 84.

Allgemeiner

naturgeschichtlicher Catalog

der

auf Schloss Jägersburg bei Forchheim in Franken

aufgestellten

v. Schlagintweit'schen Sammlungen

aus Indien und Hochasien.

Preise sind in Thalern Pr. Ct. berechnet; Bestellungen, nach Jägersburg Station Forchheim, Bayern, zu senden, werden unfrankirt expedirt, unter billigster Berechnung der Kisten und Emballage.

In der Schreibweise der indischen Namen ist als abweichend vom Deutschen nur zu erwähnen : ch = tsch; j = dsch; sh = sah; s = weiches s.

I. Anthropologische Gegenstände.

1. Menschenskelette, aufgestellt.

(Zahl zur Rechten der Racennamen bedeutet Anzahl der Individuen; Zahl in Klammern bedeutet Kind darunter; Zahl in Klammern mit * bedeutet Frau darunter.)

Preis per Skelett : 50 bis 60 Thaler, je nach Beschaffenheit und Seltenheit.

Kayath-Hindu	1	Singhalese, Ceylon	2	Kashmir-Mussalman	1	Bhot, Ladak	1
Gorkha-Hindu	2	Gond, Indien	1	(in Sack verpackt.)		Lepcha, Sikkim	2
Sikh, Panjab	3	Naga	1	Bhot, Bhutan	3	Turk, Yarkand	1
				Bhot, Sikkim	2		

Ueberdies verschiedene Skelettheile von Hindu-Leichen aus dem Hugli-Flusse, nicht vollständig aufgestellt.

2. Menschenschädel.

Preis : 15 bis 20 Thaler per Schädel, je nach Seltenheit.

Brahman, Indien	3 (1)	Thakur-Hindu	4	Bhil	2	Kashmir-Mussalman	2
Brahman, Nepal	1	Bais (Vaisia)	3	Khol	1	Bhot, Bhutan	2
Rajput	4	Kahar	1	Daphla	1	Bhot, Central-Tibet	2
Gorkha	1	Assamese	1	Garro	1	Bhot, Ladak	1
Nepal-Rajput	1	Sudra	4	Khassia	2	Lepcha, Sikkim	2
Gurung-Hindu	1	Singhalese	4 (2)	Mussalman, Indien	2	Tibet, Mussal.	3 (1*)
Newari-Hindu	1	Gond	1	Pathan-Mussalman	2	Sandwich Isl.	2 (1*)

3. Ethnographische Racentypen.

(Facsimiles in Metall und Gyps, nach hohlen Gypsmasken über Lebende; vordere Hälfte des Kopfes. Detaillirter Catalog geliefert, wenn gewünscht.)

Preis der ganzen Sammlung von 275 Individuen :

a) in Metall = 2000 Thaler; in Gruppen von wenigstens 25 Individuen = 8 Thaler das Stück.
b) in Gyps (je nach Race verschieden getönt) = 348½ Thaler; von 100 Individuen = 133⅓ Thaler.

Brahmans	14 (2*)	Sudras	12 (4*)	Tibeter : Buddhisten	32 (9*)
Rajputs, reine Race	5	Vereinzelte Hindustämme	6	Tibeter : Mussalmans	24
Himalaya-Rajputs	74 (20*)	Aboriginer	26	Turks aus Central-Asien	10
Bais (Vaisias)	5 (1*)	Mussalmans, Ind. u. Hochas.	54 (2*)	Fremde Racen in Indien	13

Figure 7.4 Hermann and Robert Schlagintweit, 'Allgemeiner naturgeschichtlicher Catalog …', Gießen, 1864; under the heading 'Anthropological Artefacts', potential buyers found lists of objects with prices of 'Human Skeletons', 'Human Skulls', and 'Ethnographic Racial Types'.
© BSB, SLGA, VI.5.1.3.

considered truthful representations.[129] The ethnographic heads proved to be a great success in this regard, with the whole series being exported to London, Paris, St Petersburg and other imperial museums in Calcutta and Madras.[130]

Apart from their commercial value, the Schlagintweits' ethnographic heads are significant in another, more fundamental regard, in relation to the history of science and the European appropriation of the world in the nineteenth century. As established in earlier chapters, different members of the Schlagintweits' expedition were simultaneously indispensable partners *and* themselves objects of research. There thus exists an unresolvable tension between the plaster casts' depiction of individual human characteristics and the purpose of wider ethnographic representation in the masks of Nain Singh and other former guides and translators, who agreed to have their heads plastered. That is, their facial casts also eventually became part of the brothers' series of 275 ethnographic heads. In their intended entirety, these heads appear less like an ensemble of differentiated individual portraits. Rather, they were intended to represent the widest possible spectrum of anthropological variety in south and central Asian 'races' and 'tribes'.[131]

To better understand this tension, it is extremely revealing to take into consideration past exhibition practices with the Schlagintweit heads. Some ethnographic collections purchased only a few specimens. Further, a number of anthropological and geographical exhibitions and congresses in the 1860s and 1870s presented only a selection of masks. That allowed more intimate encounters with the heads, as during the Second International Congress of Geographers in the Palais des Tuileries in Paris in 1875, where five masks were put on a table positioned at the centre of the 'German section' (Figure 7.5).[132]

---

[129] 'Monthly General Meeting', *JASB*, 28 (1859), 266–7.

[130] Other museums purchased at least parts of the series, including museums in Bombay and Sager (India), Milan and a few institutions in the United States, *Allgemeine Zeitung* (Augsburg), 1 April 1866, 1484; and E. Schlagintweit, 'Besuch', 1115.

[131] Itinerant anthropologists of the early twentieth century already experienced and reflected this tension; see Sysling, 'Racial Science', who argues that 'the anthropologists hoped to convey the idea of a [specific] "race" to their European audience. The casts' detailed beauty and individuality, however, make them so flexible they can also function to counter the typecasting activities of anthropologists', ibid., 128. The question of individuality and ethnographic representation is also touched on in Feldman, 'Contact Points', 248; and Driver, 'Face to Face'.

[132] After the defeat in 1870–1, the Congress assumed a higher significance for French officials and was intended as a stage to show off France's still 'high scientific rank' among its rival nations. Letter from Camille Clément de La Roncière-Le Noury, president of the Parisian Société de Géographie, to the French minister of agriculture and trade, 8 April 1875; Archives Nationales [AN], Paris, F/12/4980, Expositions universelles, internationales et nationales (1844–1921); Répertoire méthodique provisoire, folder 'Exposition des sciences geógraphiques au palais des Tuiliers, en 1875'.

Figure 7.5 Selected view of the 'German section' with several Schlagintweit objects being centrally displayed at the exhibition accompanying the Second International Geographic Congress in the Palais des Tuileries, Paris, 1875. Photographer: Alexandre Quinet.
© BnF, Société de Géographie W2/37.

However, some of the most important museums in Europe and India bought and exhibited (almost) the entire Schlagintweit series. At the South Kensington Museum, one visitor experienced, with a morbid sense of humour, the full ambivalence of the hundreds of 'Indian physiognomies': 'Some of these faces are very curious. Sometimes the expression of disgust at the feeling of the plaster mask on the living patient's face is most amusing. Sometimes a broad smile is rendered. Sometimes the sunken features betray the fact that the cast has been made from a dead subject.'[133] In addition, the Muséum national d'Histoire naturelle in Paris purchased the entire set and displayed roughly two-thirds of the heads in a vast exhibition hall, together with other natural history and anthropological artefacts, including whole animal and human skeletons. The wall of Schlagintweit masks, symmetrically hung without clear individual frames, emphasises this purpose of the casts: their collective character as physically measurable and comparable objects of research (Figure 7.6).[134]

For the Schlagintweits themselves, the casts functioned as ethnographic visual material, in some sense similar to the earlier series of three-dimensional profiles the brothers had produced of Alpine mountain formations in the 1850s. While those had captured the morphology of a specific landscape, the plaster casts of human faces now precisely documented individual physiognomic traits, but they also conveyed the sitter's identifiable momentary mood: pain, stress or fear. A set of 149 anthropometric drawings gives further clues to how the ethnographic heads were used. The Schlagintweits made those drawings after the casts. With the help of symmetrical lines, any differences from an 'ideal type' of physical form were made visible, especially by marking any deviation or irregular shape with a dotted line (Figure 7.7).

The transmission of the proportions of three-dimensional heads into two-dimensional drawings of front and profile views seems to have made it possible to take further anthropometric measurements. However, pencilled notes directly on the drawings – in the case of Nain Singh 'nose more to the left', for Mohammad Amin 'corners of the mouth higher' – indicate how difficult it was to achieve the exact equivalent of the masks' proportions on paper. At the same time, the Schlagintweit drawings confirm an important function of the masks as scientific (raw) material for further physical-anthropological research. In sum, the brothers produced photographic portraits of different

[133] 'The India Museum', *Saturday Review*, 48, 18 October 1879, 476.
[134] Daugeron and Le Goff, *Collections scientifiques*, 60f.

Figure 7.6 The brothers' collection of human faces is placed on the back wall, underneath the caption 'Collection Schlagintweit' and above showcases of human skulls and entire skeletons. Section of a photograph by Auguste Dollot, 'Galerie de Paléontologie, Eléphant de Durfort', showing the 'Galeries d'anatomie comparée et de paléontologie', taken 23 September 1908. © MNHN, IC 4514.

individuals, took anthropometrical measurements of the bodies and made plaster casts of their faces, hands and feet (Figure 7.8), only to later reproduce those by drawing them for anthropometrical purposes. It thus becomes clear how the bodies of individuals they had encountered in the flesh in Asia were recreated through multiple media, with the Schlagintweits' different anthropological techniques being closely intertwined.

And yet, for the Schlagintweits the masks always played a more ambivalent role. The two surviving brothers spent a couple of years at Castle Jägersburg, which was most intimately decorated with Indian silks, paintings, Buddha statues and other personal memorabilia, including the entire series of plastered faces. In this semi-private, semi-public space, the whole ambivalence of the ethnographical collections became visible: On the one hand the heads, as 'portraits' of individuals actually encountered, served their personal memory of some of the indigenous

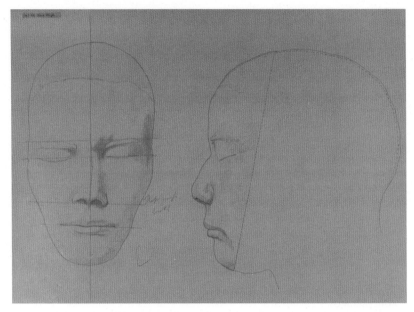

Figure 7.7 Sketch after the facial cast of Nain Singh, front and profile view. The numbers of the sketches correspond to those given in their formal list of 275 ethnographic heads.
© BSB, SLGA, IV.2a.75.

partners in south and central Asia to whom the brothers had become deeply attached, including Murad, Mohammad Amin and Nain Singh. While Hermann and Robert would never see their travel companions again, they tenderly commemorated these former partners also in writing. On the other hand, the heads were intended as working material for the planned eighth volume of the *Results* on Indian 'Ethnography', which was to include several prints of the plaster casts.

The Schlagintweits sought to capitalise on any positive remarks about their collections that appeared in private letters or newspaper articles in their quest for social and scientific recognition. The importance they attached to their international press coverage materialised in a remarkable document: the 'Collectanea Critica'. In this, Hermann Schlagintweit assembled a vast amount of printed articles and a few letters in several languages about their travels and artefacts. Whereas some of the negative voices raised against their enterprise – especially from Britain – were preserved, the Schlagintweits included, above all, positive statements – and especially about their collectables. Such praise from the press was

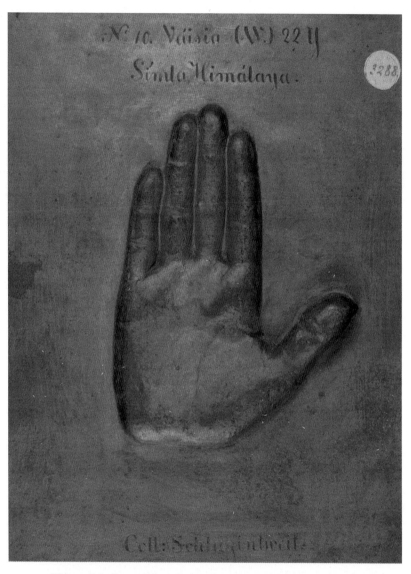

Figure 7.8 Cast of female hand, originally in plaster, reproduced in bronze. It was taken from a twenty-two-year-old woman named Váisia, from Simla in the Himalayas, whose face and foot were both modelled in plaster. Beneath the engraving, 'Coll: Schlagintweit'. No. 10 from a series of thirty plaster hands for anatomical studies.
© MNHN, Paris, HA-3288.

then quoted in their printed catalogues to enhance the perceived value of the Schlagintweit duplicates, replications and originals for sale.[135] While the artefacts were thus used as a source of income, they also served the Schlagintweits as private objects of memory and as tools to furnish an external image of themselves as 'great explorers'. That is, all visitors to Monbijou and later the Jägersburg were confronted with the aura of authority that the brothers generated through the display of their scientific trophies.

However, the Schlagintweit case also throws into sharp relief how difficult it was to transform the material artefacts of expeditionary science into permanent and unified museological collections. After a decade at the Jägersburg, the brothers again ran into financial strains, not least because the analysis of their 40,000 objects consumed much of their British and Prussian salaries. While plans to sell the Jägersburg castle did not materialise, Hermann started to look for other localities. An application to found an India museum in Munich failed. The brothers were merely provided with space in the Neue Pinakothek for their 'watercolours and sketches, photographs, maps', and their library of Asiatic scholarship and collected manuscripts, especially on Tibet.[136] The next station of the Schlagintweit collections became Nuremberg, the kingdom's second largest city. Hermann first approached its mayor in February 1876 with the suggestion that 'a collection from the discipline of ethnography in particular' would fill a gap in the already assembled 'series of works in the arts, science and technology' of the city.[137] Indeed, as he continued, 'this new material would not only have a positive effect on the many numerous foreign visitors to Nuremberg', but the Schlagintweit collections would 'also facilitate production for the Orient' by the diverse local branches of craftsmen and 'manufacturing plants'. Finally, their artefacts could also help to improve the city's 'already admirable technical education' (*technisches Schulwesen*).

In the end, the brothers were again to profit from royal patronage: In 1877 the Bavarian king, Ludwig II, offered them the use of facilities in Nuremberg Castle (the 'Bilder-Saal' and the 'Himmelsstallung') to

[135] The brothers, for instance, quoted a laudatory passage by Joseph Barnard Davis, a famous British anthropometrist and racial scientist, from his private letter in their published *Prospectus of Messrs. de Schlagintweits' Collection of Ethnographical Heads*, iv.

[136] HS and RS to the king, 4 February 1861, BayHStA, 'Findbuch über Akten der zoologischen Staatssammlung', VN 67, Abschrift 1182. I thank S. Kleidt for the reference. H. Schlagintweit, 'Bericht über die ethnographischen Gegenstände', 337.

[137] HS Mayor Karl Otto Freiherr von Reichenbach, 24 February 1876, Nürnberger Stadtarchiv [NSA], C7/I, GR, no. 9976. Subsequent quotations, ibid.

display their collections at a more convenient and centrally located site. However, the large hall of the castle remained hung with another art collection, including many Italian masterpieces. This led to a strange mixture of European oil paintings surrounded by Indian curiosities, including 'a great number of minerals, plants and seeds, types of wood, skeletons, stuffed quadrupeds and birds, and also reptiles preserved in spirits'.[138] As was critically noted in the press, 'the value of this collection would be much increased through a systematic arrangement, which had only begun', while 'object labels and explanatory texts' would further improve visitors' understanding of 'Indian craftsmanship'.[139]

Yet, even this third location for the Schlagintweits' India museum proved only temporary. In 1882, the year of Hermann's death, all collections had to be packed up again, since Prince Luitpold needed the facilities to stage the first Bavarian National Exhibition. The repeated relocation of the collections, and the brothers' futile attempts to preserve them in their own commemorative museum, increased their willingness to sell off different parts. Emil became the heir of all the Schlagintweit's artefacts at the time of Robert's death in 1885. Once the last unified display of the different branches of their collections came to an end, he spent the remainder of his life selling off various portions to private collectors, royal benefactors, scientific societies and state museums.[140]

To be sure, such a dispersal of scientific specimens is by no means disadvantageous, since it increases the likelihood of different naturalists and ethnographers engaging with the material legacies of an expedition. However, the brothers' personal aim to secure the memory of their travels in a unified collection, permanently displayed at a specific institution, never materialised. Here, as elsewhere, there thus is an element of failure. As with their only fleeting India museums, the brothers also failed to complete most of their publication projects, and could secure the publication of only a tiny fraction of their striking visual archive of exploration. It is thus perhaps as much through what the brothers did achieve as through their many disappointments, that we learn much about the occasionally unfulfilled promises of *empires of opportunity*.

---

[138] There were also: 'Leather- and glassworks, ceramics and works in stone, instruments, vessels and metal weapons. Also woodcarvings and miniature pieces made from semi-precious stones and ivory can be found here in significant quantity', 'Schlagintweit Sakünlinski'sche Sammlung', *Kunst und Gewerbe. Wochenschrift zur Förderung deutscher Kunst-Industrie*, 12 (1878), 133.

[139] Ibid.

[140] A detailed overview is given by Kleidt, 'Sammlungen', 129–37.

The Social Fabric of Science: Collecting Honours and Rejections

This final section will show that their collections ultimately played an important role in shaping the brothers' social advancement – especially at a time of much private and public vilification. Following the press defamations in England between 1857 and 1859, the brothers' reputation in the British Isles was on a downward spiral. Even the very positive reports about the display of their collection in London did little to alter their status within the scientific community. This growing estrangement of British men of science from the Munich-born travellers looms large in Joseph Hooker's correspondence. Here, a language of exclusion was ever more openly employed to stress the Schlagintweits' lack of respectability. Expressing ever-greater doubts over their integrity, Hooker wrote to the RGS president, Murchison, in 1859:

I do not say that the S[chlagintweit]'s have no results, or that their Geograph[ical] discoveries are not meritorious & useful; but this is not all we have to deal with; look at the cost, at the withholding of their Scientific results from our Societies, to the sending of all their materials to Prussia ... & I ask you whether such conduct would be tolerated for a moment in an Englishman.[141]

Whereas Hooker, at this point, grounds his criticism of the Schlagintweits in the favouring of German scientific institutions with their Asian collections and results, he continued his polemic by arguing that their proposed publication would effectively betray the works of British travellers: 'What have our public Scientific Establishments got for this Expenditure of £18000? &, what is worse, what have they not lost by the withdrawal of Every Shilling & Every Sympathy on the part of the Indian' government from British travellers, 'whose results are Either destroyed by neglect, or are now to be placed at the disposal of these men, who have not the feelings of gentlemen in such matters, & far less ability'.[142] In consequence of the brothers' purported failure to acknowledge their British predecessors, the Kew botanist concluded that their expedition should be regarded as nothing less than 'the grossest job that was ever perpetrated in this country under the name of Science'.[143] To secure British funding for British subjects, even well-respected and widely travelled naturalists such as Joseph Hooker employed an exclusionary

---

[141] JH to Murchison, 19 July 1859, RBG, DC 96, 406.
[142] Ibid. Hooker's critique was shared by other British travellers, see e.g. Torrens, *Travels*, 212.
[143] JH to Murchison, 19 July 1859.

rhetoric that stressed categories of national belonging and assumed supposedly rigid cultural differences.

This exclusion was assured not only through rhetoric, as the Schlagintweits were increasingly treated in Britain as social outcasts. The fact that they were unwelcome in the meeting places and institutions of science in London became obvious by 1859, when the brothers made an appearance at the prestigious Athenaeum Club, 'the resort of almost everyone of note in the literary, scientific, and artistic world'.[144] After a controversial invitation to this exclusive establishment, it was again Hooker who voiced severe disapproval:

> With regard to the Athenaeum ... An Englishman is blackballed if known to be personally disagreeable only to members of his own profession, & doubly so if anything having even a taint of dishonour & ticks to his name. Had the S[chlagintweit']s been Englishmen they would have been blackballed a hundred times over.[145]

These exclusionary practices testify that the Schlagintweits' authority was challenged not least for social reasons, as their behaviour provoked criticism that was clearly linked to issues of class and to the presumed lack of an appropriate, gentlemanly conduct.[146] The criticism was further fuelled by the portrayal of their travels and publication schemes as driven by greed for fame and profit – hence the antithesis of respectable, disinterested science. To the extent that the two dimensions were inextricably linked in Victorian scientific circles, the social exclusion of the Schlagintweits thus also implied the symbolic denial of a reputation as honourable gentlemen scientists by their British peers.

While the Schlagintweits were, at least to a degree, treated as imperial outsiders even prior to their departure (following the early critique of this foreign recruitment), this was even more the case *after* their return.[147] Their fall from grace in Britain should not, however, be considered as the inevitable outcome of transnational scientific collaborations, as the successful assimilation of other German scientists and administrators into the high establishment of the British Empire showed at the time.[148] It was the Schlagintweits' bold conduct that impacted considerably on their chequered careers.

---

[144] Godsall, *Burton*, 217.
[145] JH to Murchison, 19 July 1859.
[146] Even their supporter Roderick Murchison dismissed their 'obtrusive manners' in a letter to Hooker, 20 July 1859, RBG, DC 96, 409.
[147] See for early opposition, *The Athenaeum*, 1378, 25 March 1854, 376; Torrens, '*Travels*', 211–12.
[148] Weigl, 'Acclimatization'.

In the midst of conflicts with the British press and Prussian officials, the Schlagintweits came to believe that one important way their respectability could ultimately be restored was through royal recognition – and at best through a title of nobility, as if the latter would automatically grant them with an air of dignity. After their requests for further Prussian medals had failed over the thwarted museum plans, the Schlagintweits now went full circle and turned again to the Bavarian monarch in search for support and glory.[149] To pave the way, the brothers first gifted '80 seed samples and 38 ethnographic objects' to King Maximilian II in February 1859, which soon found their way into Munich's Ethnological Museum.[150] A few months later, the brothers informed the monarch that his sympathy for their cause would render it now a 'special duty during our upcoming visit to England to try to get permission' to enrich 'Bavaria's collections' with as many Schlagintweit artefacts as possible.[151]

The travellers continued their royal submission by requesting 'a favour, which perhaps can only partly be excused on the ground that its granting would ... be of the highest importance for our official relations with England'.[152] They maintained that 'nothing else could be more supportive' to their future plans than if the king would bestow on them the title of nobility from 'our fatherland Bavaria'. This, it was argued, would decisively improve their negotiating position with their British employers. Yet, the granting of such a title depended on the existence of a considerable private fortune. The brothers therefore provided a list of their current possessions. These included '21000 fl.' gained *during* the Indian expedition, hence pocketed from British and Prussian funds, which were supplemented by inheritances from their deceased parents, including Adolph's share. Altogether, the brothers declared to possess 'total assets ... of 60000 fl.'.[153] All this money, they said, was deposited at a 'Prussian bank, predominantly in the form of Prussian bonds [*Wertpapiere*]'. It is noteworthy in view of their supposed incapacity to repay the Prussian loans that they were sitting on such fortunes, soon used to purchase the Jägersburg. King Maximilian II sanctioned their unusual personal application for hereditary ennoblement on 28 August 1859.[154]

---

[149] In 1854, Frederick William IV had already bestowed on Hermann and Adolph the Eagle of the Red Cross, fourth class.
[150] See Hermann and Robert's submission to the king, Munich, 1 June 1859, BayHStA, Adelsmatr. Adelige, S 156, doc. 1.
[151] Ibid.
[152] Ibid.
[153] Ibid.
[154] 'Ministerialrath Dr. Rappel' to the king, 7 July 1859, BayHStA, S 156, doc 3.

Soon after, the brothers started to use their new title as a tool for negotiating higher salaries in England. In a letter to the India Office, Hermann stated: 'We allow ourselves to communicate to you that recently the King of Bavaria, whose subjects we are, has conferred upon us the title of nobility ... Perhaps we may be allowed to profit of this occasion for requesting again ... the adjustment of preliminary expenses and increase of pay.'[155] In their request for nobility, several layers of the Schlagintweit controversy were bundled together, including the brothers' deep-rooted quest for social recognition, their secretive communication strategies to make multiple sets of promises to different patrons and their seemingly insatiable appetite to consume even more British means – despite the ongoing debate in Britain over their financial debauchery. While different sums had been leaked to the British press, it was soon openly speculated that their schemes had already cost up to £40,000.[156]

Ever the makers of their own fortunes, the Schlagintweits did not leave the design of the noble family crest to the herald painter. They already possessed a crest (Figure 7.9), but now insisted on the incorporation of references to their Asiatic expedition that helped to lay claims to scientific achievements. The brothers therefore personally provided specific symbols from their journey to the painter.

While the old heraldic figures of the sword and the feathered arrow (which represent the name Schlagintweit) were retained, new elements were added.[157] The modified shield, now decorated with an upright sword flamant and a more impressive arrow, possessed a specific bordure as the brothers' mark of difference. The inspiration for the design was taken from 'the seal of the Mangnang monastery in Tibet', a highly exclusive place visited by Adolph Schlagintweit in 1855. Even 'the typical [Tibetan] colours for ornamental borders, red and silver, were kept'.[158] The crest furthermore featured two large Bengal tigers, in yellow and black (Figure 7.10).

Lastly, the new family motto, 'Deo duce ferro et penna' was described by the crest painter as containing 'a hint of the said travel, undertaken under the protection of God, with sword and feather in the name of Science'.[159] The Schlagintweits' 'double conquest' of having opened up unfamiliar territories through their learning and scientific approach,

---

[155] Hermann to India Office, Berlin, Schloss Monbijou, 29 September 1859, BSB, SLGA, IV.6.1. See dozens of similar requests to East India House, later the India Office, ibid., and BL, L/Mil/2/1477, e.g., Schlagintweits to Melvill, 1 October 1859.

[156] Or roughly equivalent to £4,500,000 today. *The Athenaeum*, 1768, 14 September 1861, 342.

[157] The new crest is described at length in BayHStA, S 156, doc. 15.

[158] Ibid.

[159] Ibid.

Figure 7.9 Old Schlagintweit family crest.
© DAV, NAS 12 SG, 8.4–8.7.

symbolised by the feather, only made possible by their assertive penetration, characterised by the sword, was powerfully expressed in their modified coat of arms.[160] The crest forcefully captured the way the brothers

---

[160] While the brothers were indeed armed during the expedition, they relied less on swords than on 'double and single barrelled guns', together with '1500 caps, 40 lbs of Powder' provided to them by the Military Department, Government of India, BL, IOR, L/Mil/3/ 587, coll. 26, from letter no. 30, 2 February 1856; Hermann to secretary to the Military Department, 2 December 1855; also NAI, Military Department, Quarter Master General Branch, Consultation, 28 May 1856, ref. 148.

Figure 7.10 Schlagintweit noble coat of arms.
© BayHStA, Adelsmatrikel Adel S 156.

sought to portray themselves, as courageous scientific explorers, 'pushing back the frontier of ignorance and resistance' for the sake of the advancement of science into the dangerous unknown.[161]

Yet, there existed a striking contrast between the private motives versus the public appearance of the Schlagintweits' road to ennoblement. The brothers forged a false image of their achievement of noble rank in the printed *Prospectus* (1860) for the *Results*. While the brothers had approached the king with strong promises, which in turn encouraged Maximilian II to provide the honour, their published account gave a noticeably different impression, with the monarch as the prime mover of the scheme: 'The attention of his Majesty the King of Bavaria, was, from his natural predilection for science, soon attracted by the great success of these distinguished travellers, who are his subjects; and he has accordingly been pleased to confer upon them the titles of hereditary nobility.'[162] The Schlagintweits thus strove to reassert their authority by depicting their ennoblement as supposedly objective proof of their achievements.

Their *Prospectus* also tried to capitalise on Adolph's death to create a sense of anticipation for their projected oeuvre among the German reading public. Indeed, the document appealed directly to their patriotic feelings towards the brothers and the sacrifices they had brought to the altar of science:

The public, during their absence, manifested a lively interest in the reports which, from time to time, reached Europe of the successful progress of the mission, but this interest deepened into a painful feeling of universal regret and sympathy, when the long doubtful fact was established, that one of these enterprising travellers, Adolph, had fallen a victim to his zeal by the hands of barbarous tribes in Turkistan (Central Asia).[163]

In the mid nineteenth century, the false dichotomy of European civilisation versus extra-European barbarism clearly had a strong place in the public imagination.

Soon, German, French, and British papers reported on the brothers' ennoblement. Indeed, it sometimes seemed as if this proved to be their redemption. The London *Art Journal*, for instance, informed its readers that: 'We learn that the King of Bavaria has conferred titles of nobility on the survivors, as a mark of his appreciation of the services rendered by them to the science of ethnology.'[164]

---

[161] See on this powerful notion Driver, 'Missionary Travels', 166.
[162] H., A. and R. Schlagintweit, *Prospectus: Results of a Scientific Mission*, 4.
[163] Ibid., 3.
[164] *Art Journal*, 21, 1 November 1859, 350.

News of the Schlagintweits' ennoblement was complemented by reports that they had received other scientific medals and honours. In 1859, the brothers received the gold medal of the Parisian Geographical Society for their explorations in high Asia, especially of Tibet. This was one of the highest possible distinctions – the British imperial hero David Livingstone had enjoyed the same prestige only two years earlier. It may not, however, be surprising that the brothers had lobbied for this distinction with French geographers, and had undertaken an extensive correspondence with de la Roquette, an influential member of the society since their departure in 1854. They even had laudatory German newspaper articles about their achievements specially translated into French and annotated for Roquette, thus helping to pave the way for this important award.[165]

Their active pursuit of numerous royal medals and distinctions from German, European and ultimately non-European benefactors, in which the gifting of Asiatic objects repeatedly played a crucial role, soon led to the image in Berlin that the Schlagintweits were indeed 'medal-hunters' – 'ein bisschen Ordensjäger'.[166] British naturalists remained utterly unconvinced, and saw through their carefully orchestrated image campaign. While Murchison pleaded for clemency for the two foreigners, stating that 'I really thought that there was a limit to the dislike of them by a few of my friends', his efforts proved futile.[167] When Murchison alluded to their having been granted the 'Grand Prix de la Société Geógraphique de Paris', Joseph Hooker drily replied:

I am a thorough respecter of rank & honor, but with me they are less than nothing against the opinion of candid men of Science, and I have mingled enough myself in that upper sphere, & seen too much of how Gold medals are got & given, to attach any importance to such things when unaccompanied by results & the unanimous testimony of Scientific men. These ... baubles may serve their turn in science now & again, but they will be forgotten when the scandal of this Schlagintweit affair will be a familiar episode in the history of British Science.[168]

---

[165] See the large volume of letters between the Schlagintweits, Humboldt, Murchison and Roquette; BiSG, Ms. 3611, 2631, esp. 1–4.

[166] Reizenstein to Malortie, 10 October 1859, KHA, Depot 103, 320. See also Kalka, 'Miscellanea'. For a full list of foreign honours, including from South American monarchs, see BayHStA, Ordensakten 9056. For the period 1850s–1871, Robert received thirteen medals, Hermann eleven; more on Robert's strategic medal-hunting in SLGA, V.10; material on Robert's public lectures and 'Korrespondenz zur Erlangung von Orden und Auszeichnungen'.

[167] Murchison to JH, 13 July 1859, RBG, DC 96, 404.

[168] JH to Murchison, 19 July 1859, ibid., 406.

In the history of the Schlagintweit collections, private memory of exploration, commerce, science and the brothers' personal quest for social advancement all colluded in myriad ways. But their story also sharply elucidates the vicissitudes of royal patronage-seeking in the mid nineteenth century and reflects the gradual shift of support and judgement towards more public forms of organisation. After the untimely decline of the enthusiastic Prussian king Frederick William IV, and the subsequent breakdown of the brothers' communication monopoly with their Anglo-German benefactors, their visionary plans for a permanent India museum were ultimately defeated by professional jealousies, emerging models of disciplinary order and ever-more confident administrators eager to shape Prussia's and Bavaria's museological landscape against the Schlagintweits' own institutional ambitions.

8 Asymmetric Reputations: Memories of Exploration and German Colonial Enterprise

This chapter deals with the asymmetric reputations and long-term impact that the Schlagintweit brothers had in Britain, continental Europe and India during the decades after their return. It analyses how the scientific and popular memories of this instance of expeditionary science diverged, and what ideological meanings later became attached to the Anglo-German undertaking by different national publics and administrations. By focusing exclusively on metropolitan polemics and transnational frictions, the existing literature on the Schlagintweit enterprise has so far overlooked the lasting influence that the brothers also had in British India.[1] During the expedition, which had taken the Schlagintweits over a distance of 29,000 km, the brothers and their many employees had carried out scientific observations and measurements with great vigour, filling over forty volumes of field notes. The German travellers had also successfully appropriated a wealth of indigenous knowledge and thus acquired unique insights into the geographical characteristics of India and the trans-Himalayan region, including parts of Tibet, Nepal and Chinese Turkestan. Hermann Schlagintweit's knowledge of these barely accessible territories was such that years after his return to Europe, British colonial officers and government departments still sought his advice on their surveying projects, and asked for the then Munich-based geographer's evaluation of ongoing imperial surveys. Hermann then integrated these results into his publications on central Asia's geographical formations, which were in turn consulted by British servants of the Raj. Travellers and naturalists in India were so eager to receive Hermann's judgement on their own latest theories and findings that they even provided him with materials officially marked 'confidential'.[2] As late as 1880, the German explorer was still assured by an imperial surveyor that: 'The valuable

---

[1] A valuable exception is Lüdecke, 'Untersuchungen'.
[2] Robert Gordon to HS, 5 October 1880, BSB, SLGA, IV.6.1.19, 2.

contribution to a knowledge of Tibet & the neighbouring countries made by your brothers & yourself are often quoted by our highest authorities.'[3]

The Schlagintweit Mission and the Foundation of the Indian Meteorological Department

Through their extensive data gathering, publications and correspondence with administrators on the subcontinent, the Schlagintweits ultimately shaped scientific practices and the history of institution building in India. This applied especially to Hermann Schlagintweit's innovative work in the field of meteorology. As he knew, this branch of science had major implications for colonial agriculture, the establishment of sanatoriums and questions of European settlements and troop stationing in the hills.[4] Hermann provided the first comprehensive summary of weather conditions in both India and high Asia, an ambitious work grounded in measurements taken at 250 different stations.[5]

While the brothers and their assistants had personally collected a vast amount of climatic observations, the empirical basis of their work was significantly enriched by meteorological data provided to them by British India's medical and revenue departments. The latters' own measurements extended further back and forward in time, and were systematically obtained from numerous Indian regions, allowing for comparative insights. Hermann complemented his analysis with a thorough examination of dozens of Indian parliamentary reports and scientific articles that touched upon issues such as the varying patterns of rainfall, radiation and evaporation in different Indian regions, as well as on the conundrum of the shifting positions of the perpetual snow in the Himalayas. His extensive engagement with Indian meteorology, published in almost twenty works between 1863 and 1883, implied a considerable effort since, at the time, hardly any specialist literature existed on meteorology, and no scientific journal was as yet devoted to this discipline.[6]

The Schlagintweits, in some ways, helped to lay the groundwork for modern Indian meteorology, providing results that became a central

---

[3] Ibid. All volumes of the *Results* were then integrated into the Imperial Library; see the stamped originals at the National Library of India, Kolkata.

[4] H. Schlagintweit et al., *Results* IV, *Meteorology of India*, xi; Sikka, 'India Meteorological Department', 387.

[5] Markham, *Memoir*, 207; also Lüdecke, 'Untersuchungen'.

[6] *Symons's Monthly Meteorological Magazine* was first published in 1866. See on the emergence of this branch of science, Anderson, *Meteorology*. Hermann's series of publications started with H. Schlagintweit, 'Elements'.

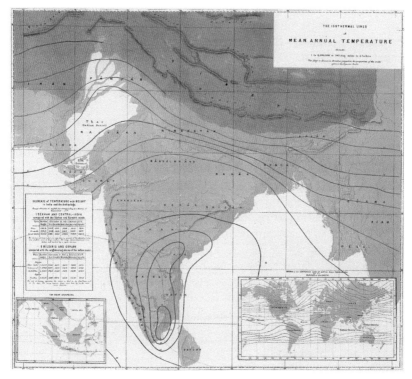

Figure 8.1 Hermann Schlagintweit, 'Illustration of the Meteorology of India'; 'The Isothermal Lines of Mean Annual Temperature'; with a smaller 'Sketch of the Isothermal Lines of Annual Mean Temperature for the Surface of the Earth', and a table on the 'Decrease of Temperature with Height in India and the Archipelago'; Hermann Schlagintweit et al., *Atlas*, Meteorological Maps and Tables 2, extract. © DAV.

reference point for decades to come.[7] Hermann worked meticulously with his data sets to calculate mean values of the temperature for each month and compared the results, as well as the length of the different Indian seasons, with those prevailing in Europe (Figure 8.1). One of his findings was that no Indian region experienced on average such cold temperatures as those in England (Greenwich). He further explained

---

[7] This instance suggests a different finding from Dirk van Laak's assumption that the German 'contribution to the scientific opening up of India was limited to Romantic philology', Laak, *Welt*, 51.

that the daily variation of temperatures differed greatly between the western and eastern Himalayas, and also noticed a climatic anomaly over the Tibetan plateau, which experienced an extremely cold winter and an extremely hot summer.[8] Hermann explained this through the influence of height and the considerable extent of the Himalayan range, thus seeking to formulate a more theoretical model on the climatic features of this exceedingly complex world region.

Always taking British colonial preoccupations into account, Hermann's approach also led him to distinguish specific places in the Himalayas that seemed particularly suitable for the recreation of Europeans, away from the tropical heat prevalent in the Indian plains. In such elevated spots, he considered the temperatures to be agreeably moderate and the air healthy and refreshing. Yet such qualities had to be balanced against the dangers that accompanied foggy periods in various mountain areas, which were believed to pose a real threat to the European bodily constitution.[9] In providing a hierarchy of the most appropriate zones, Hermann judged in particular the large valley of Kashmir as among the best regions on the entire globe for future European settlements (Figure 8.2). This colonial imagination was shared by Robert Schlagintweit, who described Kashmir in numerous public lectures as a

valley, which has been already highly celebrated amongst all educated nations from the oldest times, and Nature has poured out over it an abundance of such charms and gifts in the most lavish manner, as perhaps over scarcely any other landscape in the world. For everything that is delightful to the eye of man, everything that is pleasurable and exciting to his senses, is here found united in the most beautiful harmony.[10]

The Schlagintweits also singled out other south Asian regions for white settlement or troop convalescence, which – given the high mortality rate among European soldiery – greatly concerned imperial authorities at the time.[11] Apart from different lower parts in the Himalayas, Hermann drew attention to the Nilgiri Hills in particular, whose landscapes the brothers alluringly depicted in their *Atlas* (Figure 8.3). Of this range of hills in southern India, he declared: 'The decrease of temperature with

---

[8] H. Schlagintweit et al., *Results* IV, 468–71. Robert Schlagintweit, too, later took an interest in Indian meteorology, BSB, SLGA, V.2.2.2, 49.

[9] Lüdecke, 'Untersuchungen', 214.

[10] BSB, SLGA, V.2.2.1, 64. The two brothers decisively helped to establish Kashmir as a 'worldly paradise' for the German middle classes: 'Ein tibetanischer Tempel', *Über Land und Meer*, 17 November 1866, 140–2. On the role of such alluring descriptions of overseas regions for imperial ventures, see Laak, *Welt*, 31–2.

[11] On military inquiries into the health of European troops in India, and the medical opinion that came to greatly favour highland sanitariums, Kennedy, *Stations*, 25–6.

Figure 8.2 Watercolours by Adolph (above) and Hermann (below) Schlagintweit, October 1856. I. 'The Gardens of Shálimar and the neighbouring Mountains'; II. 'The Fort of Srinagar with the Chain of the Pir Panjál', together forming a 'Panorama of the Lake and the Gardens near Srinagar, Kashmir'; Hermann Schlagintweit et al., *Atlas*, 18; the lithographs were widely reproduced in the press.
© DAV.

height is here more rapid (quite isolated peaks excepted), and therefore more favourable to the erection of sanitaria than any of the mountainous regions to the south of the Himalaya.'[12] What was more, 'not only the higher parts [of the Nilgiri Hills], but also steps of minor elevation, are topographically well adapted to settlement and colonization'.[13] Yet the Schlagintweits not only suggested the colonisation of heights, but also the development of 'marine sanitaria on islands', off the coast of British India, distinguishing between 'eligible' and 'established stations'.[14] Hermann singled out a number of 'places ... where, without the means being cooler, the breezes are more regular and more refreshing and the atmosphere decidedly purer' than on the coast of India. An added

[12] H. Schlagintweit et al., *Results* IV, 366.
[13] Ibid.
[14] H. Schlagintweit, 'Mountain and Marine Sanitaria of Southern India, from Dr. Macpherson's Reports', *Atlas*, 'Meteorological Maps and Tables', 4.

The Kúnda Range, in the Nílgiris, Southern India.

Figure 8.3 'The Kúnda Range, in the Nílgiris, Southern India', with a striking aesthetic resemblance to the European Alps. *Aquarelle* by Adolphe Schlagintweit, March 1856, printed in oil colours by Storch & Kramer, Berlin; Hermann Schlagintweit et al., *Atlas*, 8. © DAV.

advantage of such imperial outposts was that 'malaria, if existing, can much more easily be removed where all around is the open sea'.[15]

That Hermann Schlagintweit should be so deeply concerned with European health in India is hardly surprising. He was not only the son of a medical doctor, but had also started his university studies in medicine. Even before their departure, and later when the brothers negotiated a second contract with the East India Company for transforming their travel observations into published accounts, they had also sought to accommodate the interests of their imperial paymasters in their scientific programme. This applied above all to the influential Colonel William Henry Sykes. As Hermann noted, Sykes's own works had been 'of great importance in directing the attention of Government to the relations connecting meteorology with the various questions of military and civil settlements, and soon after his reception into the Court of Directors the plan for an extensive series of meteorological observations to be distributed all over India, in charge of the medical establishment, was put in execution'.[16] Before their embarkation, Sykes had briefed the brothers to make observations on meteorology and health issues in particular. Later, all the brothers commented to him and other British authorities on their state of health whilst moving through many different climatic zones and environmental extremes.[17] Finally, Hermann's personal fate exemplified the great personal cost of exploration. He remained preoccupied with the fragility of European bodies in Indian climates not least because he contracted malaria in south Asia and suffered from its effects for the rest of his life.

Hermann Schlagintweit's studies of the climatic features of regions both within and beyond British formal rule also led him to propose in the last volume of the *Results* the foundation of an Indian meteorological service. Following the system then established in England, he recommended the establishment of one or two 'Meteorological Offices' in each Indian province, which should be connected to the lines of the Indian telegraph for quick communication, especially important in the case of threats posed by violent storms. Each office (or station) should furthermore be staffed with scientific personnel trained in the use of modern instruments, who should also be able to make calculations and climatic comparisons with the data they obtained.[18] These posts would

[15] Ibid., 163–4.
[16] H. Schlagintweit et al., *Results* IV, 7.
[17] See Schlagintweit letters to Sykes, e.g., from AS, 26 February 1856; and again 25 April 1857, FBG, SPA, 353/2.
[18] H. Schlagintweit et al., *Results* IV, 155.

report back to a 'Central Office', ideally located in Agra or Ambala, hence not too close to the coastline, 'where the marine type of climate predominates too extensively'.[19] This central office was then to pool the regional observations and turn into a veritable centre of scientific accumulation. The German explorer was furthermore convinced that:

Altogether Indian climate is so much more regular than that of Europe, that some twenty to thirty offices would already prove very valuable, if well distributed over the peninsula and gradually including the eastern shores of the Bay of Bengál, some localities from the Islands of the Indian Archipelago to the south-east, Aden, and some of the telegraphic stations recently erected in the west.[20]

Since very few meteorological departments existed even in Europe in the 1850s and early 1860s, this presented an innovative scheme. Other natural disasters and famines demonstrated the usefulness of such a department, yet its realisation would take some further time. While the organisation of 'Meteorological Departments was commenced in India in 1867' for a first few regions, a fully-fledged system of stations was only established in 1875.[21]

Some British contemporaries perceived of the Schlagintweits as being among the prime movers of this state institution, together with Anglo-Indian naturalists and surveyors such as the Stracheys and Henry Francis Blanford. This becomes clear in a series of *Despatches of the Organization of a Meteorological Department in India, 1871–73*, collected by Robert H. Scott, then director of the meteorological department in Britain. In April 1871, Scott informed the secretary of state for India 'that the sources of information which' those involved in the planning of an Indian Meteorological Department 'have found most useful ... have been Herr von Schlagintweit's Report, vol. iv, with the plates accompanying it', together with 'Mr Glaisher's Report on the Meteorology of India'.[22] The brothers' role in establishing this government department was still acknowledged in Indian scholarship well into the twentieth century.[23] German subjects were thus not merely the 'impatient observers' (Hans Fenske) of the colonial expansion by other European powers in

---

[19] Ibid.
[20] Ibid., 155–6.
[21] Royal Commission, *Minutes*, 61, account by J. Cosmo Melvill.
[22] 'Report of the Meteorological Committee of the Royal Society, 11 April 1871', in Markham, *India*, 10–12, 10. Scott refers to H. Schlagintweit et al., *Results* IV, and the accompanying *Atlas*; on earlier use of Schlagintweit data, see also Ranking, the 'Sanitary Commissioner for Madras', *Report upon the Military and Civil Station*, 35.
[23] Ghatak, 'Introduction', 3.

the nineteenth century; rather, they sought at times to be the driving forces in the processes of foreign institution building.[24]

Intergenerational Encounters: Indian Pundits and the Schlagintweits

The Schlagintweits also had an inconspicuous influence as regards the renewed training and dispatch of indigenous surveyors, or pundits, to explore the Asian highlands. Adolph's brutal murder at Kashgar, whilst on a mission to gather meteorological and magnetic data and to survey different routes into Chinese Turkestan, convinced colonial officials that any further use of European officers would be too risky in view of the political and military turbulence that wrecked this part of central Asia in the years after the Great Rebellion.[25] While far from being the only Company employee to have lost his life during a journey that combined scientific instrumentation with intelligence gathering, Adolph's murder further tipped the balance in favour of sending out 'native' surveyors over the next few decades.[26]

Yet, there was more to the relation between the Schlagintweits and the Indian pundits. The three travelling brothers had first trained the later most famous of these, Nain Singh, in the art of surveying and observation in 1855. Nain Singh, after his own remarkable career as an independent and secret explorer of Tibet in British service, then instructed, until 1879, the next generation of pundits in the use of sextants, prismatic compasses and boiling-point thermometers – precisely those instruments whose handling he had first learnt from the German travellers.[27] The last pupil that Nain Singh, this 'instrument-reproducing instrument', formally taught in secret exploration was Sarat Chandra Das, 'an Indian of a different order from the other pundits of the Survey of India'.[28] Das proved to be a highly learned scholar himself, whose many talents and interests led him to become a 'traveller, and explorer ... a linguist, a lexicographer, and ethnographer and an eminent Tibetologist'.[29]

In an unusual series of personal and textual encounters between the Schlagintweits and Indian pundits over two generations, Sarat Chandra Das took up his interest in Tibetan scholarship thanks to the inspiration

---

[24] Fenske, 'Zuschauer'.
[25] Mason, 'Explorers', 430.
[26] Johnson, *Spying*, 100–3; the pundits are attracting increasing attention, see Kreutzmann, *Exploration*; Brescius, 'Cultural Brokers'; Driver, 'Face to Face'.
[27] Waller, *Pundits*, 124, 196.
[28] Ibid., 193; Raj, *Relocating Modern Science*, 199.
[29] Mahadevprasad Saha, quoted in Das, *Autobiography*, iii.

he drew from Emil Schlagintweit's work on Buddhism. Emil had been among the first continental Europeans to describe the itineraries and secret surveys in high Asia by Das's teacher, Nain Singh, for larger readerships in Europe.[30] Yet his own work on Tibetan Buddhism, largely written on the basis of manuscripts and artefacts that his older brothers had collected during their explorations (not least through the help of intermediaries like Mani and Nain Singh), also found eager readers in India and the Himalayas.[31] The learned Das and Emil Schlagintweit later engaged in direct correspondence with each other, in which the retired pundit claimed: 'I must candidly confess that my first impression on Buddhism were due to your excellent handbook on that subject. You have been the pioneer in that untrodden field of research. I have only fallowed [sic] your footsteps.'[32]

Like the pundit Das, who authored *A Tibetan–English Dictionary, with Sanskrit Synonyms* and *A Journey to Lhasa and Central Tibet*,[33] the German orientalist Emil Schlagintweit also combined deep religious, historical and philological interests in India and Tibet with more contemporary concerns about Asia (and, in Emil's case, later Africa, too).[34] While the relation between these two works remains unclear, it is notable that Emil had already produced in 1870 a 258-page unpublished manuscript: a 'Tibetan–Sanskrit Dictionary'.[35] For his wide-ranging interests, Emil built up an extensive network of correspondents, ranging from scholars and officials in London and British servants of the Raj to Russian orientalists and German missionaries active in the Himalayas.[36] Among them were the Moravian missionaries Heinrich August Jäschke and August Hermann Francke, who lived close to the borders of Tibet and proved valuable local informants.[37] Building on these vital

---

[30] E. Schlagintweit, 'Forschungsreisen'.

[31] Huber and Niermann, 'Tibetan Studies', 101.

[32] Letter from Darjeeling, 4 September 1901, reprinted in E. Schlagintweit, 'Adresse an den Dalai Lama', 665–7.

[33] Das (Rai Bahadur), *Journey to Lhasa*; Das, *Tibetan–English Dictionary*.

[34] Once a unified Germany acquired its own formal colonies in Africa, Emil also collected a wealth of materials about German colonial enterprise, held among his private papers, BSB, SLGA, VI.3.24.1, 'Kolonialwesen in Afrika'; he also followed other imperial affairs, gathering data on 'China's relation with foreign countries, especially Russia and the European powers', ibid., VI.3.17.10.

[35] BSB, SLGA, VI.1.1.

[36] For Emil's insistent applications to be supplied with ongoing administrative reports on British India and 'all Native States', NAI, Foreign Department, 1882, General B, Progs. June 1882, nos. 2/5–6, doc. 1. While it was the rule that 'no [rare] books should be given away at all to outsiders, such as Dr. Schlagintweit', Emil's efforts proved successful, not least because the Foreign Department's library had received and cherished his own work; ibid., doc. 3.

[37] BSB, SLGA, VI.6.1-2. See also Neuhaus, 'Tibet-Forschung', 225.

information channels, Emil became a prolific writer throughout his life on religious and cultural issues, but also often with a keen eye on contemporary British affairs and wider imperial developments. He addressed, in a series of articles, the 'question about the routes, on which a Russian invading army from Russian Turkestan could advance against India, and the dangers, which such a military clash must likely have for the survival of the English Empire in India'.[38] He also wrote about the lack of settlement by Europeans in the Himalayas (as a buffer against Russian invasions), the prospects of British trade and political expansion into central Asia and the results of the British war with Persia.[39] His status as an oriental scholar in India and beyond was considerable, so that Emil was able to mobilise important scientific and official support in his attempt to request information on Buddhist manuscripts in Tibetan temples from the Dalai Lama in 1902. Das personally translated Emil's address to that high authority.[40] While there existed also critical voices about Emil's scholarship,[41] he was nonetheless an internationally known Tibetan expert, who outlived his older itinerant brothers by almost a decade and never tired of keeping alive the memories of their exploratory feats among international audiences. Emil was the only Schlagintweit to have his own entry in the *Dictionary of Indian Biography*, where, however, his older siblings are mentioned as 'great explorers'.[42]

Dry Technicalities, Supreme Landscape Views: The Schlagintweits' Mixed Legacy in Britain

No matter how favourably colonial officials in India received the Schlagintweits' scientific findings and geographical discoveries, it did not significantly change the reception of their published work in Britain. On the contrary, it seems as if some of Britain's most influential journals had already reached their verdict on their oeuvre before the evidence was presented. When the first of nine projected volumes of the *Results* appeared, the reaction of the London *Athenaeum* was unsurprisingly harsh. However, one remarkable aspect of the review was that Berthold Carl Seemann, a German botanist working at Kew, had submitted this critical piece.[43] Previously, Seemann had been working as a botanist in

---

[38] E. Schlagintweit, 'Indiens Grenznachbarn', 105.
[39] E. Schlagintweit, 'Uferstaaten'; 'Völker Ost-Turkistans'; 'Zustände in Bhutan'.
[40] E. Schlagintweit, 'Adresse an den Dalai Lama'.
[41] On the critical appraisal, Martin, 'Translating Tibet', 104.
[42] Buckland, *Dictionary*, 377.
[43] Seemann, 'Review'. For the authorship, http://athenaeum.soi.city.ac.uk/athall.html; I thank Ulrich Päßler for this information.

Hanover, but had moved to London in 1844 to receive further training as a plant hunter. Owing to William Hooker's patronage, he was sent out only two years later as the official naturalist for a British survey mission to the Pacific.[44] In 1860, Seemann climbed even higher within Britain's imperial establishment, securing an appointment by the Colonial Office 'to report on the Fiji Islands, before the British government accepted their cession'.[45] Unlike the Schlagintweits, Seemann had thus thoroughly assimilated into British metropolitan society. He had become a fellow of the Linnean Society in 1852, acted as vice-president of the Anthropological Society, enjoyed RGS membership and was also later married to an Englishwoman.[46]

A close comparison of Seemann's review with unpublished letters written by Joseph Hooker reveals a remarkable similarity in the content and style of their Schlagintweit critique. Seemann's published piece echoed Hooker's complaints, stating that a 'slight' had been cast upon British naturalists and Company servants, since not only 'one, but *all* members of a foreign family' had been appointed for completing the geomagnetic survey of India.[47] The assessment also repeated Hooker's catchphrase that the brothers' 'appointment was one of the most gigantic jobs that ever disgraced the annals of science'.[48] It therefore seems likely that Hooker had briefed his Kew colleague for his polemic.

To be sure, *The Athenaeum* did acknowledge that the brothers had succeeded in pioneering excursions beyond 'the chains of the Karakorum' in central Asia. On the magnetic survey, they also conceded that as an 'important fact' of their researches 'may be mentioned the particular modification of the lines of intensity where they pass through the interior of India Proper, and all along the northern parts of the Himalayas'. Lastly, the journal lauded in particular the *Atlas*. It contained 'beautiful maps' and 'different views of the higher districts of Asia', said to be 'beautifully executed', and which 'will be acknowledged to be faithful representations by those who have traversed these charming districts'.[49]

However, this fleeting praise was largely outweighed by the critical thrust of the review, which attacked the brothers on different levels. For one thing, the brothers had made the rare gesture of prominently

---

[44] Seemann published *Botany of the Voyage* with the help of William Hooker.
[45] Boulger, 'Seemann'.
[46] Ibid.
[47] Seemann, 'Review', original emphasis. Also quoted in Finkelstein, 'Mission', 199.
[48] Hooker had written to Murchison that the 'Schlagintweit affair … is the grossest job that was ever perpetrated in this country under the name of Science I do not hesitate to believe', 19 July 1859, RBG, DC 96, 406.
[49] Seemann, 'Review'.

introducing European readers in the first volume of *Results* to their most important indigenous assistants. Felix Driver has argued that rather like 'experimenters faithfully describing their apparatus at the start of a laboratory report, the Schlagintweits were presenting for inspection the moral, racial and intellectual characteristics of their intermediaries. As instruments of science, the equipment had to be tested and calibrated, and any weaknesses examined, before an account could be given of the data collected in the field.'[50] While this is true, the brothers' long description of key members of their establishment was nonetheless unusual as their many guides, translators and other companions received open praise, friendly mention and, for those who had died, even words of personal commemoration. In describing formed intimacies and friendships, the brothers noted that the death of their lauded assistant Abdul 'is a subject of sincere regret to us'.[51] This positive acknowledgement of indigenous contributions, sometimes even of partnership, was, in fact, more prominent than the brothers' references to any British naturalists or Company officials in their published account.

The brothers' appreciation of their relations with south Asian informants and co-workers in the field provoked a scathing reaction from the British journal:

There are actually biographical sketches, written in the most matter-of-fact style, of all the observers, interpreters, collectors and servants, filling seven quarto pages. Some of these sketches rather remind us of the contents of the pieces of paper which on our arrival at the Indian ports natives force into our hands, recommending their services as washermen, valets, or something worse. Those bold-hand writers who are so accommodating in giving the poor natives a character when they have none might be glad of the following model![52]

The scorching review then mockingly reprinted an entire passage from the book, in which the Schlagintweits described the rise of their indigenous assistant Cheji. The twenty-five-year-old Lepcha from Sikkim rose from a 'plant collector' to the 'extremely useful' main translator in the expedition group of Hermann's Himalayan travels. Decried at the time as 'absolutely useless matter', these biographical fragments presented a clear break with Victorian conventions on expeditionary publishing.[53]

After ridiculing the Schlagintweits' indigenous sympathies, *The Athenaeum* went as far as to challenge the idea that the brothers may possess *any* higher scientific credentials at all:

---

[50] Driver, 'Face to Face', 454.
[51] H. Schlagintweit et al., *Results* I, 37.
[52] Seemann, 'Review', 216.
[53] Brescius, 'Forscherdrang', 85.

[W]e hold the Brothers de Schlagintweit quite incapable of taking a comprehensive view of any given subject; and we presume we are stating the general opinion of the scientific world correctly when we say that they can *take* observations, but not *make* observations. Place good instruments in their hands, and they will take astronomical, magnetic and meteorological observations with accuracy; but ask them to furnish a comprehensive account, founded upon their observations ... and they will ... thoroughly disappoint you.[54]

Regardless of the brothers' achievements as Alpine explorers, it was even insisted that 'Messrs. de Schlagintweit would have made good subordinates in a larger expedition, but they were remarkably ill-chosen for undertaking the lead of a great scientific mission'.[55] Faced with hundreds of pages of scientific computation, the review lamented their supposed 'absolute worthlessness' – indeed: 'Dry technicalities will never pass off for the results of abstruse science.'[56] To be sure, for contemporaries, abstract or 'abstruse' science described the highest level of natural philosophical research.[57] The rejection of the Schlagintweits' treatise as dry lists of instrumental readings was grounded in the purported lack of a higher natural-philosophical theory of the Himalayan and central Asian mountain chains. It was argued that their entire oeuvre would fall short of achieving such a synthesis of their knowledge of numerous local phenomena. Instead of a philosophical treatise like the one Humboldt had offered on the physical geography of the Andean mountains, the Schlagintweits failed, at least in the eyes of many of their British peers, to extrapolate any novel theories from their 'mountains of data'.[58] What further undermined their authority was the fact that the Schlagintweits later overconfidently sought to correct results obtained from the long-established and well-staffed and extremely accurate Great Trigonometrical Survey of India, whose achievements the brothers tried to refute in broad brushstrokes by presenting alternative coordinates 'determined from march-routes alone'.[59]

Certainly, the loss of Adolph Schlagintweit's expertise was a crucial factor in their perceived failure to meet the high expectations the brothers themselves had raised. However, even without his knowledge, there was no compromise over the publication project.[60] The goal was to

---

[54] Original emphasis, Seemann, 'Review'; also quoted in Finkelstein, 'Mission', 200.
[55] Seemann, 'Review', 216.
[56] Ibid.
[57] Reeve, *Portraits*, 81–6; Johnson, *Dictionary* (1839), where 'abstruse' is defined as 'hidden', 'difficult', as 'remote from conception or apprehension. It is opposed to obvious and easy', 6.
[58] Finkelstein, 'Berge'.
[59] Golubief, 'Observations', 46.
[60] HS and RS to Sir Charles Wood, 21 July 1859, BL, IOR, L/Mil/2/1477.

compensate for Adolph's absence by hiring further assistants to complete their monumental task nonetheless. Driven by Humboldtian ambitions, the two surviving brothers proved incapable of adjusting the eventual shape of their work to their time, energy, scientific expertise – and financial means. By comparison, Humboldt had published his *Voyage aux régions équinoxiales du nouveau continent* between 1805 and 1839, letting his insights mature into a masterful oeuvre that combined scientific innovation with cultural observations to provide a holistic treatment of South and Central America's human and natural worlds. The Schlagintweits, by contrast, proposed to finish *nine* volumes, equally aspirational in scope, within only *three* years – a megalomaniac plan.[61] In both a spatial and epistemological sense, their reach undoubtedly exceeded their grasp.

A further reason why the brothers did not secure a lasting legacy in Britain was their unfortunate choice of genre, which frustrated the expectations Victorian audiences had of an epic journey of exploration. No part of the *Results* was intended to be a pleasing and more general travel account – even if their extraordinary story, full of tensions, death and undoubtable exploratory and mountaineering achievements, could have made for a great literary success. Yet, in their serious attempt to take scientific understandings of India and high Asia to new heights, the Schlagintweits' highly technical treatises failed to grip anyone's imagination. This deprived them of wider readerships, not least since at the time 'exploration appealed as much to emotion as it did to reason'.[62] The brothers even conceded that '[i]t was with great difficulty and chiefly owing to the lively interest of our late friend Baron Humboldt, that we have been able to find a publisher for this work' at all.[63]

The Schlagintweits' work lost further appeal due to its considerable price of £4 4s per volume (roughly £450 today). This was especially true for the otherwise well-received *Atlas of Panoramas and Views*, which had the aesthetic and scientific qualities to become an iconic oeuvre (Figure 8.4). Yet, its elaborate production entailed spiralling expenses both for the India Office and the brothers themselves.[64] Initially, they had planned to publish their views in a work that 'will contain ... 700 Views pretty equally distributed over the countries of India and High Asia'. In 1859, the brothers were still optimistic that the 'work will appear in 20 charts (livraisons) each containing on an average 35 coloured drawings,

[61] J. D. Dickinson (secretary of East India House) to HS and RS, 8 July 1858, BL, IOR, E/1/309, 3393.
[62] Kennedy, 'Reinterpreting', 10.
[63] HS and RS to J. C. Melvill, 20 June 1859, BL, IOR, L/F/2/230.
[64] Anon. letter to the Perthes publishing house, FBG, SPA, ARCH PGM 353/1, 117, no date.

and will be completed in about 15 months', with each view being accompanied by 'an explanatory sheet'.[65] In seeking a broad readership, the Schlagintweits stated that in addition to private consumers, three copies of the work 'could be sent out directly to the seats of govt. at Calcutta, Bombay, & Madras'. They suggested the remaining number would 'find its full use in the India House, since it will be a work particularly adapted for presentation in India and in the Colonies.'[66] However, this magnificent work was never realised. The brothers were also forced to abandon their plan to produce, *en masse*, a popular, commercially successful, luxury volume [*Prachtband*] with a set of 125 selected views.[67] They only completed 43 of 120 planned views and maps from India and high Asia for their stunning *Atlas*.

To understand, however, how the brothers could secure a reputation as such outstanding landscape artists in Britain, India and the European continent requires a brief sketch of the views' various production stages. Such an analysis reveals much about the wider artistic and technical circles that sustained the making of such visual legacies of expeditionary science. The publication demanded a great amount of preparatory work in close cooperation with other artists. While this assistance was never formally conceded, the brothers enlisted, in fact, the service of at least *nine* different Munich landscape painters after their return.[68] Their spectacular, partly iconic views of little-known landscapes in the Himalayas, of central Asian steppes or their depiction of specific meteorological phenomena across the subcontinent, were thereafter regarded as objective scientific images – purportedly in the vein of the empirically grounded, instrument-based and true-to-life praxis of Humboldtian science. Hence, despite their undoubtable aesthetic qualities, the Schlagintweit views have so far been understood as accurate naturalistic landscape panoramas. However, as new scholarship on European expeditions has demonstrated, what scientific travellers wrote and sketched during their trips 'often went through multiple revisions before the works appeared in print, a process that substantially altered the texts' and images, thus 'eroding their authenticity as a record of immediate observations and experiences'.[69] And indeed, while the brothers began their views *in situ*

---

[65] HS and RS to J. C. Melvill, 20 June 1859.

[66] Ibid.

[67] Hermann to the Smithsonian Institution; Smithsonian, *Annual Report*, 84–5. An incomplete copy exists in BSB, SLGA, IV.1.26, entitled *Coloured Photographs from India and High Asia* (Munich, 1860).

[68] Kleidt, 'Kunstwerk'.

[69] Kennedy, 'Introduction', 9; Barrett-Gaines, 'Silences'; Keighren, Withers and Bell, *Travels into Print*.

in Asia, their artistic assistants in Munich, among them such well-known names as Wilhelm Scheuchzer, Fritz Bamberger and August Löffler, were subsequently asked to modify or complete them, or – after having been presented with the original *aquarelle* – to even produce entirely new versions.[70] For this process of modification and completion, the brothers glued their original images on large pieces of cardboard, on which they then inserted numerous marginalia clarifying the desired alterations.[71]

Since none of the Munich artists ever set foot in Asia and were trained in the European tradition of landscape painting, many of the Schlagintweit images became aesthetically 'Alpinised'. The views from the Asian highlands – the Himalayas, the Karakoram and the Kunlun – above all ultimately conformed to European conventions of high-altitude scenery. It is indeed possible to identify which Schlagintweit *aquarelle* was later modified by which Munich-based painter, since individual styles of painting cloud formations and other natural phenomena at times dominate the original sketches. However, this was not the last stage of modification. Rather, around a hundred completed *aquarelles* were then photographed and subsequently coloured by half a dozen different colourists employed by the famous royal photographer Joseph Albert in Munich.[72] Each coloured photograph looked different, and in a further selection process, the brothers ultimately chose their favourite version for the *Atlas*.[73] These were then finished by another set of intermediaries, including the lithographers Storch & Kramer (Berlin), the artist, lithograph and printer Wilhelm Loeillot (Berlin), the lithographer Sabatier and the printer Lemercier, both from Paris.

In other words, many of the *Atlas* views do not reflect immediate impressions of intimate encounters by the Schlagintweits with foreign landscapes and cultural formations. On the contrary, they were revised multiple times by artists and technicians, the results then dismissed or accepted. However, at least in the perception of the Schlagintweits themselves, who carefully watched over every step of alteration, the printed

---

[70] The other artists employed were Karl Millner, August Geist, Ludwig Meixner, one of Bamberger's pupils called Müller and another as yet unknown painter; Kleidt, 'Kunstwerk', 151.

[71] A set of around 300 Schlagintweit views with annotations survives in the Alpine Museum (Munich). The cardboard frames of roughly the same number of Schlagintweit views held in the Staatliche Graphische Sammlung (Munich) have been removed.

[72] Some eighty-seven motifs, differently coloured on 163 sheets by Albert's team (sometimes with four different versions of the same view), are held at the Staatsbibliothek Bamberg (StBB), HVG, 47/19–182. Another set of five coloured views is held at the Wellcome Library, London.

[73] A list of sketches provided to Albert and his team (with the dates 21 February 1859 and 1 April 1859) is at BSB, SLGA, VI.5.11.1.

Figure 8.4 Hermann, Adolph and Robert Schlagintweit, *Atlas of Panoramas and Views*, 1861–6.
© DAV.

views were nonetheless authentic: they were believed to truthfully depict the specific *qualities* of flora, light and colours of different scenes. This reveals another crucial function of the Schlagintweit views as aides-mémoires. In the brothers' attempt 'to see, to know, and to remember',[74] the finished *aquarelles* and sketches, just like their photographs and eastern artefacts (which were likewise preserved and sometimes modified for display in peculiar ways), promised to capture the true nature of those tropical and high-altitude landscapes:

[T]o the traveller his painting, if sufficiently large not to exclude those minor details of tint and form which he cannot leave unobserved, recalls the *total impression* with more power than any length of description could do; the effect of the aspect is instantaneous, and, for the author particularly vivid too, by the variety of the recollections called up. Perhaps also the reader in a distant country may find them effective enough to assist him in combining the impression of landscape and climate, of the charms of tropical scenery with the prose of glare and heat and steam.[75]

---

[74] Leibsohn, 'Introduction', 12.
[75] H. Schlagintweit et al., *Results* IV, 116–17, my emphasis.

Indeed, despite all criticism of the Schlagintweits as self-serving travellers, whenever their Asian views were presented to British popular and scientific audiences, the praise was univocal.[76] British geographers and imperial administrators alike enthusiastically commented on them, including EIC officials in London, who noted that: 'The whole number [of drawings] ... is 700, all of which with very few exceptions are entirely new, and of which the far greater portion represent scenes hitherto unvisited by European travellers. They are therefore of peculiar topographical and geographical value whilst the Panoramic views of the Himalayas of the principal Peaks, Glaciers and Lakes are also of geological importance.'[77]

When judging the necessary subsidies to print and publish the Schlagintweits' *aquarelles* and sketches, India Office associates therefore only remarked that even if this 'involves a considerable cost', the benefit of reproducing these rare 'Panoramas and views' would nonetheless be considerable. Copies of their views 'may be advantageously distributed to the Library and the Principal authorities in this country and in India'.[78] The usefulness and novelty of the drawings was readily acknowledged precisely because they made visible important regions outside formal British possessions. In the words of British officials, the Schlagintweit views were indeed 'forming a permanent and authentic record of the characteristic features of regions either imperfectly known or wholly unknown to European research'.[79] The geographer Roderick Murchison likewise described their views with great approval, pointing in particular to the precious and rare knowledge recorded in them. 'When I reflect that these brothers have penetrated farther into Thibet and Tartary from the plains of India on the south, than any other European, that their physical, geological, and geographical observations are of the highest value, and that they have even made photographic sketches at heights of 20,000 feet above the sea, I cannot but rejoice.'[80]

Despite universal praise for their visual materials, the Schlagintweits almost descended into oblivion in Britain after the heated debates from 1857 to 1861. Even if much more positive reports appeared on the later volumes of their *Results*, their reputation in Britain had already been

---

[76] 'New Contributions to the India House Museum', *Illustrated News of the World*, 2 July 1859; *Star*, June 1859, copy in ÖAV, CC, 109.
[77] Note by the India Office Library, 24 June 1859, BL, IOR, L/F/2/230.
[78] Ibid.
[79] Ibid.
[80] Murchison, 'Address', 1857, clviii–clix.

lastingly tarnished. Hermann bitterly noted this lack of acknowledgement in 1874:

It may be of a more general interest that the later [British] expeditions ... have confirmed the Karakorum chain as the central watershed, as it appeared to us in 1856 and 1857 ... Our precedence as regards scientific results remains often entirely unmentioned among the English, even if Adolph's unhappy fate found a lot of commiseration, but this admittedly due to the political circumstances.[81]

Especially when compared with the situation in the German states, the Schlagintweits' slide into obscurity can, however, also partly be explained by the fact that there existed no need for *German* scientific heroes in Britain. The brothers' travels and publications overlapped with a wealth of sensationalised stories of British overseas exploration and achievements. This unavoidable competition with other heroic travellers in the marketplace was captured in a letter from Murchison to Humboldt. Writing about the unbounded public excitement of the African explorer's recent journeys, Murchison wrote that '[t]he great African Lion Livingstone is likely to be smothered by public ovations. I have strenuously urged him to go to work & make his book which will sell exceedingly well.'[82] It did. When Livingstone published his *Missionary Travels*, its first 12,000 copies sold immediately, leading to nine further editions within a few years.[83]

For many British contemporaries, the African traveller and missionary scholar Livingstone personified an assumed set of British characteristics, and seemed to exemplify his nation's civilising influence on the 'primitive' world outside of Europe. As Tony Hopkins has argued about the relationship between individual scientific heroism and national identity in Victorian Britain:

[H]eroic myths are invented for purposes that are specific to the society concerned ...The hero personifies the national ideal by his exemplary actions. He disappears into remote, primitive lands because the contrast they offer underlines the superiority of his own advanced, progressive society. His exploits create opportunities, through feats of endurance and discovery, to reaffirm and strengthen the values that underpin the greatness of the motherland.[84]

The deeds and reputations of British expeditionary heroes and martyrs of empire easily outshone the fate and contested achievements of the Schlagintweits as recruited foreigners. Livingstone and the vanished Arctic explorer Sir John Franklin are just two cases in point. Britons'

[81] HS to unknown recipient (likely A. Petermann), Munich, 22 December 1874, FBG, SPA, 353/1, 31–2.
[82] Murchison to Humboldt, 17 January 1857, copy of letter at BBAW.
[83] On this sensational success and its reasons, see Driver, 'Missionary Travels', 164.
[84] Hopkins, 'Explorers' Tales', 671; incisive also is Sèbe, *Heroic Imperialists*.

feats and failures in the opening up of *terrae incognitae* in not only Africa's interior, but also in central Asia, Australia (as with Burke and Wills's expedition, 1860–1), South America and the Polar regions drew such interest at home that the resulting travel accounts, and gripping stories of 'rescue expeditions', often created sensational publishing successes.[85] By contrast, even if usually quoted by British specialists, not a single volume of the Schlagintweits' *Results* was ever reprinted, never threw off a profit for its authors nor gripped the public imagination.

The Triumph of 'German Science': The Schlagintweits as Scientific Heroes in a Pre-colonial Land

One can only be struck with the almost antithetical reputation that the same itinerant naturalists received from peers and the wider publics on the European continent – especially in pre-1871 Germany. Here existed specific historical constellations in which the Schlagintweits gradually assumed an outstanding reputation as scientific 'heroes'.[86] Yet, there was no sense of inevitability about their idealisation. On the contrary, the ways in which the Schlagintweits ultimately came to be regarded as exemplary figures of German national virtue and scientific achievement can tell us much about the manner in which expeditionary heroes were both manufactured and marketed. The scientific travellers manipulated this process of increasing public adoration, as the brothers communicated their exploratory feats in specific ways to selected intermediaries and audiences, recounting them in skilfully woven narratives. In doing so, the Schlagintweits actively intervened in the process of myth building in Germany's pre-colonial era.

The first factor that decisively shaped the brothers' diverging reputation was the existence of a substantial *knowledge gap* between audiences in Britain and those in the German states. It is misleading when historians assume a purportedly shared 'European knowledge of Asia' (or of any other continent) in the past.[87] Such homogenising assumptions of a purportedly diffuse knowledge across European communities rather invites us to critically ask who possessed such privileged knowledge, but also what groups may have been excluded from such access? What

---

[85] See, however, the case of the German-born hero of colonial society, Ludwig Leichhardt in Australia: Veracini, 'Expeditions'.

[86] In German, the word 'Held' still maintained a strong military, chivalric connotation in the nineteenth century, denoting above all an 'excellent warrior', but it also acknowledged those individuals who had achieved outstanding feats; Grimm, *Wörterbuch* X, cols. 930–5.

[87] Cf. Whitfield, *Maps*, 127; Ballhatchet, 'Relations', 218.

possibilities did the varying degrees of familiarity with India's history of exploration present for the brothers and their self-fashioning in front of different readerships?

Expert geographers of the standing of Carl Ritter, Alexander von Humboldt or August Petermann were closely following the scientific progress on Indian and high Asian geographies. Likewise, German encyclopaedias, specialist journals and great synthesising books in the fields of geography and European exploration were usually up-to-date.[88] However, given their small print runs and high prices, such works were largely reserved for a small section of society. Even German journalists found themselves at a loss when reporting on the Schlagintweit expeditions, conceding that if they were to report on their topographical details, 'then we must assume the erudition of an entire geographical society'.[89] RGS president Roderick Murchison also presumed a much lesser familiarity with overseas geography among the German reading classes. He thus stated that regarding 'foreign traveller[s] in the British service' such as the Schlagintweits, their compatriots 'should feel a just pride whether in perusing or in publishing the writings sent home to them in their vernacular freshness from remote corners of the earth, with which they are necessarily less familiar than the people of a maritime country like our own'.[90]

The depth of British accumulated knowledge on south and high Asia was indeed lost on German popular readerships. Since the latter were ignorant of travellers like Hodgson, the Stracheys, Cunningham, Thomson or Hooker, there was plenty of room in the German imagination for national scientific heroes to emerge for this world region.[91] Above all, Robert Schlagintweit's public lectures, hundreds of them given across Germany between 1864 and the early 1880s, played a crucial role in cementing the brothers' public standing as pioneers, great explorers and colonial improvers, not least because Robert again silenced the brothers' many British predecessors.[92] The Schlagintweits

---

[88] See, e.g., the serial production of articles on Asian geography and ethnology in the archive of *PGM*; Meyer (ed.), *Konversations-Lexikon*, under 'Himalaya', 1016–21.

[89] 'Schlagintweit's Vortrag in der Gesamtsitzung der kaiserlichen Akademie der Wissenschaften zu Wien', *Neue Münchener Zeitung*, 7 January 1858, 22.

[90] Murchison, 'Address', 1857, cxlix.

[91] This was a long-lasting myth, which sometimes even extended to India proper; see the propagandistic work by Mindt, *Der Erste war ein Deutscher!*, 1; or Terra, *Humboldt*, 129.

[92] When writing to the influential US scientist and editor Benjamin Silliman about a planned review of their *Results* in his *American Journal of Science*, Hermann, at least, added that '[t]he only wish we may add perhaps is that the [names] of the Englishmen', i.e. former Himalayan travellers, 'might be also occasionally quoted in your journal, as ... this may, as Mr. Brockhaus duly thinks, much contribute to exclude any provocation

were thus believed to be 'true trailblazers' in more respects than deserved.[93]

Yet, exploration and its imprint on the public imagination was always a collaborative enterprise. It involved numerous actors and 'a larger network of domestic institutions and interest groups that gave exploration much of its shape and purpose'.[94] Humboldt, for one, distributed the brothers' reports from Asia to be printed in scientific publications, sometimes providing catchy titles himself 'to excite the interest'.[95] Humboldt also acknowledged the Schlagintweits in his *Cosmos*. The old patron stressed the brothers' pioneering explorations as the first Europeans to have crossed the Kunlun, and cited the empirical data collected by his longstanding protégés in his sections on the Himalayas.[96] This accolade meant that numerous compilations of geographical discoveries and popular treatises were subsequently published in the German states that prominently praised their achievements.[97]

The authority of the German explorers was also negotiated and displayed in several learned journals. In the archive of *Petermann's Geographical Observations*, there are traces of numerous interventions by the Schlagintweits seeking to mould their representation. Hermann once directly informed Petermann that the Schlagintweits' ethnographic heads had been 'deliberately not mentioned' in Gordon Latham's work *Descriptive Ethnology*. While Hermann said he wished to write 'entirely privately' about the disregard, he nonetheless told the editor that 'I leave it entirely to you if you believe you are able to consider' this slight in an upcoming article.[98] At another point, Hermann wrote to Petermann with characteristic modesty: 'Perhaps the fact may also be mentioned that before us, no European has crossed the crest of the Karakorum in a northward direction, and [also] the Kuenluen from north [to south], which even Marco Polo never managed to achieve.'[99] Petermann, in turn, demanded 'free copies' of all the expensive volumes of the *Results* and any other publication, adding: 'Surely, no scientific journal has more persistently discussed your great mission than mine, which has followed ...

of feelings of rival animosity in England'. Johns Hopkins University (JHU), Special Collections, Papers of Benjamin Silliman (MS 30), 6 April 1861.

[93] 'Neue englische Expedition nach Inner-Asien', *Globus*, 1 (1862), 94.
[94] Kennedy, 'Introduction', 9; Flüchter, 'Identität', 164–5.
[95] Humboldt to Ritter, 18 December 1856, in Päßler, *Briefwechsel*, 184–5.
[96] Humboldt, *Cosmos* V (1858), 438.
[97] Among over seventy identified works, Zimmermann, *Völkerkunde*, 285–90; Sievers, *Asien*, 30–2.
[98] Cf. Latham, *Ethnology*; H. Schlagintweit to Petermann, 13 April 1859, FBG, SPA, 353/ 1, 69–70.
[99] HS to Petermann, 17 April 1861, ibid., 77.

with empathy and great interest all your writings for ten years, and could not for a moment be muzzled by English judgements.'[100] Traditionally, historians of exploration have focused on 'the mechanisms of hero-making' within distinct national communities of commemoration.[101] Yet, attention has recently turned to competitive hero-worship among different European imperial powers.[102] As regards British and French African travellers as heroes of their respective empires, it has become clear, for instance, how French glorification of Colonel Marchand was spurred by the British elevation of Kitchener to hero status after the confrontation of the two figures during the Fashoda Crisis in 1898.[103] Yet, what has been less explored is the phenomenon of how public degradation in one society spurred the furious glorification of the same *personae* in another. In consequence of the mutual observations between Britain and Germany within the landscape of popular print culture, the Schlagintweit case opens up an important new dimension for studying the transformation of European travellers into imperial heroes – in formally non-colonial societies, too.

There existed a distinct pattern and chronology in the way the Schlagintweits were depicted by the German press over the course of their careers. At first, after they had already established their names as daunting Alpine naturalists, numerous articles – sometimes accompanied by maps – appeared in daily papers and scientific magazines during the Indian expedition that made them household names amongst the greater part of the German reading classes. During this first stage, little attention was paid to foreign critique, whether stemming from the British Isles or India itself.[104] After the completion of the mission, when the first reviews appeared in *The Athenaeum* in late 1857, German commentaries changed. British criticism was now openly registered. But there still existed de-escalating voices. The journal *Bonplandia* noted, for instance, that 'for a few years, certain London papers have sought to debase the merits of German travellers'.[105] Yet, the paper continued to stress in a conciliatory tone that 'all national feeling should cease in the field of science'.[106]

In the third stage, however, the rhetoric of German papers sharpened. This mirrored the fact that by the early 1860s, the affair over these

[100] Petermann to HS, 9 May 1864, ibid., 95.
[101] Sèbe, *Heroic Imperialists*, 19; see among many, Berenson, *Heroes*.
[102] MacKenzie, 'Introduction'.
[103] Sèbe, *Heroic Imperialists*.
[104] *The Englishman* (Calcutta), 3 April 1855, copy in ÖAV, CC, 93.
[105] *Bonplandia*, 21, 15 November 1857, 332.
[106] Ibid.

lucratively employed 'foreigners' had become heavily politicised in England. The longstanding supporter of the brothers in Britain, Edward Sabine, had been portrayed as a man whose judgement could not be trusted, since *The Athenaeum* demanded that 'scientific men will be consulted' to judge their mission 'who have not received any Prussian decorations'.[107] When *The Athenaeum*'s campaign against the Schlagintweits had forced even respected British men of science to publicly defend their role in the scheme, this shift in tone was closely noted among German papers.[108] Parts of Edward Sabine's published reply to charges of favouritism and corruption against him were thus reprinted in German papers, although in a modified way. One claimed, for instance, that Sabine had supposedly praised the Schlagintweits' *Results* as an oeuvre 'that will do credit to [the brothers] themselves and to the Indian Government in all times to come'.[109] To fight fire with fire, the German papers thus selectively reprinted positive statements from British scientists and papers in order to directly contradict the more critical responses.[110]

In the increasingly vicious fight over the travellers' reputation, other German papers employed even more manipulative strategies by twisting the words of British papers. When the Schlagintweit mission was said to have been 'universally condemned' in England (*Saturday Review*), this was taken as an expression of an ill-informed nationalist frenzy, not a qualitative judgement on their scientific results.[111] But more than that: when the brothers' met with veritable storms of abuse, the British critique was bluntly reinterpreted in the German press as a symptom of outright 'jealousy' of their achievements.[112] In this way, foreign defamations could be presented as veiled acknowledgement. Conversely, when a welcoming commentary appeared in Britain papers, such praise was taken literally.

The public critique of lauded national figures was not a one-way phenomenon. On the contrary, German journals also openly challenged

---

[107] Seemann, 'Review', 215.
[108] *The Athenaeum*, 1767, 7 September 1861, 320–1, for Sabine's and Murchison's public statements about their role in the Schlagintweit mission; ibid., 1768, 14 September, 1861, 342; also, review of 'Chesney, Narrative of the Euphrates Expedition', *The Athenaeum*, 2201, 1 January 1870, 18; Alcock, 'Address', 314.
[109] *Deutsche Allgemeine Zeitung*, 214, 18 September 1861, 73.
[110] Specific British papers were singled out and lauded for their supposed unbiased reviews of the German travellers' work – always positive in approach. This included the *London Review*, the *Illustrated News of the World*, *The Reader* and others; see 'Das Werk der Brüder Schlagintweit', *Allgemeine Zeitung München*, 284, 11 October 1863, 4707.
[111] Ibid.
[112] 'Das grosse Reisewerk der Gebrüder Schlagintweit', *Illustrierte Zeitung*, 919, 9 February 1861, copy in ÖAV, CC, 74–5.

iconic British travellers such as David Livingstone, and devalued their findings when compared with those achieved by the African explorer Heinrich Barth. Humboldt thus wrote to Carl Ritter that 'Petermann, in his last issue, took some vengeance against the dwarfed Livingstone'.[113] Earlier, Petermann had set Livingstone's travels against Barth's exploratory feats, claiming that the former's expedition had provided 'only meagre scientific results'.[114] By contrast, Barth's journeys were said to have significantly expanded the geographical knowledge of Africa, not least by tracing a number of caravan routes, which he had linked on maps to his own itinerary.[115] Yet, such foreign criticism did not alter the public veneration Livingstone then received in Britain. Whatever the German attempts to shrink this quintessential British heroic traveller down to human size, his celebrated sixteen-year African travels had already catapulted Livingstone to Olympian heights.[116]

Conversely, the serious British critique of the Schlagintweits provoked substantial reactions among German papers and the wider public. Above all, it led to diametrically opposed reactions in the national press. In fact, only one Prussian satirical paper, the *Kladderadatsch*, whose mockery of the Schlagintweits was not taken up by any other publication, was a single critical voice in what was otherwise an ocean of praise. As the analysis of thousands of newspaper articles published in the 1860s and 1870s has shown, the Schlagintweits were commonly seen in a positive, if not glorifying, light. There was general consensus that the brothers' 'travels ... must be considered among the greatest scientific undertakings of modern times', that their 'name will for all times be ... a bright meteor in the scientific heaven'.[117]

The Schlagintweits' diverging reputations were forged under almost reverse conditions in Britain and the German states. In Britain, an Indian or Himalayan expedition was no outstanding event, whereas in the German-speaking world, such an enterprise had the potential to be perceived as a unique undertaking: as the first grand mission under German leadership into the supposed unknown of high Asia. Moreover, Britain had seen thousands of Company officers and servants die in India and other parts of the empire, and several also in the course of scientific exploration and political missions beyond south Asia's northern frontier. By contrast, Adolph Schlagintweit's brutal execution as a British

---

[113] Humboldt to Ritter, 21 May 1857, in Päßler (ed.), *Briefwechsel*, 201–2.
[114] Ibid., 202, n. 5.
[115] Petermann, 'Livingstone's Reisen'; Petermann, 'Kartenskizze'.
[116] Lewis, *Empire of Sentiment*.
[117] 'Hochasien und sein Handel', *Vorwärts!*, 16 (1866), 36.

employee in central Asia was a singular event. The mystery surrounding his death added to the dramatic appeal of the episode and soon found its place in numerous popular papers. Lastly, in direct contrast to the British public, there existed in the German press a decisive willingness to create, in the brothers, explicitly *German* heroes.

In this context, the work of Benedict Anderson on modern nations as 'imagined communities' provides important insights. As Anderson has shown, 'print-capitalism', in particular, 'made it possible for rapidly growing numbers of people to think about themselves, and to relate themselves to others, in profoundly new ways'.[118] Through an emerging mass media, people came to imagine themselves as part of a larger community – even though 'the members of even the smallest nation will never know most of their fellow-members, meet them, or even hear of them, yet in the minds of each lives the image of their communion'.[119] In the case of the politically fragmented German states, this sense of belonging to a shared cultural unit beyond such political divisions is particularly striking. The dream of a future German nation-state had been a central reference point for German liberals with decisively expansionist agendas since the 1840s.[120] In the press coverage of the Schlagintweits, an imagined Germany, glued together by shared cultural-linguistic features and scientific achievements, was indeed repeatedly evoked.

Language undoubtedly played a significant role in the imaginary bonds between the fragmented German political landscapes.[121] Even before the first volume of the Schlagintweits' *Results* appeared, it was critically noted: 'The fact that this work appears in the English language has indeed something embarrassing for our national feeling [*Nationalgefühl*]; we thought until now that those times are gone when even an Alexander von Humboldt saw himself forced to have his great travel work published in French.'[122] Such linguistic patriotism emerged in relation to other German works as well. *Petermanns Geographische Mitteilungen* also objected to the fact that 'even the report on the Novara expedition, the first scientific voyage around the world sent out by a German state, had

---

[118] Anderson, *Imagined Communities*, 36.
[119] Ibid., 6.
[120] Fitzpatrick, *Liberal Imperialism*; see, however, also the valuable consideration of particularisms in nineteenth-century German states in Green, *Fatherlands*. Her work reminds us that there were both strong particularist and nationalist cultures, which, however, did not need to be mutually exclusive; ibid., 147.
[121] See also Fahrmeir, *Deutschen und ihre Nation*, for the importance of language for a nineteenth-century German identity.
[122] 'Notizen', *Deutsches Museum*, 32, 9 August 1860, 214; *Literarisches Centralblatt für Deutschland*, 18, 24 April 1869, 513.

first to be published in London in English'.[123] (*PGM* thus considered even Austria, which had sent out the Novara expedition, still part of the broader imagined German nation.) Publishing German scientific works in foreign tongue, the journal continued, 'offends our feeling as Germans, and must throw a curious light on our highly praised scientific genius in the eyes of other countries'.[124]

In direct response to 'abominable' British polemics against their fellow German travellers, branded as a 'dishonourable game' by German magazines, the latter issued ever-more sharp attacks in reply. The counterblasts sought to ridicule those British men of science who had publicly dared to denigrate the Schlagintweit brothers. As the widely read journal *Globus* stated, 'there are in London a number of scientists, who regard with a narrow-minded jealousy all progresses in the sciences that have not been made by John Bull'.[125] In singling out the Schlagintweits' longstanding opponent, the journal continued its attack that 'even a man like the botanist Hooker has the effrontery [*entblödet sich nicht*] to claim that he would know English officers, "whose studies and magnetic surveys would be of infinitely higher value than those of the Schlagintweits"'.[126] After Murchison and Sabine had been pressured by *The Athenaeum* to explain their role in the Schlagintweit appointment, Hooker, too, spoke out and openly distanced himself from the brothers. He stated that the whole affair and the neglect of British talent 'excite my deepest indignation'.[127] By rubbing salt into the wound that many British officers had failed to secure Company grants for scientific research, the German paper asked why those English officers that Hooker had praised had failed to accomplish the brothers' monumental tasks, and 'rather ceded the glory to three German men'?[128]

With British papers downplaying or even woefully neglecting the Schlagintweits' 'colossal works', German editors increasingly adopted a chauvinistic stance.[129] This became especially apparent in the notion of 'German science' and its claimed grandiose achievements.[130] The brothers became increasingly portrayed as its very embodiments.[131] What

[123] The work discussed was Scherzer, *Circumnavigation*; 'Geographische Literatur', *PGM*, 7 (1861), 202–3.
[124] Ibid.
[125] 'Der Streit über den Gorilla und du Chaillu', *Globus*, 1 (1861), 121.
[126] Ibid.
[127] Hooker, 'The Messrs. Schlagintweit', *The Athenaeum*, 1769, 21 September 1864, 374.
[128] 'Streit über den Gorilla und du Chaillu', 121.
[129] 'Das grosse Reisewerk der Gebrüder Schlagintweit'; 'Ostindien', *Magazin für die Literatur des Auslandes*, 38/11 (1869), 158–9.
[130] Ibid.
[131] Ibid.

is most important about the trope of 'deutsche Wissenschaft' was the fact that the sciences were singled out as a realm in which Germans could compete with, and purportedly even outclass, other European nations during the nineteenth century. While many contemporaries lamented Germany's relative backwardness in the mid nineteenth century and looked to the industrial achievements and global expansion of the British state and other empires around them with envy, the triumph of German culture and science became a source of unifying pride and identification in the German states.[132] Indeed, in many press reports on expeditions to Africa, the supposed *eminence* of German travellers was frequently assumed and favourably compared with the British exploration of the continent's interior. While its claim to German grandeur again implied a reference to the British Empire as the measuring stick, a popular German paper proudly noted that '[t]he participation of Germans in overseas explorations has never been greater than during the current times. In Africa, we even outperform the English.'[133]

Since the age of enlightenment, European expeditions of discovery had become closely linked with notions of the modern.[134] To present the German states as a vital participant in the 'unveiling' of the interior of the world's continents, the 'German nation' (beyond its political fragmentation) could explicitly be imagined as a progressive force of humanity. For nineteenth-century contemporaries, the higher purpose of overseas exploration was bound up with western technological and medical advancements,[135] which allowed European voyages of exploration to circumnavigate the globe and reach deep into remote regions, leading explorers 'to map out the geographical coordinates of their routes as guides to those who followed'.[136] In the light of this idea of perpetual progress, the Schlagintweits' explorations in central Asia – unfinished not least because of Adolph's sudden death – came to be perceived as a national 'task' that subsequent German itinerant naturalists were to realise.[137] German papers established a veritable genealogy of German central Asian exploration. Thus, the travels of people such as Gustav Radde and Ferdinand von Richthofen were incorporated into a German

---

[132] The notion of a superior 'German science' was also celebrated in a global overview of German overseas achievements in Scherzer, 'Deutsche Arbeit', a lecture widely reprinted at the time.

[133] 'Neue deutsche Reisende', *Westermanns illustrierte deutsche Monatshefte*, 12 (1862), 334.

[134] Stern, 'Exploration'.

[135] Headrick, *Tools*.

[136] Kennedy, 'Introduction', 2.

[137] 'Neue deutsche Reisende', 334: 'Die Aufgabe, die Adolph Schlagintweit zu erfüllen durch einen Mörder gehindert wurde, haben zwei andere Reisende wieder aufgenommen.' See also Gollwitzer, 'Expansionsideologien'.

genealogy of high Asian exploration, which the brothers were claimed to have begun.[138]

This rhetoric of a specific German 'task' to be completed overseas for the sake of the fatherland's glory applied not only to the Schlagintweit brothers and central Asia. Petermann argued in 1860 that 'the German nation' was also loudly called upon to act after the disappearance of the German traveller Eduard Vogel in Africa's interior, while he was 'in the service of German science'.[139] In sending out a search expedition to find the 'relics' of Vogel – 'to save the last notes from his hand, his collections, the results of his strenuous labours [and] the price of his sacrifice' – Petermann pressed for public support for further German-led expeditions into those same African regions. Such follow-up expeditions were intended to 'finish his work, to solve … the task he had set himself, and thereby to erect an honorific monument not only to his memory, but also to German science and German spirit'.[140]

Other German papers went beyond central Asia and Africa, and unfolded a *global panorama* in claiming German achievements in having opened up the interiors of *all* continents. The *Illustrirte Zeitung* (Leipzig) displayed remarkable national hubris in the realm of science:

Again, it is Germans who enrich the sciences in a magnificent manner. We do not possess one foot of land in foreign continents, but we know those more thoroughly than any other people. Beyond doubt, it is our travellers who take the first rank of scientists [and] explorers … the shining deeds speak for themselves, and cannot be denied. The Schlagintweits are equal to the greatest names in their field of science, and high posthumous reputation is secured for them for all ages to come.[141]

German explorers were thus not considered only as *primi inter pares*. Rather, the Schlagintweits and other itinerant geographers were portrayed as superior. Regardless of the fact that petty 'jealousy' would deny their acknowledgement, the paper claimed that it was indeed Germans who had scientifically triumphed in every corner of the globe – and often in foreign imperial service.[142]

---

[138] 'Neue deutsche Reisende', 334. Hermann and Adolph Schlagintweit knew Richthofen personally. While they all had certainly met in Berlin (where Richthofen studied from 1852 to 1856, at the time Hermann was lecturing at Berlin University), Hermann and Richthofen were also later involved in the 'German section' of the exhibition that accompanied the second international geographical congress in Paris, in 1875. Richthofen, 'Bericht' (1875), 186.

[139] Petermann, 'Expedition nach Wadai'.

[140] Ibid.

[141] 'Das grosse Reisewerk der Gebrüder Schlagintweit'.

[142] This finding significantly diverts from Ulrike Kirchberger's conclusion (*Aspekte*, 462) that German itinerant scientists employed in British overseas service were often

The English, just as little as the Russians or the Dutch, have no right to lament about the services that German men have afforded them. The best description of *Guyana* we have through Schomburgk; *New Zealand* has been made more thoroughly familiar to us through Dieffenbach, and partly through Hochstetter; Leichhardt lost his life in *Australia*; Barth brought clarity to the geography of *Inner Africa*. The Asiatic part of *Russia* has primarily been, we can indeed say, discovered by Germans ... Kaempfer and Siebold gave us the best descriptions of *Japan*; for the region of the *Nile*, [the works of] Burckhardt, Russeger, Werne are classics; Niebuhr opened up the Mahomedan *Orient*; in *America*, there is no region that would not have been travelled through and described by Germans ... [All of them] aggrandise the glory of the German name in scientific research. Whoever realises what they have achieved, cannot refuse his admiration.[143]

In this global panoply of heroic explorations, the 'political resonances of science' are strikingly evident.[144] National celebrations for the discovery of overseas territories became a central part of the 'constitutive stories' of imagined German nationhood.[145] 'German science', with its inherent claim of a specific German predilection for thorough scientific exploration, became an important space for German self-fulfilment. The quotation above on the global achievements of German exploration also demonstrate that the Schlagintweits' history became part of a wider process in later nineteenth-century Germany that saw the creation of a pantheon of national 'scientific heroes'.[146] This is well captured, among other symbolic acts, in the integration of Hermann and Robert Schlagintweit into the 'gallery of eminent naturalists of modern times', a series of portraits of a hundred of (as was claimed) Europe's most influential scientists (Figure 8.5).

While this may not have applied to all German travellers enumerated above, the public glorification of the Schlagintweits, at least, served clear political ends. At a time when German merchant houses already enjoyed a global presence, quickly reacting to ever-new trading opportunities overseas, the brothers were openly praised for having identified new avenues for European trade and settlement in Inner Asia.[147] While the military value of the intelligence the brothers provided was believed to be most

---

'ignored' in their homelands whilst they found acknowledgement overseas and in Britain.

[143] 'Das grosse Reisewerk der Gebrüder Schlagintweit', emphasis mine. See also Richter, *Fortschritt*, 67–8. This trope still echoed in the early twentieth century, see Paletschek, 'Modernisierungsleistungen', 49.

[144] Jordanova, 'Science and Nationhood'. This refutes the contemporary idea that in the fields of 'science and technology, nationalisms appeared absurd' in nineteenth-century Europe, as discussed by Laak, *Welt*, 44.

[145] I borrow the term from Colley, *Acts of Union*.

[146] Cf. Torma's fine work on the hero status of Sven Hedin, *Turkestan-Expeditionen*, 34.

[147] Scherzer, 'Deutsche Arbeit'.

Figure 8.5 Hermann Schlagintweit, lithograph, 'Gallerie ausgezeich-
neter lebender Naturforscher', G. A. Lenoir (1856).
© Universitätsbibliothek HU Berlin [HUB], 'Porträtsammlung: H. v.
Schlagintweit-Sakülünski', photographer: Helffter, image 12863.

useful to the British,[148] the information on settlements and trade opportunities, by contrast, appealed directly to the German middle classes.
Hermann Schlagintweit's evaluation of the 'possibility of colonisation
by Europeans' of 'Assam' and the 'forelands of the Himalayas' was,
for instance, noted and discussed in the German press.[149] Commercial
journals such as *Forward! Magazine for Merchants* – with the expansive
thrust of German trade captured in its title – readily picked up the intelligence the Schlagintweits provided.[150] For the Himalayas, the trade
journal emphasised in particular the existence of 'much building material,

---

[148] 'Der dritte Band des Schlagintweit'schen Werks über Indien und Hochasien', *Beilage
der Deutschen Allgemeinen Zeitung*, 365, 8 August 1864, 1551–2, 1551.

[149] 'Der Krieg der Engländer gegen Bhutan, im östlichen Himalaya', *Bayer. Zeitung*
[unknown date], copy in ÖAV, CC.

[150] *Vorwärts!* was edited by the German Eduard Amthor (1820–84), a learned orientalist,
publisher and director of a commercial school [*Handelsschule*], which he opened, after
travels to England and France, in Hildburghausen in 1849. Lülfing, 'Amthor'.

fine woodland, and metals in abundance'.[151] Tibet was said to offer, by contrast, an 'abundance of salt, excellent horses and extensive sheep-farming'. The fact that the products of both regions seemed mutually complementary reinforced the paper's conviction that there was indeed a 'real need for trade' in those areas, just awaiting German intervention.

> It is surely a matter of extraordinary practical importance for Europe to facilitate … this great trade in high Asia through appropriate infrastructure, so that it will gain in both breadth and size. No expensive railways will serve this trade of high Asia … but rather, according to the personal observations of Professor Schlagintweit, bridges over the bigger rivers, and of such an extent that they are able to carry heavy beasts of burden.[152]

Such technological interventions were, however, only the first stage of a more thorough future penetration of these regions. Hence, once the 'lines of communication' are realised, and 'perhaps Europeans have settled there, then high Asia can look forward to a bright future'. Indeed, the 'commerce' of this world region 'will flourish up to a hitherto unimaginable level, guaranteed by its abundance of metals, animals and plants'.[153] The journal's call for trade expansion into high Asia, based exclusively on the Schlagintweits' accounts, aptly demonstrates to what considerable extent the brothers came to be seen as vanguards for commercial outthrust.

The Schlagintweits' careers are therefore relevant to an increased historiographic interest in the permeation of imperial ideologies into European domestic societies. Recent scholarship has predominantly rejected Bernard Porter's assumptions that empire had no meaningful impact on British society and culture, even in the nineteenth and twentieth centuries, when it reached its most global scope.[154] A number of works have rather highlighted the popular repercussion of empire on domestic debates and cultural imaginations, argued for both Britain and France.[155] While the literature on Germany's colonial history has traditionally focused on the late nineteenth and early twentieth centuries, a proliferating scholarship has turned attention away from this short period

---

[151] 'Hochasien und sein Handel', 37.
[152] Ibid. All subsequent quotations, ibid.
[153] Ibid.
[154] An important book that provoked numerous follow-up studies was Porter, *Absent-Minded Imperialists*.
[155] Take, as examples for the so-called New Imperial History, Hall, *Civilising Subjects*; Nechtman, *Nabobs*; Berenson, *Heroes of Empire*; Sèbe, *Heroic Imperialists*; MacKenzie (ed.), *European Empires*; decisive was the early intervention by Cooper and Stoler, *Tensions of Empire*.

of actual empire, yet sometimes with a problematic teleological tendency. The literary scholar Susanne Zantop, for instance, argued that: 'Imaginary colonialism anticipated actual imperialism, words, actions. In the end, reality just caught up with the imagination.'[156] Such a view, however, ignores the fact that Germans had travelled under the protection of foreign empires or had actively participated in their imperial projects long before 1884, and that some of those travellers then channelled imperial practices, objects and ideologies back home.[157] Recently, Bernhard Gißibl has argued that the careers of German missionaries, mercenaries, scientists and entrepreneurs might be regarded as a form of 'vicarious imperialism'.[158] Yet, as the popular reception of the Schlagintweits' achievements in – as was often critically noted – British imperial service makes clear, such an ersatz empire was not considered sufficient.[159] Rather, the brothers' services and sacrifices to the British imperial cause were appropriated to fervently demand *German* overseas possessions.

The two surviving brothers achieved the 'popularisation of empire' in the non-colonial German states above all by adopting more popular forms of science communication. In contrast to the technical nature of their *Results*, the explorers exploited a huge market for popular versions of their travel accounts in the German-speaking world. For one thing, Robert and Hermann Schlagintweit provided a constant stream of more accessible articles in daily newspapers and bourgeois magazines. In addition, it was Hermann who embarked on another huge publishing effort, and this time finished four volumes of a popular German travelogue.[160] His *Travels in India and High Asia* was the only publication by the travellers that was reissued in a further edition, and became acclaimed as a work that had purportedly inspired numerous German fellow travellers

---

[156] Zantop, *Colonial Fantasies*, 9. Other significant works on the pre-colonial period of German history include Kundrus, *Phantasiereiche*; Fitzpatrick, *Liberal Imperialism*, Müller, 'Imperialist Ambitions'; and, with particular attention to various and changing relationships between German-speaking central Europe and the Middle East over a long chronology, Berman, *German Literature*. For a perceptive reading of anti-colonial sentiments in German culture and thought of the enlightenment, see, however, Berman, *Enlightenment*. A good overview of current post-Saidian interests in German orientalism is given in Hodkinson et al. (eds.), *Deploying Orientalism*.

[157] Conrad, 'Rethinking German Colonialism', 550, on Richthofen's imperial projects in China and on the German exploration of Africa. See also the rich empirical analysis in Naranch, 'Beyond the Fatherland'.

[158] Gißibl, 'Imagination', 164; the author usefully draws attention to the fact that German imperial sojourning in other empires happened both long before and also after the short-lived German colonial Empire.

[159] Werner (ed.), *Ostindien*, v.

[160] H. Schlagintweit, *Reisen*.

and merchant firms to turn to the opportunities opened up by the Schlagintweits in Asia.[161]

Yet, even more than the printed word, it was the hundreds of public lectures that Robert Schlagintweit delivered in front of middle-class German audiences of scientific, amateur, manufacturing and commercial societies that had the greatest impact. Despite holding a chair in geography and statistics at the University of Giessen since 1864, Robert turned into a widely celebrated public orator and science populariser (Figure 8.6). Over two decades, he delivered his accounts on the Indian mission (and later his two North American travels in 1868–9 and 1880) in many corners of the German states (190 cities in total), regularly drawing audiences of several hundred listeners.[162] In doing so, Robert popularised numerous 'opportunities of empire'. For instance, he always reiterated in front of commercial and bourgeois spectators that once 'European colonists … have but once settled there [in the Himalayas] in considerable numbers, then difficulties will no longer present themselves as regards extension of tea plantations, laying out vineyards, erection of tobacco manufactories and industrial establishments of all kinds, and bringing into use the immeasurable woods now laying waste [sic]'.[163]

What is indeed striking is the performative aspect of the Schlagintweits' later careers, which seemed to effortlessly combine 'science and circus'.[164] Robert's public presentations and lecture tours demonstrated that expertise was something that could be performed in a carefully orchestrated and entertaining show. In each town, Robert had specific supporting frames manufactured in advance to put up various illustrations.[165] He always travelled with selected artefacts and visual legacies from his former explorations, ranging from maps and printed *aquarelles* to minerals and human remains, including skulls and plaster casts, which were passed through the seated audience. While his listeners stared at the striking images the Schlagintweit enterprise had produced, and touched with their hands the material memorabilia from Asia, Robert delivered his gripping account of their travels and travails, which he had

---

[161] Werner (ed.), *Ostindien*, esp. v–vi.
[162] See the incredibly detailed memorabilia and notes Robert kept on each of his over 1,300 lectures, in BSB, SLGA, V.1.
[163] BSB, SLGA, V.2.2.1, 105–6.
[164] Such public performances stood, of course, in a much longer tradition, see Bensaude-Vincent and Blondel (eds.), *Science and Spectacle*; also thought-provoking is Shreider, 'Circus'. On the intertwining of entertainment and learning in racialised ethnographic performances see Zimmerman, *Anthropology*; also Liebersohn, *Music and the New Global Culture*.
[165] RS to unknown recipient, 29 November 1872, Koninklijke Bibliotheek, The Hague [KB], 130 G 17.

Figure 8.6 Advertising poster for a series of Robert Schlagintweit's lectures in Germany, in 1878, 'Ueber den von seinen Brüdern und ihm während mehrjähriger Reisen nach den verschiedenen Richtungen durchzogenen HIMALAYA (erläutert durch landschaftliche Ansichten, Photographien & Karten)'; and 'Ueber die von ihm zweimal ihrer ganzen Ausdehnung nach bereiste Pacific-Eisenbahn Nordamerika's'. © BSB, SLGA, V.1.11, 101.

fully memorised to appear more lively and enthralling. His performances followed a carefully developed and continuously improved ritual – a popular scientific spectacle that reached his consumers through different senses: sight, touch and sound. It allowed German and wider European audiences to undertake imaginary travels to lands whose 'exotic' qualities Robert painted in vivid colours.

Above all, in exemplifying the popular culture of imperialism, Robert's series of lectures was a money-spinning business. While many popular historical accounts of expeditionary enterprises stress the explorers' eccentric characters, and portray them as larger-than-life figures solely devoted to discoveries and willing to pay the ultimate price for their ambitions, Robert and Hermann Schlagintweit's case strongly reminds us that nineteenth-century explorers were also ingenious entrepreneurs, keen to sell their knowledge, lived experiences and the material archives of exploration to multiple audiences in a transnational arena of consumers.[166] Robert's professional management of his lecture tours does indeed shed fresh light on 'the economy of expeditions and their affinity with other capitalist initiatives that use models of subscription'.[167] The Giessen professor turned science populariser generally used subscription lists in advance to guarantee commercial success for his attempts to widely disseminate the results of the Schlagintweit expedition across central Europe, Russia (where his lectures yielded the highest profits) and North America.[168] Robert's two tours to the USA, where he crossed the entire North American continent from coast to coast via railway, led to additional lectures on his American travels and awaiting opportunities in the 'New World'. This helped to attract further German emigration to the States, with Robert's published works on the USA becoming central sources of information for willing German emigrants.[169]

If there is a danger in mistaking the published opinions of journalists as the views of the public at large, any doubts about the increasing popularity of Robert Schlagintweit are put to rest by his large audiences voting with their feet.[170] His considerable talent for rhetoric, univocally praised by amateur, learned and mercantile *Vereine*, at the same time points to the missed opportunity and ill-chosen genre of the Schlagintweits' dry

---

[166] See on Ludwig Leichhardt's entrepreneurialism for realising his Australian expeditions, Thomas, 'Expedition'.
[167] Thomas, 'What is an Expedition?', 9.
[168] BSB, SLGA, V.1.1, 6, 45.
[169] Jaehn, *Germans in the Southwest*, 15–18.
[170] Franu Leibing, 'Wandervorträge', *Der Bildungsverein*, 33, 19 August 1874, copy in BSB, SLGA, V.1.17, 96.

Results.[171] Surviving evidence from newspapers and diaries also shows that there was a strong gender dimension in the appeal of these male 'heroic' travellers. While next to nothing is known of their intimate life, in Europe or Asia, neither Hermann, Adolph nor Robert Schlagintweit ever married, and Robert was only once briefly engaged (their younger sibling Emil compensated for this fact by having six children with his wife). But Robert eagerly noted the consistently large presence of women at his public spectacles, and claimed that the gripping depiction of his adventures had female listeners in the packed audiences occasionally lose consciousness.[172]

In France, too, the Schlagintweits became widely known travellers, and their story was interwoven into some of the most successful popular works of literature at the time. If there was one nineteenth-century writer who best captured the heightened interest in mobility and adventures, it was Jules Verne. In his first successful novel, *Five Weeks in a Balloon* (1863), Jules Verne's protagonist travelled 'during the years 1855 to 1857' through 'the west of Thibet, in company with the brothers Schlagintweit'.[173] After their initial appearance, the author with an out-spoken geographical passion memorialised the brothers' Asiatic travels in several other novels, helping the Schlagintweits become household names in France also. Like Robert Schlagintweit in his lectures, Jules Verne also immortalised the tragic death of Adolph Schlagintweit as a pioneering traveller to central Asia. Verne informed his readers that since the time of Tamerlane, 'there have been fierce sultans [in central Asia], it is true'; 'among others that Ouali Khan Toulla [sic], who in 1857, strangled Schlagintweit, one of the most learned and most daring explorers of the Asiatic continent'.[174]

The brothers also became celebrated explorers in Russia, where Robert Schlagintweit had given a series of lectures in 1867–68. Robert was even invited by Russian officials to explore parts of central Asia, but this time in the service of Britain's supposed imperial competitor. Testifying to the shifting loyalties of these imperial opportunity seekers, Robert only declined this offer because he had already planned his first

---

[171] The 'motto' for his lectures was: 'Überall bin ich zu Hause// Überall bin ich bekannt,// Macht mein Glück im Norden Pause,// Ist der Süd mein Vaterland.', capturing his sense that all the German lands were his fatherland and he himself by now a national icon: BSB, SLGA, V.1.1, 2.

[172] On the gender appeal of 'heroic' exploration, with women sometimes presenting 'the majority' of listeners at the public lectures, see BSB, SLGA, V.1.1, 10.

[173] Jules Verne's works with references to the Schlagintweits include: *Five Weeks in a Balloon* (first published in French in 1863); *In Search of the Castaways* (1868); *The Steam House* (1879–80); *Robur the Conqueror* (1886).

[174] Verne, *Bombarnac*, 111.

lecture tour to another aspiring empire: the westward-expanding United States.[175] Against many contemporary and current assumptions about an Anglo-Russian antagonism in central Asia, a perspective that often problematically focuses merely on these two powers and less on central Asian polities and their own agendas, Robert returned from Russia convinced that no risk existed of any potential Russian invasion of British India.[176] Rather, he welcomed Russia's supposed 'civilising' influence in Inner Asia, and saw in China the greatest foe of western powers (among which he included Russia).[177] Although Robert Schlagintweit had rejected the Russian offer, a few celebrated Russian explorers of central Asia, including Nikolai Mikhailovich Przheval'skii, *the* celebrated Russian explorer of central Asia, became deeply moved by Adolph's fate, and pushed for a memorial to be erected at the place of Adolph's beheading in Turkestan.[178]

Yet, while the initiative for the Schlagintweit monument came from outside, some prominent German officials sought to become involved in its realisation. One of them was the diplomat and East Asian expert Max von Brandt, a former Prussian officer who had become minister resident in Japan in 1872, only to be promoted to the rank of consul in China from 1875 to 1893.[179] Brandt never lost sight of German commercial and colonial interests in Asia. He had accompanied the Prussian expedition to east Asia between 1860 and 1862, and had actively advanced the idea in the mid 1860s of taking the Japanese island of Hokkaido as a formal German colony.[180] Later, he supported the establishment of the German Steamship Line during the Kaiserreich to encourage German trade with its colonies in Africa and the Pacific, and was involved in the foundation of the Deutsch-Asiatische Bank in Shanghai in 1889.[181]

In consequence of his eagerness to advance German interests and prestige abroad, Brandt saw the creation of Adolph's projected memorial as an important opportunity. From 1887 onwards, he stayed in close correspondence with the German Foreign Office and Russian officials such as the consul in Kashgar, Mr Petrovskij, to bring the plan about. He was

---

[175] The idea of a further exploration into central Asia on Russian 'state expenses' was proposed to Robert by the Russian privy councillor Nicolai von Stein, BSB, SLGA, V.1.5, 8.

[176] As recounted in summaries of his lectures, ibid., V.1.1, *Bromberger Zeitung*, 15 April 1868, 281.

[177] BSB, SLGA, V.1.3, *Der Freischütz*, 14 February 1867, 43.

[178] Petrovskij, *Turkestanskie Pis'ma* [*Turkestan Letters*], no. 89, Petrovskij to N. M. Przheval'skii, Kashgar, 30 January 1887, 195. On his life, Waldron, 'Asia and Empire'.

[179] Franke, 'Brandt', 531.

[180] Brandt, *Ost-Asien*, vol. I; Wippich, *Japan als Kolonie*.

[181] Franke, 'Brandt'.

also responsible for interventions with the Chinese government in Peking to secure the right to erect the monument.[182] With the energetic support of the German consul in Peking, and the vice-president of the Petersburg Society, Baron von Osten-Sacken, the Schlagintweit monument could be unveiled in June 1889.[183] While no German economic interests were at stake, for German officials the monument in the difficult-to-reach interior of central Asia potently symbolised the extension of German overseas exploration.[184]

For this transnational act of commemoration, Emil Schlagintweit, other officials of the Bavarian government and even the German emperor Wilhelm II personally wrote letters to the members of the Petersburg Society, thanking them for the gesture of recognition for the brothers and – through them – German overseas accomplishments.[185] It is indicative of how the Schlagintweits had indeed become German heroes that the secretary of the Paris Geographical Society, Edouard Blanc, defended the French plaque on the monument, which he had personally visited in 1890:

The German government had full right to appropriate the glory of the Bavarian Schlagintweit for the sake of the whole Empire in the name of German unity. We, in our capacity as geographers … appropriate the glory of the deceased for the whole of Europe, even for the whole world in the name of science. On the field of science, there should be no enemies, only collaborators.[186]

However, in contrast to such claims about the universal value of scientific achievements, Adolph's death still assumed a different political meaning in Germany – where this victim of overseas exploration became incorporated into a nationalistic culture of imperialism.[187]

---

[182] Letters from Brandt from Peking to the board of the Berlin Geographical Society, 8 July 1887, and to a Russian minister, 9 July 1887, at PA AA, Peking II 891.

[183] E. Schlagintweit, 'Bericht'.

[184] See the 'Subscription List' for donations to the monument, which was to be erected 'to the glory of Adolph von Schlagintweit'; report by Ludwig von Ammon, in 1889, at DMM, HS08683; Sto: NL 12/171. In 1886, the German geographer Friedrich Ratzel, who developed the theory of *Lebensraum*, had supported the installation of a 'commemorative plaque' to the Schlagintweits at their birthplace. ES to Hüther, 25 September 1886, Stadtarchiv, Munich [StM], no. 123.

[185] BayHStA, MA 53157, doc. 1, 15 January 1890, 'Kgl. Bayerische Gesandtschaft in St. Petersburg'; for the German emperor's personal letter of gratitude to Petrowskij, ibid., doc. 3, 'Kgl. Bayr. Gesandtschaft in St. P.' to 'B. Staatsministerium des Kgl. Hauses und des Aeusseren', 10 August 1891.

[186] *St. Petersburger Herold*, 8 August/27 July 1887; BayHStA, MA 53157, appendix to doc. 3.

[187] See the communication by the 'Kgl. Bayr. Gesandtschaft' to the 'B. Staatsministerium des Kgl. Hauses', 10 August 1891.

Even though Schlagintweit's murder had also been mourned in Britain, the loss of a young and talented German explorer (and the disappearance of other German overseas travellers in Africa and Australia) assumed a specific importance for contemporaries. It was emphasised, time and again, that Adolph's early demise symbolised both the contributions and the personal sacrifices that *Germans* also made for the sake of European overseas discoveries. It was precisely because the German states were neither unified as a nation until 1871, nor possessed any colonies abroad before 1884, that the Schlagintweit expedition and the personal suffering of the brothers could feed into a German colonial imagination and ideology. Adolph Schlagintweit, like several African explorers, became not only a hero, but also a 'martyr of German science'.[188] As they had often fallen in the service of a foreign empire, their glorification served clear political ends. That is, the involvement of German subjects in opening up non-European landscapes to western science and commerce was used to assert German claims to enter the colonial race, too.[189] A case in point is the publication by a colonial agitator that appeared in 1867 under the title *The Foundation of Prussian-German Colonies in the Indian and Great Ocean, with a Particular Focus on East Asia*. In it, the author drew this ideological connection between 'German sacrifice' and colonial right:

From a scientific standpoint, what a great opportunity finally opens up to us with the colonisation of Formosa [Taiwan]! A country long since called one of wonders and mysteries, whose geographical characteristics, natural resources and population are as yet hardly known, it offers in many respects a virgin field for explorations of all kinds. And who is better qualified to exploit it than the German people, which long ago acquired the entitlement to its own colonial possessions with the health and life of his most noble sons – and especially Prussia, which has undertaken with the greatest liberality so many scientific and practical expeditions into the remotest countries. Humboldt, Leichhard, Schlagintweit ... Barth ... are names, of which even the greatest maritime and colonial states are jealous, and we will justifiably call any project of German colonisation ... an act of piety towards our martyrs of science who are buried in foreign soil.[190]

---

[188] The list of contemporary works that described Adolph as a German 'martyr' of science and exploration is endless; see, e.g., *Landshuter Zeitung*, 103, 7 May 1859, 416; Adolph's well-attended funeral service is also described in the language of martyrdom, *Bayerischer Kurier*, 125, 7 May 1859, 846. Even for the orientalist and traveller Ármin Vambéry, Adolph led the list of the most eminent 'martyrs of the science of geography' in having opened the interior of Asia for European rule. Vambéry, *Westlicher Kultureinfluss*, 2. On German African explorers and the commercial implications of their work: Gesellschaft für Erdkunde zu Berlin, 'Erster Aufruf', 170–1.

[189] Connected to these discursive themes, but also putting forward an important intervention against a teleological view that assumes the realisation of colonial 'fantasies' into formal rule, Conrad, *German Colonialism*, ch. 2: 'Colonialism before the Colonial Empire'.

[190] Friedel, *Gründung preußisch-deutscher Colonien*, 82–3.

The loss of 'German lives' in the service of other empires thus became a justification for German politicians, merchants, liberal journalists and colonial agitators to demand an enhanced German presence in the extra-European world. Hence, such pre-colonial experiences of German subjects were *not* only 'retrospectively adopted by the colonial movement during the 1880s and ... pronounced to be the pre-history of contemporary expansionist plans'.[191] The 'vector of continuity' went not merely backwards into the past to justify the acquisition of colonies in the late nineteenth century.[192] Rather, the Schlagintweits' pursuit of scientific careers in a foreign empire would have important ramifications in the country where these imperial outsiders originated. It fed a sense of prospective colonial entitlement in the German states as well.

[191] Conrad, *German Colonialism*, 17.
[192] Ibid.

Conclusion

Research on modern European colonial empires has undergone significant change. While historians had long recognised the transnational nature of empires in the early modern period, where a multinational management of overseas possessions was often the norm rather than a deviation, modern colonial formations continued to be frequently represented as more distinct national enterprises, believed to be internally homogenous and self-reliant and externally competitive.[1] For Eric Hobsbawm and many other historians of the later twentieth century, the 'Age of Empire' was 'essentially an age of state rivalry', and therefore of mutual antagonism that seemed to leave little space for collaborative cross-border ventures and transfers.[2] Yet empires were always embedded in larger environments – and many were unable, or indeed unwilling, to shield themselves from external ideological, political or scientific influences.[3] It has been a key concern of this book to demonstrate that the emergence of global perspectives on the past allows a fundamental reconsideration of the inner workings of empire. The work forms part of a wider endeavour by historians in recent years to take seriously the remarkable permeability and openness of imperial systems for indigenous and other groups of actors, who originated from beyond the boundaries of the respective imperial motherland.[4] A retrieval of their life stories, it has been argued, illuminates how transversal mobility could lead to significant movements of skills, ideologies and expertise across imagined and political boundaries often too quickly taken for granted.

Through recovering in detail the experiences of a set of German scientific travellers in and beyond the borders of 'British' India, this work has examined how the imperial exploration, documentation and exploitation

---

[1] Kumar, *Visions of Empire*, 149; Arnold, 'Contingent Colonialism'.
[2] Hobsbawm, *Age of Empire*, 51.
[3] For example, on Anglo-German colonial cooperation in Africa, Linder, *Koloniale Begegnungen*.
[4] Alavi, *Muslim Cosmopolitanism*.

of territories opened up opportunities for a range of enterprising foreigners – among them many experts in the fields of expeditionary and natural sciences and surveying. Until the establishment of a colonial empire by a then unified and expansive nation state in 1884, German naturalists were confronted by the lack of formal overseas possessions of their politically fragmented native countries in which they could have realised scientific ventures. However, as they underwent progressive university training in the German states and were thus equipped with internationally recognised expertise in fields as diverse as physical geography, meteorology, botany, geology, agricultural chemistry and scientific forestry, German scientists faced what I have called multiple *empires of opportunity*. These were the colonial systems and infrastructures that other European powers had established abroad – with the extensive and geographically complex Indian subcontinent attracting particular attention from a large series of curious non-nationals.[5] The British East India Company, itself a hybrid institution with commercial and state-like qualities acted, in fact, as one of the most significant patrons for scientific positions outside of Britain and continental Europe from the mid eighteenth century until the end of Company rule in 1858.

Since the early 1800s, British imperial authorities had increasingly relied on a mobile workforce of German and north European naturalists and scientific travellers, who were at times regarded by British patrons as superior to their countrymen in practice and outlook. Yet, such foreign recruitments also occurred because there were simply not enough qualified people amongst the home population willing or able to fill the increasing number of scientific, administrative and technical positions in India and at the fringes of the expanding Company Raj (Chapter 2). Expert migration thus proved to be a two-way street. In effect, British officials and German scientific specialists utilised each other for different motives, but with mutual benefit.

The influential careers of imperial outsiders from the German states (and Scandinavia, as in the case of the Danish botanist Nathaniel Wallich and others) demonstrate that the history of scientific activity, material extraction and institution building in India was much more than the result of British imperial agency and ambition alone. This crucial finding informs the wider point that in this period European imperialism was, to a degree, a collaborative project. It involved numerous actors of different western nationalities and, as this work has particularly highlighted

---

[5] Already contemporaries were aware of the transformative impact of non-British and especially continental European outsiders for western science in south Asia; Liebig, 'Natural Science'.

(Chapters 4 and 5), a wide range of people from south Asian polities who entered the service of the Company or the Raj as surveyors, informants or scribes out of diverse interests and motivations.

Yet, while I have proposed this more inclusive model of European imperialism, the book has also demonstrated that foreign influences and the large-scale reliance on external expertise and manpower for British rule in south Asia never went uncontested. On the contrary, this work meticulously explored how imperial dependence on outside staff and know-how provoked fierce debates among the scientific community and the wider publics in Britain and India alike. These mid-nineteenth-century debates were part of the rhetorical, albeit not actual, 'nationalisation' of empire at the time. Through a symptomatic reading of a wealth of manuscript and printed materials associated with the Schlagintweit enterprise, the work traced this hardening of mental rigidities in reaction to a perceived wave of overly privileged outsiders in British service at a moment of great social and political crisis in the wake of the Crimean War and the Indian Rebellion. Going beyond simple assertions that there existed diffuse British antipathies towards German recruits at the time, the book has deployed a rich contextual analysis around the conduct and scientific programme that the brothers launched in India before and during the political conjunctures of the EIC's suppression and the subsequent transfer of authority from Company to Crown. By narrating the violent controversy that for years surrounded the travellers' careers from as many standpoints as the surviving documentation allows, I have sought to provide a more nuanced picture of their contested exploration, comportment and legacy. This multi-sided analysis was also necessary to capture the complex interactions and manipulations between scientific circles in Berlin and London, imperial authorities in Britain and India and various reading publics in Europe.

One of the work's principal concerns was to establish the precise role of colonial, national and indigenous interests in the career and fate of the Schlagintweit mission. As a succession of chapters on their multinational sponsors, eminent scholarly patrons and their equally notable and vocal critics have shown, what made this transnational scheme worthy of intense study is precisely the fact that its key motives seemed to shift according to the particular agenda of the diverse agents who held a stake in it – either as benign promoters or fierce opponents. Some British men of science, such as the Royal Society magnate Edward Sabine, initially thought their project was to form part of a dominant model of magnetic surveying across the Indian subcontinent. The brothers, driven by great personal ambition, however, almost instantly transformed this geomagnetic focus and definition of their programme into a Humboldtian enterprise of global physics, large-scale climatology and an interdisciplinary geographical exploration of south and central Asia. Their inspiring mentor and tireless supporter Alexander von Humboldt, in turn, threw his entire weight behind their scheme precisely because he sought to defend his own model of more holistic empirical engagements with extra-European natural and cultural worlds at a moment of increasing challenges by more narrow specialists in diverse branches of the sciences.

The brothers' fierce and noteworthy opponents, by contrast, used the Schlagintweit appointment as an occasion to expose and attack the Court of Directors' erratic support for scientific activities, which they felt was lacking commitment to meritocratic considerations and a willingness to also support more abstract, non-utilitarian goals. More fundamentally, however, the critics also sought to establish or consolidate their own disciplinary and cultural expertise by criticising the Schlagintweit enterprise as either lacking the detail and reliability of restricted enquiries in a more narrow field, as shown with the imperial botanist Joseph Hooker and the Prussian mineralogist Weiss; or, as in the case of the naturalist Berthold Seemann, by denouncing the brothers because of their over-attention to local precision and supposed lack of a general theoretical overview. Their expeditionary project was thus open to somewhat contradictory attacks, which brought into the open obvious and tangible ambiguities around disciplinary specialism. Indeed, rigid scientific disciplines, when understood as 'well-institutionalized homogeneous systems of formal behavior', hardly existed in the mid nineteenth century, when the status of fieldwork and of precise sets of standard practices were violently debated among metropolitan and colonial naturalists and travellers alike.[6] Rather, by deploying a contextual approach to a large set of critical reactions, this work has shown how the private and public responses to the Schlagintweit mission by some of Europe's leading men of science helped to shape disciplinary politics and order.

An interpretative theme that has run through this book is that of authority – its social logics and multifarious connections with the work of trans-cultural scientific exploration. As the detailed analysis established, authority as a form of power and influence in a given social or scientific setting was never universal or stable, but was rather the product of constant negotiation and conflict – in the Schlagintweits' case in an explicitly transnational arena. Taking the always volatile nature of authority seriously – its establishment, challenge, eventual loss and recovery in other contexts and in front of different audiences – helps to make sense

---

[6] From the incisive reflections by Schaffer on 'Disciplines', 58.

of the brothers' diverse trajectories, and how they ultimately proffered their results and legacies of exploration, often in a carefully orchestrated manner, to target specific consumers.

Although Alexander von Humboldt by no means possessed an unchallenged monopoly over the disciplinary cartography of the first half of the nineteenth century, his work still cast a monumental shadow over expeditionary science in the 1850s. The brothers' scientific careers were launched, as Chapter 1 has demonstrated, when the international authority of 'German science' was high and Humboldtian models of enquiry, despite growing specialist attacks, still found followers and admirers across continental Europe, Britain, North America, India and elsewhere. The Schlagintweits' early exploration of the Alps, then widely considered the least-known region within Europe, mirrored the larger outthrust of German naturalists and scholars into various corners of the globe – often, though by far not exclusively, under the protective umbrella of other states and their military and scientific services. This intrepid *Welterfahrung* coincided with a burst of philological and other scholarly researches in the German states that allowed German oriental scholarship and scientific enquiry to enjoy high status and credibility in relation to distant lands, their cultures, histories and religions amongst other scientific communities.[7] It is indicative of the international authority of German humanist learning that the leading orientalist in mid Victorian Britain was Max Müller, professor at Oxford, who, among others, decisively shaped the comparative study of religions and Indo-European languages.[8] The elite standing of both German scholarship and science also found tangible expression in the frequent appointment of German ethnographers and naturalists by British, Russian and Dutch authorities and patrons of science. The Schlagintweits' recruitment into the Company Raj in India through efficient patronage networks that were cleverly exploited by their mentors Humboldt and Carl Ritter only exemplified a much older and larger migration of German talent into foreign employment.

At the same time, the intense connections between Humboldt, Ritter and other Prussian geographers and British colleagues, Indian naturalists and officials demonstrated that there were, in fact, multiple 'centres' in Europe that engaged in significant ways with south and central Asian societies and natural environments. By putting the focus on Berlin and nearby Gotha as one such scientific hub, and by examining a number of collaborative enterprises, this work has shown how some German

---

[7] Marchand, *German Orientalism*.
[8] Davis, 'Max Müller'; Halbfass, *India and Europe*.

states became thoroughly integrated into imperial knowledge networks. Eminent German practitioners of geographical sciences received and synthesised observational data from numerous colonial servants, travellers, officers and medical and military institutions.[9] While such transversal exchanges relied partly on a polite culture of scientific correspondence and instruction, travelling naturalists like the Schlagintweits were equally crucial as mediators in such cross-border transfers of knowledge, yet the ramifications that their careers in foreign service had for their non-colonial homelands have previously hardly been studied. The work has demonstrated that the specific forms of glorification and public memory around their careers in British employment later fuelled a national-imperial discourse that claimed since the 1860s that the exploratory feats of German overseas travellers in the past would justify a formal colonial outthrust of their fatherland in the future (Chapter 8). At the same time, the significant presence of German scientific travellers and specialists in the East India Company and later the Raj makes clear that the German contribution to the history of European imperialism was never limited to the carving out of spheres of 'informal empire' (from the Middle East and central Asia to South America) or the establishment of formal colonial possessions. It also found expression in the development and despatch of manpower and scientific and technical *Produktivkräfte* and know-how into the service of foreign powers.

As the Schlagintweits' project of expeditionary science has forcefully shown, the authority of 'German science' also had a strong political dimension. This became especially pronounced in published German responses to the increasingly hostile British attacks on their Asiatic scheme and its early published results. 'German science' – as a discourse on alleged German thoroughness, profundity and skilful interdisciplinary enquiry, rather than as a specific set of practices – now became a topos for articulating 'national' solidarity with the brothers from Prussia to Bavaria. But 'German science' turned not only into a realm of reassurance but also of national hubris, in which a supposed German superiority over other powers was confidently claimed and widely expressed. Yet there also existed, with equally strong political connotations, a discourse on 'British science' and its historical achievements in India – and the two spectacularly collided in the Schlagintweit enterprise and its media reception. Precisely because the Munich-born brothers seemed to 'colonise' British India, in the sense of declaring scientific discoveries already

---

[9] This book therefore expands the existing literature on imperial capitals as, in Latour's phrase, 'centres of calculation'; see, e.g., Miller, 'Joseph Banks, Empire, and "Centers of Calculation"'.

claimed by British surveyors, officers and naturalists, could the public reactions reach such fever pitch across the Channel. The reason was that British authority over the colonised peoples of south Asia was widely believed to partly depend on the imperial nation's scientific triumphs, and on the sense of the superiority of British intellect that was supposedly shown to the 'native mind' by the pre-eminence of the Company's British officers and their longstanding projects, such as the monumental Great Trigonometrical Survey. This close connection between 'British science' and the alleged legitimacy of colonial rule on the subcontinent became especially pronounced in the wake of the Indian Rebellion, when the Company's domination was fundamentally challenged, militarily in India, and publicly in Europe (Chapter 6).

It was in the midst of this large crisis, when Britain's trust in their 'Indian subjects' was deeply eroded after the outbreak and violent suppression of the Sepoy Rebellion, that the returned Schlagintweits published their first scientific works on Asia. While sometimes ridiculing the religiously informed worldviews of their many Hindu assistants, the brothers nonetheless openly acknowledged their at times total dependence upon the authority and guidance of their south Asian companions. But thereby hung a tale. While from today's perspective such an act of recognition appears honest and laudable, it exposed the Schlagintweits to contemporary critique and biting ridicule. The reason was that the scientific authority and trustworthiness of field accounts then largely depended on the sources of geographical and other knowledges. Since the Schlagintweits depicted their enterprise as a fundamentally collaborative and trans-cultural project – one marked by shifting hierarchies among the members of the diverse expedition party that sometimes excluded the brothers from taking key decisions about routes, sustenance and strategies – they broke with widely shared conventions of mid Victorian exploration and publishing.[10]

The relationship between trans-cultural exploration and scientific authority was thus highly ambiguous. While Indian and central Asian guides were indispensable for some of the mission's most celebrated exploratory feats, and while their independent excursions considerably widened the scope of the entire scheme, the brothers' acknowledgement of indigenous leadership and instruction in the field at the same time undermined their standing in Europe. The scathing British reception of their accounts, especially of those sections depicting the enterprise as a collective work, confirms the fact that individual authority in

---

[10] Driver, 'Face to Face', 455.

the scientific culture of the time hinged massively on considerations of training, race, gender and class.[11]

However, beyond such racist sneers against the Schlagintweits' Indian and Himalayan employees, there was an element of truth to the critique that the large and predominantly untrained indigenous workforce of the expedition accounted for some of the shortcomings of its results. This was especially the case with their rich scientific collections, whose later value depended heavily on the correct and careful practices of sketching, preparation, conservation and minute description in the field if the objects were not to lose their use for metropolitan analysis and comparison. The Schlagintweits, driven by the desire to collect enough specimens to fill their own projected India museum in Berlin, engaged the services of sometimes entire villages in the Asian highlands (Chapter 5). This involvement of amateur hands undoubtedly diminished, in turn, the scientific appeal of their artefacts – and hence also the authority of the expeditions' European 'leaders' – as Joseph Hooker's caustic critique of their botanical herbarium evinced. Equally telling is the fate of the brothers' extensive mineralogical collections. Never properly analysed because it lacked complementary data, Robert and Emil Schlagintweit later sought to sell the entire series through a mineralogical dealer in Bonn, without success. In the end, Emil received thanks for his consent to 'the removal of the worthless debris' – together with an invitation to Bonn for 'a walk over the path that is paved with it'.[12]

Yet, despite such anecdotal setbacks, this work has proven how closely related were the Schlagintweits' claim to expertise and authority over south and central Asian geographies, cultures and 'racial' varieties on the one hand, and their accumulation of large sets of material documentation on the other. While not part of their initial contract (which had only roughly established their later fields of enquiry) with the EIC's Court of Directors in 1854, the ethnographic and physical anthropological surveys of the brothers were especially well received upon their return. Britain's leading anthropometrist, Joseph Barnard Davis, even claimed that their series of 275 ethnographic heads formed from 'living specimens' would represent 'by far the most important contribution ever made to Indian Ethnology'.[13] With no prior ethnographic or anthropological training or experience, the Schlagintweits indeed became lastingly concerned with documenting social and cultural variation across India

[11] Bourguet and Licoppe, 'Voyages'.
[12] Rheinisches Mineraliencomptoir Dr A. Krantz to ES, 12 May 1887, BSB, SLGA, VI.5.2.
[13] Davis' letter was reprinted in H. and R. Schlagintweit, *Prospectus of Messrs. de Schlagintweits' Collection of Ethnographical Heads*, iv.

and the Himalayas in sketches, photography, bodily measurements and plaster-casting that made their transit expedition significant as an early example of large-scale comparative ethnography years before the official survey of the 'People of India' (1868–75) used photographic illustrations to 'represent the different varieties of the Indian races'.[14]

While the itinerant brothers were not primarily motivated by the concerns of oriental studies when they set out, it could be shown that in the wake of their Asiatic expedition, Emil's use of their manifold collections, Robert's popular lectures and Hermann's travelogues did engage with popular and scholarly orientalism. What therefore added significance to their scheme is that, due to the Schlagintweits' practices of large-scale collection, seizure and accumulation in the field, it ultimately contributed to both work in the humanities and in various branches of the sciences – from Indian meteorology to racial science. The rise of anthropology and other natural sciences was, as Andrew Zimmerman has conclusively shown, central to an antihumanist reconfiguration of the European self in the age of high imperialism, especially in Germany with its old tradition of humanist *Bildung*.[15] Ultimately, as part of the brothers' ambivalent legacy, not only leading German racial scientists but also specialists from France, Britain and elsewhere came to rely on the Schlagintweits' large sets of anthropometric data, ethnographic observations and the plaster faces of 'living specimens' displayed in numerous museums to compile, in the later nineteenth century, racial cartographies and genealogies.[16] The youngest Schlagintweit brother, Emil, profited perhaps most directly from the wealth of Asiatic objects and Buddhist artefacts his siblings had acquired. While he arranged their diverse manuscript collections, he also utilised them for philological and historical studies that secured him an international reputation as one of the leading Tibetologists among scholarly circles across Europe, Russia and British India into the twentieth century.[17]

During their lifetime, the Schlagintweits themselves magnificently exploited their collections as a source of scientific authority, income and career advancement. Given the relative scarcity of philological, natural-historical and anthropological collections from overseas in German

---

[14] Howe, 'Hooper', 713.

[15] Zimmerman, *Anthropology*.

[16] See, among many others, Welcker, 'Kraniologische Mittheilungen', 90; Peschel, *Völkerkunde*, 79, 383, 407; Quatrefages and Hamy, *Les crânes des races humaines*, 190, 324, 363. Topinard, 'Visite', 295.

[17] See the laudatory tribute to Emil's lasting achievements by the British orientalist Frederick W. Thomas, who lost with him a cherished colleague as 'one of the few Europeans interested in the study of the Tibetan language'; Thomas, 'Obituary', 215.

museums and universities in mid-century, especially when compared with those of great maritime powers such as Britain or France, the appeal of their objects and textual and visual materials was significant.[18] While the brothers' far-reaching plans for their own permanent India museum in Berlin (and later in Franconia, Munich and Nuremberg) did not materialise, in which their status was to be enhanced through their directorship, other uses of the collections proved successful. The brothers gifted parts of their collections to aristocratic and royal benefactors in return for lavish orders and distinctions, thus also obtaining further chances of future patronage. In sending out large sections to international colleagues, who often helped in the analysis and publication of the material, the brothers also established their reputation across various national borders and learned institutions. There was a strong commercial and entrepreneurial dimension, too. Hermann and Robert profited from the mechanical reproduction of some especially striking series of artefacts, leading them to sell, for instance, the entire series of plaster heads to public holdings from London, Paris, St Petersburg, to Madras and Calcutta. During public performances, including lectures or presentations at colonial and world fairs, the brothers also made sure to always prominently represent their names and achievements in the history of European exploration through the careful display of Asian artefacts. These were complemented with stunning iconographic works by the brothers, who used the most up-to-date technologies and lithographic devices to depict – with the unacknowledged assistance of numerous German artists – various tropical and high mountain scenes and traces of indigenous culture for western audiences.

The Schlagintweits' turn towards more public forms of scientific display and instruction was analysed in this work as a strategy to compensate for their more controversial standing among British and German subject specialists – even if the two surviving brothers received prominent scientific distinctions and membership in prestigious academies. Hermann became a member of the German Academy of Naturalists Leopoldina in 1863 and of the Bavarian Academy of the Arts and Sciences in 1881, a year before his death. Robert Schlagintweit managed to secure a university chair in geography and comparative statistics in Gießen in 1864. Yet, academic circles and modes of enquiry had lost much of their appeal in comparison with public spectacle and popular entertainment, where the defamed traveller found a rich new field of self-fashioning as purportedly one of the greatest travellers and scientific popularisers

---

[18] Manias, *Race*, 23.

of his age. However, while leading British men of science in particular denied the brothers' recognition as scientific equals, the Schlagintweits enjoyed, at the same time, unbroken esteem in Indian scientific circles and survey departments, who consulted the German explorers for advice and judgement even decades after their return – a striking incidence of vastly asymmetric reputations between colonial periphery and metropolitan centre.

Yet, authority played another fundamental role in the Schlagintweit enterprise. When not only taken to mean someone's commanding manner or recognised knowledge, but also understood as tangible power – as the ability to give orders, to enforce obedience or to gain rights of access to otherwise closed spaces – then 'authority' also helps to make sense of how this expeditionary project could actually be conducted in its specific colonial context. To take the example of their anthropological work, the brothers' eagerness to compile large sets of anthropometric data on 'racial types' encouraged them to make use of their privileged mobility as officers in the service of the Indian government, and to fully exploit their access to sites deeply marked by colonial power asymmetries. The Schlagintweits photographed and quantified the bodies of Indian prisoners, but also added human remains from colonial hospitals and plundered tombs to their swelling collections. Their climatic and magnetic studies across far-flung regions in south and high Asia equally relied fundamentally on the mobilisation of the EIC's various scientific, technical and medical services. These provided data for much longer periods than the Schlagintweits themselves spent in Asia, and which subsequently became the foundation of some of their most important published accounts, especially on south and high Asian climatology. Within British territory, at least, the Schlagintweits also received official authority to press porters and other helpers into service, a practice they likened to conduct during military marches (Chapter 4). The crisscrossing expeditions undertaken by the Schlagintweits with large sets of changing south Asian companions and contractual workers can thus be understood, in a first sense, as an imperial enterprise, because it was in its ultimate form heavily dependent on imperial infrastructure and support. The brothers' close cooperation with British agents and institutions, including the Surveyor-General's Offices at Dehra Dun and Calcutta, or the military stations and training centres of the Office of the Quarter Master General at Madras (from where several of their employed draughtsman and surveyors originated) confirms this point.

More ambivalent is the interpretation of the extent to which their enterprise was also an imperial expedition in the sense of being instrumental to British authorities and interests – or at least intended to be so.

Of course, outcome does not determine intention, and the eighteenth and nineteenth centuries saw numerous expeditions and other scientific schemes undertaken for imperial expansion or other metropolitan benefits that ended in disappointment.[19] Such failure does not, however, turn them ex-post into disinterested enterprises. What the in-depth analysis of the Schlagintweits' scientific programme in the field, in combination with their later negotiations with the East India Company, has unmistakably shown is that more disinterested scientific enquiries *and* imperial concerns about resource identification, route intelligence and the gathering of commercial information and objects were never mutually exclusive. This is further evinced by their subsequent publication of a substantial *Route-Book*, itself a military, not a scientific, genre, full of elaborately described routes and passes across the north Indian frontier and the Himalayas at different seasons.

The practical dimension of the Schlagintweits' programme is also revealed by their accumulation of water and soil samples to be explicitly tested with the view of improving the agricultural output of British India. Imperial concerns were also palpable in the Schlagintweits' lasting obsession with the wellbeing of Europeans in India and the Himalayas, and their incessant search for suitable sites for colonisation and the foundation of sanatoriums to restore the fragile health of 'whites'. The same quest for useful knowledge is evident in their rich series of plants and industrial collections of textiles and papers that were verifiably of wider interest to European manufacturers and producers. While far from all Schlagintweit objects had immediate industrial implication, many ethnographic and natural-history artefacts could be sold commercially. I follow Felix Driver in suggesting that much of the brothers' collecting efforts were part of a programme of *applied natural history*, 'in the sense that they were explicitly concerned with raw materials and industrial processes that could be exploited in Europe for economic benefit' (Chapter 7).[20] The Schlagintweit enterprise thus offered an exceptionally rich opportunity to explore the ambiguities around the precise relation between Humboldtian physical geography and the economic interests of colonial botany and resource imperialism. The brothers' careers, and even more so those of later German scientific specialists like Gustav Mann, Dietrich Brandis or the London-born John Augustus Voelcker active and

---

[19] Instances of imperial failures, of official schemes of colonial resource exploitation running into the ground, are attracting scholarly attention as they help to illuminate the limits of power of European empires; Mellilo, 'Global Entomologies'; Sivasundaram, 'Sciences and the Global', 155.

[20] Driver, 'Face to Face', 448.

influential in the Raj point to a significant shift in the role of science as a 'tool of empire', as it became more interventionist, indeed indispensable, for conquering vast new swaths of colonial territories since the 1870s and managing their various resources.[21]

While the mission was partly, but by no means exclusively, framed by colonial and imperial interests, these were, however, often hushed up in public descriptions of the mission. This book has carefully illustrated how flexible were indeed the definitions of the expedition's aims and scope in front of mixed audiences and patrons (Chapter 6). Yet, when the content of the Schlagintweits' later publications and public lectures are considered, produced by Hermann and Robert long after British support had ceased, it is remarkable how enduring the brothers' interests and investments in imperial issues really were. This indicates that the personal experience of the 'opportunities of empire' was changing in a more fundamental sense: whereas the Schlagintweits' Alpine studies had been free from political concerns and completely uninterested in culture, their service in south Asia had opened up new horizons and ultimately changed the brothers' perception of the imperial world around them. Robert Schlagintweit in particular, fully accepting a supposed British civilising mission, demonstrated a lifelong concern with agricultural and commercial improvement schemes and the material opportunities in south and central Asia awaiting European and North American intervention and enterprise (Chapter 8). To be sure, Robert, unlike his younger half-brother Max Schlagintweit, never sought to establish formal German colonial possessions in the east; his was not a state-centred imperial outlook, but rather one that encouraged further western exploration, private trade and resource exploitation in the Himalayas and beyond in the wake of further British territorial expansion that he forcefully promoted.[22]

To understand how the itinerant brothers could mobilise and address different publics, from specialist to popular audiences across Britain, continental Europe, India and the United States, the last three chapters have paid close attention to their intricate publication strategies, their role as orators and the transnational distribution networks that sustained popular interest in their expeditionary exploits. Here, however, a salient ambiguity prevails. Too often, general historiographic approaches assume that the appropriate mode of analysis of such major expeditions is to focus exclusively on the strategies of publicity, of propaganda and

[21] Headrick, *Tools of Empire*; Hodge, *Triumph of the Expert*; Fischer-Tiné, 'Rural Development'.
[22] BSB, SLGA, V.2.2.1, 105–6.

of distribution of exploratory results.[23] Indeed, the brothers' manipulation of their image as 'heroic' travellers and, in the case of Adolph, as a tragic 'martyr of science' proved a significant topic – especially in their native country, in view of the distinct politicised memories and repercussions of their overseas careers. Yet the brothers' case also offered a rare chance to explore the role of secrecy at work in transnationally funded schemes, and has drawn attention to a large range of techniques involved in consciously blocking communication, and in preventing peers and publics from gaining immediate access to the details of their enterprise. Especially in reaction to British criticism and scrutiny, the Schlagintweits' communication strategies involved strategic silences and the conscious exclusion of former colleagues from their correspondence – which clouded, to bewildered and furious contemporaries, their multiple British sponsorships in an aura of 'mystery'.[24]

In the Schlagintweits' careers, there was another significant aspect about the interplay of the private and the public. That is, the surviving travellers countered the widespread critique of their scientific programme by carefully managing self-constructed archives that were supposed to project a more desirable legacy to later generations. The brothers first created a public India museum in Berlin in 1857, yet one that had to be transformed in 1860 into a semi-private institution hosted in a Bavarian chateau explicitly acquired for this purpose. This was a lavishly designed space in which their wide-ranging travels were acknowledged and celebrated through displaying 'Schlagintweit objects' obtained from numerous Asiatic regions. The walls were hung with portraits of Indian rajas they had met, and with fine silks and curious objects that told anecdotes of their former adventures and cross-cultural encounters.

With regard to the newspaper war that raged over the travellers' contested achievements, a second private archive was a large collection of international articles on their expedition, Hermann's 'Collectanea Critica'. This compilation decisively favoured positive reviews and at times hagiographic responses from British, German and French scientific and artistic journals. Private correspondence completed these written memorabilia.

Finally, Robert Schlagintweit, the explorer turned public persona, sought to secure his place in history by leaving behind a meticulous

[23] Important works in this regard are Riffenburgh, *Myth*; Sèbe, *Heroic Imperialists*; and a long tail of more popular, often problematically hagiographic, works on single white men, including Jeal, *Stanley: Africa's Greatest Explorer* or Schlager, 'Vergessene Helden'. See for an insightful reflection on the different political appropriations of a single explorer's life, Livingstone, *Livingstone's 'Lives'*.

[24] *The Athenaeum*, 1768, 14 September 1861, 342.

documentation of his role as a celebrated populariser of science. The material legacy of his transformation into a scientific showman is enormous: in forty-one bound volumes of several thousand pages, it includes hundreds of his letters alongside newspaper articles, critiques and reviews, (auto-)biographical notes, advertisements, circular letters, tickets, invitations, business cards and posters announcing his public appearances. What makes this collection particularly striking is the way he was keen to carefully frame its later use. At the beginning of the first volume, Robert stated that the 'extensive materials' on his intercontinental tours, preserved 'as complete as possible', could 'perhaps later' provide a useful 'contribution to the history of public lecturing'.[25] For this purpose, he also provided 'a short historical sketch' of how he had managed to conquer considerable audiences to consume the accounts of his Asian and later American travels. Robert Schlagintweit also published and widely circulated an account of his popularising achievements.[26] In pointing to the brothers' interventions in shaping their legacies and the commemoration of their missions, Robert's astonishing archive presents another attempted self-inscription into the history of science, with him as pioneer of such travelling spectacle. Over the decades, the Schlagintweits' struggles for recognition thus turned them from Alpine naturalists into overseas explorers and artists, museological directors and finally 'heroes' and 'martyrs' of a future German empire whose experiences were rewritten many times to suit the ideological foundations of a soon unified nation with its own global aspirations.

Ultimately, this book's focus on the dynamics of inclusion and exclusion at play in schemes of transnational science addressed significant current and broader concerns about global history writing and its challenges in an age of anti-global backlash and resurgent nationalism. Since much, and significant, work in global history has tended to privilege accounts and comparisons of large-scale economic developments and macro-processes of convergence and integration, the field has sometimes (but by no means always) treated human mobility and experiences more in the mass. Yet to understand the broader phenomena of motion and orders – the opportunities created through cross-border mobility *and* local or national resentment against such connections and career-making and their attempted suppression – it is vital to consider detailed and complex human stories.[27] These can serve as starting points from

[25] BSB, SLGA, V.1.1, 'Correspondenz über öffentliche, wissenschaftliche Vorträge', 'Band I. Einleitung. –Statistische Tabellen', 3.

[26] R. Schlagintweit, *Bericht*.

[27] Significant therefore are works from the burgeoning field of global micro-histories of mostly itinerant lives of individuals taken to be illustrative of broader patterns of their

where to explore the establishment, resilience and ultimately the fragility of larger systems and cross-border networks when faced with discord and open opposition. It is precisely when mobile people ran against visible or imaginary walls, when transversal mobility did not lead to cosmopolitan outlooks but rather triggered exclusion and protective attitudes and actions in the political and social domains that the deeper issues and trends of a time are thrown into sharp relief.[28] As the case of the maligned Schlagintweit brothers suggests, it is often outsiders who are most aware of such defensive sentiments and orders, less the people 'within'.[29] Their exemplary story leads us to go beyond global history's proclivity for connectivity and integration, and rather to look critically at the contradictory legacy of growing globality during the past centuries.

---

time: Lambert and Lester (eds.), *Colonial Lives*; Subrahmanyam, *Ways to be Alien*; Davis, *Trickster Travels*; Colley, *Ordeal*; Ogborn, *Global Lives*.

[28] Conrad and Eckert, 'Globalgeschichte', 21.

[29] Nicholas Purcell, 'Seven Types of Mobility around Global Ancient Mediterraneans', unpublished paper, Inaugural Conference of the 'Global Nodes, Global Orders' Leverhulme International Network, 25–27 June 2015, Oxford.

Archives

<table>
<tr><td>AN</td><td>Archives Nationales, Paris, Pierrefitte sur Seine
F/12/4980, Expositions universelles, internationales et nationales
(1844–1921); Répertoire méthodique provisoire</td></tr>
<tr><td>APS</td><td>American Philosophical Society, Philadelphia
Archives, John Edward Gray Papers, 1783–1884
PMP, v.1196, no. 5, 8</td></tr>
<tr><td>BArch</td><td>Bundesarchiv, Berlin
R901, 37418, 'Akten betreffend: die wissenschaftliche Reise der Brüder Schlagintweit'</td></tr>
<tr><td>BayHStA</td><td>Bayerisches Hauptstaatsarchiv, Munich
Abt. II Geheimes Staatsarchiv, MA 53157 'Denkmalerrichtung für den Asienforscher Adolph Schlagintweit in Kaschgar, 1890'
Adelsmatrikel Adelige, S 156 (1859)
Ordensakten 9056
VN 67, Findbuch über Akten der zoologischen Staatssammlung</td></tr>
<tr><td>BBAW</td><td>Alexander-von-Humboldt-Forschungsstelle der Berlin-Brandenburgischen Akademie der Wissenschaften</td></tr>
<tr><td>BiSG</td><td>Bibliothèque Sainte-Geneviève, Paris
Ms. 3611, 263, 'Pièces diverses, manuscrites et imprimées, relatives aux frères Hermann, Adolphe et Robert Schlagintweit, physiciens et géologues de Berlin'</td></tr>
<tr><td>BL</td><td>British Library, London
India Office Records (IOR)
B/234
E/1/300
E/1/309
E/2/25
E/4/835
L/F/2/230
L/F/2/241
L/Mil/2/1477
L/Mil/3/587
MSS EUR F 195/5
PRO/30/12/22
T3787</td></tr>
</table>

X 366, *Technical Objects from India and High Asia, Collected by Hermann, Adolph and Robert Schlagintweit, 1854 to 1858*

BnF   Bibliothèque nationale de France, Paris
       Département Société de Géographie
BSB   Bayerische Staatsbibliothek, Munich
       Schlagintweitiana [SLGA]:
       II.1.5
       II.1.37–8
       II.1.43
       IV.2.63
       IV.2.68
       IV.2.90
       IV.2.94
       IV.2a.75
       IV.5.58
       IV.6.1
       IV.6.2
       V, 'Material zu den Vorträgen Robert von Schlagintweits'
       V.2.2.1–V2.2.2, 'Lecture notes': 'English Lectures on High Asia Delivered during the Years 1868 and 1869 in Various Towns of the United States of America', 2 vols.
       VI.1.23
       VI.5.6.1
       VI.8.3.1–11
DAV   Archiv des Deutschen Alpenvereins, Munich
       Collection of Schlagintweit views and Alpine portrait
       'Execution of "John Company"'; Or, the Blowing up (there ought to be) in Leadenhall Street', *Punch, Or the London Charivari*, 15 August 1857
       NAS 12 SG, 8.4–8.7
DLAM   Deutsches Literaturarchiv, Marbach
       Alexander von Humboldt, private papers
       Cotta, private papers
DMM   Deutsches Museum, Munich
ETH   ETH-Bibliothek, Geologisches Institut und Sammlung, Zürich
EUL   Edinburgh University Library
       Murchison Papers
FBG   Forschungsbibliothek Gotha, Sammlung Perthes Archiv, [SPA]
       SPA PGM 353/1, Schlagintweit, Adolf / Schlagintweit, Hermann v., Schlagintweit, Robert v.
       SPA PGM 353/2, Schlagintweit, Emil / Sykes, William Henry, Schlagintweit, Max v. / Schreiber, Paul
GNM   Germanisches Nationalmuseum, Nürnberg
GStAPK Geheimes Staatsarchiv Preußischer Kulturbesitz, Berlin
       I. HA Rep. 76 Kultusministerium Ve Sekt. 1 Abt. XV zu Nr. 189, Beiheft 'Acta Commissionis Ves Geh. Oberregierungsraths

Lehnert betreffend den Erwerb der Sammlungen der Gebrüder Schlagintweit' [AC]
I. HA Rep. 76 Ve, Akt Kultusministerium (Ministerium für Geistl. Angelegenheiten), Sekt. 1 Abt. XV Nr. 189, 'Wissenschaftliche Reisen der Gebrüder Schlagintweit nach Indien, Hochasien, sowie die Ausstellung und Benutzung der von denselben mitgebrachten Sammlungen' [WR]
I. HA Rep. 81, Gesandtschaften und Konsulate nach 1807 [GuK]
I. HA Rep. 89 Geh. Zivilkabinett, jüngere Periode Nr. 19767 'Acta des Kgl. Geh. Cabinets betr. Die von den Gebrüdern Dr. Hermann Alfred Robert Schlagintweit und Dr. Adolph Hugo Schlagintweit aus München, jetzt in Berlin eingereichten Schriften etc. 1852–1885' [AKlC]
I. HA Rep. 162, Verwaltung des Staatsschatzes Nr. 107, Section, 1, Pars. 4, No 17, 'Acta betreffend: der den Gelehrten, Gebrüdern Adolph, und Hermann Schlagintweit Allerhöchstgewährter Reisezuschuß, 1854'. [Reisezuschuß]
III. HA Ministerium der auswärtigen Angelegenheiten [MdA], Nr. 18929, 'Himalaya Expedition der Brüder Schlagintweit, 1853–1889' [HEBS]
VI. HA Familienarchive und Nachlässe, Bunsen, Karl Josias von; therein: Rep 92, Dep. K. J. v. Bunsen, B. No. 59 [FA]

HU	Harvard University, Cambridge, MA
	Collections of the Department of History of Science
HUB	Humboldt-Universität, Berlin
	Wissenschaftliche Sammlungen, Porträtsammlung Berliner Hochschullehrer
JC	St John's College Archive, Cambridge
	Adams Papers
JHU	Johns Hopkins University, Baltimore, MD, Special Collections, Milton S. Eisenhower Library
	Papers of Benjamin Silliman
KB	Koninklijke Bibliotheek, National Library of the Netherlands, The Hague
KHA	Königliches Hausarchiv, Hanover
	Depot 103, XX, No. 320, 'Acta, betr. Annahme verschiedener ethnographischer Gegenstände, sowie literarischer Sendungen in solchem Betreffe seitens des Herrn von Schlagintweit 1859'
LRN	Library of the Rashtrapati Niwas, Shimla
LUL	Leiden University Library
	Papers of Michael Jan de Goeje
MFK	Museum Fünf Kontinente, Munich [formerly Staatliches Museum für Völkerkunde]
MNHN	Muséum national d'Histoire naturelle, Paris
MVK	Museum für Völkerkunde, Berlin
NA	The National Archives, Kew

	BJ 3/53, 'Records of the Meteorological Office, 1849–1854 (Indian Subcontinent'
NAI	National Archives of India, New Delhi
	Department of Revenue, Agriculture and Commerce
	Cartographic collections, Historical Maps
	Foreign Department
	Military Department
	Revenue & Agriculture, Branch: Forests
NMAH	National Museum of American History, Washington, DC
NSA	Nürnberger Stadtarchiv
ÖAV	Österreichischer Alpenverein, Alpenverein-Museum, Innsbruck
	PERS 26.1/5, 'Robert und Hermann Schlagintweit, 'Collectanea Critica, 1848–65' [CC]
PA AA	Politisches Archiv des Auswärtigen Amtes, Berlin
	'Berichte der Gesandtschaft Peking': Peking II 891, fols. 77ff., 'Wissenschaftliche Bestrebungen'
PAL	Punjab Archives, Lahore
	Political Department
RASA	Royal Asiatic Society Archives, London
	BHH/1/81
	BHH/1/94
RBG	Archive of the Royal Botanic Gardens, Kew
	Directors' Correspondence (= DC)
	DC 51, 'German etc. Letters, 1841–55'
	DC 55, 'Indian, Chinese & Mauritius &c. Letters 1851–1856'
	DC 96, 'English Letters Moo–Myl, 1847–1900'
	DC 102, 'English Letters SME–SYM 1855–1900'
	Joseph Dalton Hooker (= JDH)
	JDH/1/9, 'Travel Journals and Correspondence: India, 1842–1911'
	JDH/1/9/1, 'Indian Journal 1848'
	JDH/1/10, 'Indian Letters 1847–1851'
	JDH/2/1/12, Letters to J. D. Hooker HOW–LEI c. 1840s–1900s
	JDH 2/4/4, Letters from J. D. Hooker PAL–WIL c. 1836–1908
	JDH 2/9, 'Miscellaneous Letters c. 1850–1922'
RGS	Archive of the Royal Geographical Society, London
	Correspondence CB4/279, 1854–60
	Image archive
SBB	Staatsbibliothek, Berlin
	Nachlass Ritter [NR]
	Sammlung Darmstaedter, Asien 1855, Schlagintweit, Adolph, Hermann, Robert Schlagintweit
	Sammlung Darmstaedter, Nachlass A. v. Humboldt, [NAvH]
SG	Société de Géographie, Paris
SSMB	Stiftung Stadtmuseum, Berlin
	Humboldt-Slg. Hein
StBB	Staatsbibliothek, Bamberg, Depositum des Historischen Vereins Bamberg

	HVG 47/1–200, nos. 47/5 and 47/1
	HVG 47/19–182
StM	Stadtarchiv, Munich
UAW	Universitätsarchiv Würzburg
	ARS-Akte 1589
UGL	University of Georgia Libraries, Athens, GA, Special Collections
	Joseph Hooker Collection
YUL	Yale University Library, New Haven, CT, Manuscripts and Archives
	Dana Family Papers

Bibliography

SERIALS

Archiv für Anthropologie
Art Journal
Asiatic Researches
Astronomische Nachrichten
The Athenaeum: A Journal of Literature, Science, the Fine Arts
Aus der Natur: die neuesten Entdeckungen auf dem Gebiete der Naturwissenschaften
Das Ausland
Bonplandia: Zeitschrift für die Gesammte Botanik
Bulletin de la Société de Géographie [BSG]
Calcutta Review
Comptes rendus hebdomadaires des séances de l'Académie des Sciences [CRAS]
Deutsches Museum
The Economist
The Englishman (Calcutta)
Erheiterungen, Beiblatt zur Aschaffenburger Zeitung
Gardeners' Chronicle
Gartenlaube
Geographical Journal [GJ]
Geological Magazine
Globus: Illustrierte Zeitschrift für Länder- und Völkerkunde
Im Neuen Reich. Wochenschrift für das Leben des deutschen Volkes in Staat, Wissenschaft
    und Kunst
Journal of the Asiatic Society of Bengal [JASB]
Journal of Indian Art
Journal of the Royal Geographical Society [JRGS]
Journal of the Society of Arts
Kladderadatsch. Das deutsche Magazin für Unpolitische
Kunst und Gewerbe. Wochenschrift zur Förderung deutscher Kunst-Industrie
Leisure Hour
Literary Gazette: A Weekly Journal of Literature, Science, and the Fine Arts
Nautical Magazine
Österreichische militärische Zeitschrift
Petermanns Geographische Mitteilungen [PGM]
Proceedings of the Royal Geographical Society [PRGS]
Proceedings of the Royal Society of London [PRS]
Professional Papers on Indian Engineering
Quarterly Journal of Science
Saturday Review of Politics, Literature, Science, Art, and Finance
Serapeum. Zeitschrift für Bibliothekswissenschaft, Handschriftenkunde und ältere Literatur
The Spectator

Symons's Monthly Meteorological Magazine
Transactions of the Ethnological Society of London
Universal Review
Vorwärts! Magazin für Kaufleute
Westermanns illustrierte deutsche Monatshefte
Westminster Review. American edition
Zeitschrift für Allgemeine Erdkunde [ZAE]
Zeitschrift für deutsche Industrie und Gewerbe
Zeitschrift der Gesellschaft für Erdkunde zu Berlin [ZGE]
Zeitschrift des Landwirthschaftlichen Vereins in Bayern

NEWSPAPERS

Allen's Indian Mail and Register of Intelligence for British and Foreign India
Allgemeine Zeitung (Augsburg)
Allgemeine Zeitung München [AZM]
Bayerischer Kurier
Bayerische Zeitung
Beilage der Allgemeinen Zeitung
Beilage der Deutschen Allgemeinen Zeitung
Der Bildungsverein
Illustrated London News
Illustrated News of the World
Illustrirte Zeitung
Ingolstädter Tagblatt
Kemptner Zeitung
Laibacher Zeitung
Landshuter Zeitung
Literarisches Centralblatt für Deutschland
Moreton Bay Courier
The Nation
Neue Münchener Zeitung
The Observer
Regensburger Anzeiger
Regensburger Zeitung
Rigaer Zeitung
Star
St. Petersburger Herold
The Times
Über Land und Meer: Allgemeine illustrirte Zeitung

BOOKS AND ARTICLES

Adelman, Jeremy, 'What is Global History Now?', aeon (2 March 2017), https://aeon.co/essays/is-global-history-still-possible-or-has-it-had-its-moment.
Agassiz, Louis and Joseph Bettannier, Études sur les glaciers (Neuchâtel: Jent et Gassmann, 1840).
Aguiar, Marian, Tracking Modernity: India's Railway and the Culture of Mobility (Minneapolis: University of Minnesota Press, 2011).
Ahuja, Ravi, Pathways of Empire: Circulation, 'Public Works' and Social Space in Colonial Orissa (Hyderabad: Orient Blackswan, 2009).

Alavi, Seema, '"Fugitive Mullahs and Outlawed Fanatics": Indian Muslims in Nineteenth Century Trans-Asiatic Imperial Rivalries', *Modern Asian Studies*, 45 (2011), 1337–82.

Muslim Cosmopolitanism in the Age of Empire (Cambridge, MA: Harvard University Press, 2014).

Alcock, Helga, 'Three Pioneers: The Schlagintweit Brothers', *Himalayan Journal*, 36 (1978/79), 156–61.

Alcock, Rutherford, 'Address to the Royal Geographical Society', *PRGS*, 22 (1877), 305–79.

Alder, Garry, *Beyond Bokhara: The Life of William Moorcroft, Asian Explorer and Pioneer Veterinary Surgeon, 1767–1825* (London: Century, 1985).

Alexandrowicz, C. H., 'G. F. de Martens on Asian Treaty Practice (1964)', in C. H. Alexandrowicz, *The Law of Nations in Global History*, ed. David Armitage and Jennifer Pitts (Oxford: Oxford University Press, 2017).

Allen, Charles, *The Prisoner of Kathmandu: Brian Hodgson in Nepal 1820–43* (London: Haus Publishing, 2015).

Allen, David Elliston, 'The Early Professionals in British Natural History', in David Elliston Allen, *Naturalists and Society: The Culture of Natural History in Britain, 1700–1900* (Aldershot: Ashgate/Variorum, 2001), 1–12.

'On Parallel Lines: Natural History and Biology from the Late Victorian Period', *Archives of Natural History*, 25 (1998), 361–71.

Altick, Richard Daniel, *The Shows of London: A Panoramic History of Exhibitions, 1600–1862* (Cambridge, MA: Belknap Press of Harvard University Press, 1978).

Anderson, Benedict, *Imagined Communities. Reflections on the Origin and Spread of Nationalism*, rev. edn. (London: Verso, 2006).

Anderson, Katherine, *Predicting the Weather: Victorians and the Science of Meteorology* (Chicago: University of Chicago Press, 2005).

Armitage, Geoff, 'The Schlagintweit Collections', *Indian Journal of History of Science*, 24 (1989), 67–83.

Arnold, David, 'Globalization and Contingent Colonialism: Towards a Transnational History of "British" India', *Journal of Colonialism and Colonial History*, 16 (2015), doi:10.1353/cch.2015.0019.

'Hodgson, Hooker and the Himalayan Frontier, 1848–50', in David M. Waterhouse (ed.), *The Origin of Himalayan Studies: Brian Hodgson in Kathmandu and Darjeeling, 1820–1858* (London and New York: Routledge, 2004), 189–205.

The New Cambridge History of India, vol. III.5: *Science, Technology and Medicine in Colonial India* (Cambridge: Cambridge University Press, 2000).

'Plant Capitalism and Company Science: The Indian Career of Nathaniel Wallich', *Modern Asian Studies*, 42 (2008), 899–928.

The Tropics and the Traveling Gaze: India, Landscape, and Science, 1800–1856 (Seattle: University of Washington Press, 2006).

Appadurai, Arjun, *Modernity at Large: Cultural Dimensions of Globalization* (Minneapolis and London: University of Minnesota Press, 1996).

Baack, Lawrence J., *Undying Curiosity: Carsten Niebuhr and the Royal Danish Expedition to Arabia (1761–1767)* (Stuttgart: Franz Steiner, 2014).

Baber, Zaheer, *The Science of Empire: Scientific Knowledge, Civilization, and Colonial Rule in India* (Albany: State University of New York Press, 1996).

Baigent, Elizabeth, 'Moorcroft, William (1767–1825)', *Oxford Dictionary of National Biography* (Oxford: Oxford University Press, 2004), online edn, May 2015, https://doi.org/10.1093/ref:odnb/19093.

Balfour, Edward, *The Second Supplement, with Index, to the Cyclopaedia of India and of Southern Asia, Commercial, Industrial and Scientific: Products of the Mineral, Vegetable and Animal Kingdoms, Useful Arts and Manufactures* (Madras: Athenaeum Press, 1862).

Ballantyne, Tony, 'Rereading the Archive and Opening Up the Nation-State: Colonial Knowledge in South Asia (and Beyond)', in Antoinette Burton (ed.), *After the Imperial Turn: Thinking with and through the Nation* (Durham, NC and London: Duke University Press, 2003), 102–21.

'Strategic Intimacies: Knowledge and Colonization in Southern New Zealand', *Journal of New Zealand Studies*, n.s. 14 (2013), 4–18.

Webs of Empire: Locating New Zealand's Colonial Past (Wellington: Bridget Williams Books, 2012).

Ballantyne, Tony and Antoinette Burton, 'Introduction', in Tony Ballantyne and Antoinette Burton (eds.), *Moving Subjects: Gender, Mobility, and Intimacy in an Age of Global Empire* (Chicago: University of Chicago Press, 2009), 1–28.

Ballhatchet, Kenneth, 'European Relations with Asia and Africa', in Albert Goodwin (ed.), *New Cambridge Modern History*, vol. VIII: *The American and French Revolutions 1763–93* (Cambridge: Cambridge University Press, 1965), 218–36.

Banerjee, Sukanya, *Becoming Imperial Citizens: Indians in the Late-Victorian Empire* (Durham, NC: Duke University Press, 2010).

Barrett-Gaines, Kathryn, 'Travel Writing, Experiences, and Silences: What Is Left Out of European Travelers' Accounts: The Case of Richard D. Mohun', *History in Africa*, 24 (1997), 53–70.

Barringer, Tim, 'The South Kensington Museum and the Colonial Project', in Tim Barringer and Tom Flynn (eds.), *Colonialism and the Object: Empire, Material Culture and the Museum* (London: Routledge, 1998), 11–27.

Basa, Kishor K., 'Anthropology and Museums in India', in Gwen R. Schug and Subhash R. Walimbe (eds.), *A Companion to South Asia in the Past* (Chichester: Wiley Blackwell, 2016), 465–81.

Basalla, George, 'The Spread of Western Science', *Science*, 156 (1967), 611–22.

Baud, Aymon, Philippe Forêt and Svetlana Gorshenina, *La Haute-Asie telle qu'ils l'ont vue: explorateurs et scientifiques de 1820 à 1940* (Geneva: Olizane, 2003), 52.

Bayly, Christopher Alan, *The Birth of the Modern World, 1780–1914: Global Connections and Comparisons* (Malden, MA and Oxford: Blackwell, 2004).

'Elphinstone, Mountstuart (1779–1859)', *Oxford Dictionary of National Biography* (Oxford: Oxford University Press, 2004), online edn, May 2017, https://doi.org/10.1093/ref:odnb/8752.

Empire and Information: Intelligence Gathering and Social Communication in India, 1780–1870 (Cambridge: Cambridge University Press, 1996).

Beck, Hanno, 'Georg Forster und Alexander von Humboldt. Zur Polarität ihres geographischen Denkens', in Detlef Rasmussen (ed.), *Der Weltumsegler und seine Freunde: Georg Forster als gesellschaftlicher Schriftsteller der Goethe-Zeit* (Tübingen: Narr, 1988), 175–88.

Behrnauer, W. F. A., 'Die Schlagintweit'schen Sammlungen auf der Jägerburg', *Serapeum* (1867), 374–9.

Bell, Morag, Robin A. Butlin and Michael Heffernan, 'Introduction: Geography and Imperialism, 1820–1940', in Morag Bell, Robin A. Butlin and Michael Heffernan (eds.), *Geography and Imperialism, 1820–1940* (Manchester: Manchester University Press, 1995), 1–12.

Bellon, Richard, 'Joseph Dalton Hooker's Ideals for a Professional Man of Science', *Journal of the History of Biology*, 34 (2001), 51–82.

Bensaude-Vincent, Bernadette and Christine Blondel (eds.), *Science and Spectacle in the European Enlightenment* (New York: Routledge, 2016).

Berenson, Edward, *Heroes of Empire: Five Charismatic Men and the Conquest of Africa* (Berkeley: University of California Press, 2010).

Berg, Maxine, 'Useful Knowledge, "Industrial Enlightenment", and the Place of India', *Journal of Global History*, 8 (2013), 117–41.

Berman, Nina, *German Literature on the Middle East: Discourses and Practices 1000–1989* (Ann Arbor: University of Michigan Press, 2011).

Berman, Russell A., *Enlightenment or Empire: Colonial Discourse in German Culture* (Lincoln: University of Nebraska Press, 1998).

Bhattacharya, Nandini, *Contagion and Enclaves: Tropical Medicine in Colonial India* (Liverpool: Liverpool University Press, 2012).

Biermann, Kurt-R., *Miscellanea Humboldtiana* (Berlin: Akademie, 1990).

Bishop, Peter, *The Myth of Shangri-La: Tibet, Travel Writing, and the Western Creation of Sacred Landscapes* (Berkeley: University of California Press, 1989).

Blackbourn, David, 'Germans Abroad and "Auslandsdeutsche": Places, Networks and Experiences from the Sixteenth to the Twentieth Century', *Geschichte und Gesellschaft*, 41 (2015), 321–46.

'Germany and the Birth of the Modern World, 1780–1820', *Bulletin of the German Historical Institute*, 51 (2012), 9–21.

Blanchard, Ian, 'The "Great Silk Road", ca. 1650/70–ca. 1855', in Markus A. Denzel, Jan De Vries and Philipp Robinson Rossner (eds.), *Small Is Beautiful? Interlopers and Smaller Trading Nations in the Pre-Industrial Period* (Stuttgart: Franz Steiner, 2011), 253–71.

Blanford, Henry Francis, 'On the Geological Structure of the Nilghirí Hills (Madras)', in Thomas Oldham (ed.), *Memoirs of the Geological Survey of India*, vol. I (Calcutta: Military Orphan Press, 1859), 211–48.

Bonneuil, Christophe, 'The Manufacture of Species. Kew Gardens, the Empire and the Standardisation of Taxonomic Practices in Late Nineteenth-Century Botany', in Marie-Noëlle Bourget, Chrisitan Licoppe and Otto Sibum (eds.), *Instruments, Travel and Science: Itineraries of Precision from the Seventeenth to the Twentieth Century* (London: Routledge, 2002), 189–215.

Bosma, Ulbe, *The Sugar Plantation in India and Indonesia: Industrial Production, 1770–2010* (Cambridge: Cambridge University Press, 2013).

Bossi, Maurizio and Claudio Greppi (eds.), *Viaggi e scienza. Le istruzioni scientifiche per i viaggiatori nei secoli XVII–XIX* (Florence: Leo S. Olschki, 2005).

Boulger, George S., 'Seemann, Berthold Carl (1825–1871)', rev. Andrew Grout, *Oxford Dictionary of National Biography* (Oxford: Oxford University Press, 2004), online edn, September 2004, https://doi.org/10.1093/ref:odnb/25029.

'Wallich, Nathaniel (1785–1854)', rev. Andrew Grout, *Oxford Dictionary of National Biography* (Oxford: Oxford University Press, 2004), online edn, May 2005, https://doi.org/10.1093/ref:odnb/28564.

Bourguet, Marie-Noëlle and Christian Licoppe, 'Voyages, mesures et instruments: une nouvelle expérience du monde au siècle des lumières', *Annales. Histoire, Sciences Sociales*, 52 (1997), 1115–51.

Brandt, Max von, *Dreiunddreißig Jahre in Ost-Asien. Erinnerungen eines deutsches Diplomaten*, 3 vols., vol. I: *Die preußische Expedition nach Ost-Asien. Japan, China, Siam 1860–1862* (Leipzig: Wigand, 1901).

Brantlinger, Patrick, 'Victorians and Africans: The Genealogy of the Myth of the Dark Continent', *Critical Inquiry*, 12 (1985), 166–203.

Bray, John, 'A History of the Moravian Church in India', in Moravian Church (ed.), *The Himalayan Mission: Moravian Church Centenary, Leh, Ladakh, India 1885–1985* (Leh: Moravian Church, 1985), 27–75.

Brescius, Moritz von, 'Cultural Brokers: Nain Singh und das Innenleben der Schlagintweit-Expeditionen in Asien, 1854–58', *Jahrbuch für europäische Überseegeschichte*, 17 (Wiesbaden: Harrassowitz, 2018), 75–119.

'Empires of Opportunity: German Scholars between Asia and Europe in the 1850s' (PhD thesis, European University Institute, Florence, 2015).

'Hochstapler, Kolonial-Gehilfen, Helden. Die kontroverse Rezeption der Schlagintweit-Expedition', in Brescius et al. (eds.), *Über den Himalaya*, 251–80.

'Humboldt'scher Forscherdrang und britische Kolonialinteressen: die Indien- und Hochasien-Reise der Brüder Schlagintweit 1854 bis 1858', in Brescius et al. (eds.), *Über den Himalaya*, 31–88.

Brescius, Moritz von, Friederike Kaiser and Stephanie Kleidt (eds.), *Über den Himalaya. Die Expedition der Brüder Schlagintweit nach Indien und Zentralasien 1854 bis 1858* (Cologne: Böhlau, 2015).

Brewster, David, *A Treatise on Magnetism* (Edinburgh: A. & C. Black, 1837).

Brockliss, Laurence, *Calvet's Web: Enlightenment and the Republic of Letters in Eighteenth-Century France* (Oxford: Oxford University Press, 2002).

Brockway, Lucile H., *Science and Colonial Expansion: The Role of the British Royal Botanic Gardens* (New Haven, CT and London: Yale University Press, 1979).

Brogiato, Heinz Peter, Bernhard Fritscher and Ute Wardenga, 'Visualisierungen in der deutschen Geographie des 19. Jahrhunderts: die Beispiele Robert Schlagintweit und Hans Meyer', *Berichte zur Wissenschaftsgeschichte*, 28 (2005), 237–54.

Brown, Richard E., 'Public Health in Imperialism: Early Rockefeller Programs at Home and Abroad', *American Journal of Public Health*, 66 (1976), 897–903.

Bruhns, Karl, *Life of Alexander von Humboldt*, 2 vols. (London: Longmans, Green and Co., 1873).

Buckland, Charles E., *Dictionary of Indian Biography* (London: Swan Sonnenschein and Co., 1906).

Bührer, Tanja, 'Intercultural Diplomacy and Empire: French, British and Asian Intermediaries at the Court of Hyderabad, *c.* 1770–1815' (unpublished Habilitation, University of Bern, 2019).

Bührer, Tanja, Flavio Eichmann, Stig Förster and Benedikt Stuchtey (eds.), *Cooperation and Empire: Local Realities of Global Processes* (New York and Oxford: Berghahn, 2017).

Burke, Peter, *What is the History of Knowledge?* (Cambridge: Polity Press, 2016).

Burnett, D. Graham, '"It Is Impossible to Make a Step without the Indians": Nineteenth-Century Geographical Exploration and the Amerindians of British Guiana', *Ethnohistory*, 41, 1 (2002), 3–40.

Masters of All They Surveyed: Exploration, Geography, and a British El Dorado (Chicago and London: University of Chicago Press, 2000).

Camerini, Jane R., 'Wallace in the Field', in Henrika Kucklick and Robert E. Kohler (eds.), *Science in the Field* (Chicago: University of Chicago Press, 1996), 44–65.

Campbell, Archibald, 'A Register of the Temperature of the Surface of the Ocean from the Hooghly to the Thames', *JASB*, 27 (1859), 170–5.

Cañizares-Esguerra, Jorge, 'How Derivative Was Humboldt? Microcosmic Nature Narratives in Early Modern Spanish America and the (Other) Origins of Humboldt's Ecological Sensibilities', in Londa Schiebinger and Claudia Swan (eds.), *Colonial Botany: Science, Commerce, and Politics in the Early Modern World* (Philadelphia: University of Pennsylvania Press, 2005), 148–65.

Cannon, Susan Faye, *Science in Culture: The Early Victorian Period* (Folkestone: Dawson, 1978).

Castells, Manuel, *The Informational City: Information Technology, Economic Restructuring, and the Urban-Regional Process* (Oxford: Blackwell, 1989).

Cawood, John, 'The Magnetic Crusade: Science and Politics in Early Victorian Britain', *Isis*, 70, 4 (1979), 493–518.

Chakrabarti, Pratik, *Bacteriology in British India: Laboratory Medicine and the Tropics* (Rochester, NY: University of Rochester Press, 2012).

Chapman, Sydney, 'Alexander von Humboldt and Geomagnetic Science', *Archive for History of Exact Sciences*, 2 (1962), 41–51.

Chaudhuri, Kirti N., *The Trading World of Asia and the English East India Company: 1660–1760* (Cambridge: Cambridge University Press, 1978).

Cheek, Martin, 'Gustav Mann', online resource, Royal Botanic Gardens, Kew, http://apps.kew.org/herbcat/gotoMann.do.

Christie, Manson and Woods International Inc., *Sale Catalogue: Fine Watches, Clocks, Scientific Instruments and Related Books, FASCIA-6172* (New York: Christie's, 1986).

Cittadino, Eugene, *Nature as the Laboratory: Darwinian Plant Ecology in the German Empire, 1880–1900* (Cambridge: Cambridge University Press, 1990).

Coghe, Samuël, 'Inter-imperial Learning and African Health Care in Portuguese Angola in the Interwar Period', *Social History of Medicine*, 28 (2015), 134–54.

Cohn, Bernard S., *Colonialism and Its Forms of Knowledge* (Princeton, NJ: Princeton University Press, 1996).

Colley, Linda, *Acts of Union and Disunion* (London: Profile Books, 2014).

The Ordeal of Elizabeth Marsh: A Woman in World History (London: HarperPress, 2007).

Conrad, Sebastian, *German Colonialism: A Short History* (Cambridge: Cambridge University Press, 2011).

Globalgeschichte. Eine Einführung (Munich: Beck, 2013).

'Rethinking German Colonialism in a Global Age', *Journal of Imperial and Commonwealth History*, 41 (2013), 543–66.

Conrad, Sebastian and Andreas Eckert, 'Globalgeschichte, Globalisierung, multiple Modernen: zur Geschichtsschreibung der modernen Welt', in Sebastian Conrad and Ulrike Freitag (eds.), *Globalgeschichte. Theorien, Ansätze, Themen* (Frankfurt am Main: Campus, 2007), 7–49.

Cooper, Alix, 'From the Alps to Egypt (and Back Again): Dolomieu, Scientific Voyaging, and the Construction of the Field in Eighteenth-Century Europe', in Crosbie Smith and Jon Agar (eds.), *Making Space: Territorial Themes in the History of Science* (London: Macmillan, 1998), 39–63.

Cooper, Frederick and Ann Laura Stoler, *Tensions of Empire. Colonial Cultures in a Bourgeois World* (Berkeley: University of California Press, 1997).

Cornish, Caroline, 'Curating Science in an Age of Empire: Kew's Museum of Economic Botany' (unpublished PhD thesis, University of London, 2013).

Crary, Jonathan, *Techniques of the Observer: On Vision and Modernity in the Nineteenth Century* (Cambridge, MA and London: MIT Press, 1992).

Crosbie, Barry, *Irish Imperial Networks: Migration, Social Communication and Exchange in Nineteenth-Century India* (Cambridge: Cambridge University Press, 2012).

Darwin, John, *The Empire Project: The Rise and Fall of the British World-System, 1830–1970* (Cambridge: Cambridge University Press, 2009).

'Imperialism and the Victorians: The Dynamics of Territorial Expansion', *English Historical Review*, 112 (1997), 614–42.

Das, Sarat Chandra, *Autobiography: Narratives of the Incidents of My Early Life* (Calcutta: K. L. Mukhopadhyay, 1969).

Indian Pandits in the Land of Snow (Calcutta: Baptist Mission Press, 1893).

Journey to Lhasa and Central Tibet, ed. William Woodville Rockhill (London: John Murray, 1902).

Tibetan–English Dictionary with Sanskrit Synonyms (Calcutta: Bengal Secretariat Book Depot, 1902).

Daston, Lorraine, 'The Humboldtian Gaze', in Moritz Epple and Claus Zittel (eds.), *Science as Cultural Practice*, vol. I: *Cultures and Politics of Research from the Early Modern Period to the Age of Extremes* (Berlin: Akademie, 2010), 45–60.

Daugeron, Bertrand and Armelle Le Goff, *Penser, classer, administrer. Pour une histoire croisée des collections scientifiques* (Paris: Publications scientifiques du MNHN, 2014).

Daum, Andreas W., '*Wissenschaft* and Knowledge', in Jonathan Sperber (ed.), *Germany 1800–1870* (Oxford: Oxford University Press, 2004), 137–61.

Wissenschaftspopularisierung im 19. Jahrhundert: bürgerliche Kultur, naturwissenschaftliche Bildung und die deutsche Öffentlichkeit 1848–1914 (Munich: Oldenbourg, 1998).

Davies, R. H., *Punjab (India): Report on the Trade and Resources of the Countries on the North-Western Boundary of India* (Lahore: Government Press, 1862).

Daviron, Benoit 'Mobilizing Labour in African Agriculture: The Role of the International Colonial Institute in the Elaboration of a Standard of Colonial Administration, 1895–1930', *Journal of Global History*, 5 (2010), 479–501.

Davis, John R. 'Friedrich Max Müller and the Migration of German Academics to Britain in the Nineteenth Century', in Stefan Manz, Margit Schulte Beerbühl and John R. Davis (eds.), *Migration and Transfer from Germany to Britain, 1660–1914* (Munich: Saur, 2007), 93–106.

Davis, John R., Stefan Manz and Margrit Schulte Beerbühl (eds.), *Transnational Networks: German Migrants in the British Empire, 1670–1914* (Leiden: Brill, 2012).

Davis, Joseph Barnard, 'On the Method of Measurements as a Diagnostic Means of Distinguishing Human races', *Transactions of the Ethnological Society of London*, n.s. 1 (1861), 123–8.

Davis, Natalie Zemon, *Trickster Travels: A Sixteenth-Century Muslim between Worlds* (New York: Hill and Wang, 2006).

Dejung, Christof, *Commodity Trading, Globalization and the Colonial World: Spinning the Web of the Global Market* (New York: Routledge, 2018).

Denzel, Markus A. (ed.), *Deutsche Eliten in Übersee (16. bis frühes 20. Jahrhundert)* (St. Katharinen: Scripta Mercaturae, 2006).

Derix, Simone, 'Vom Leben in Netzen. Neue geschichts- und sozialwissenschaftliche Perspektiven auf soziale Beziehungen', *Neue Politische Literatur*, 56 (2011), 185–206.

Desmond, Ray, *The European Discovery of the Indian Flora* (Oxford: Oxford University Press, 1992).

The India Museum, 1801–1879 (London: HMSO, 1982).

Dettelbach, Michael, 'Global Physics and Aesthetic Empire: Humboldt's Physical Portrait of the Tropics', in Daniel Philip Miller and Peter Hanns Reill (eds.), *Visions of Empire: Voyages, Botany, and Representations of Nature* (Cambridge: Cambridge University Press, 1996), 258–92.

'Humboldtian Science', in N. Jardine, J. Secord and E. C. Spary (eds.), *Cultures of Natural History* (Cambridge: Cambridge University Press, 1996), 287–304.

Dickens, Charles Hildesley, *A Project for Canals of Irrigation and Navigation from the River Soane in South Behar* (Calcutta: O. T. Cutter, 1861).

Dorn, Harold and James E. McClellan III, *Science and Technology in World History: An Introduction* (Baltimore, MD: John Hopkins University Press, 1999).

Drayton, Richard, 'Knowledge and Empire', in W. Roger Louis (ed.-in-chief), *The Oxford History of the British Empire*, vol. II: *The Eighteenth Century*, ed. P. J. Marshall (Oxford: Oxford University Press, 1998), 231–52.

Nature's Government: Science, Imperial Britain, and the 'Improvement' of the World (New Haven, CT and London: Yale University Press, 2000).

Drew, Frederic, *The Jummoo and Kashmir Territories* (London: E. Stanford, 1875).

Dritsas, Lawrence, *Zambesi: David Livingstone and Expeditionary Science in Africa* (London: I. B. Tauris, 2010).

Driver, Felix, 'Face to Face with Nain Singh: The Schlagintweit Collections and their Uses', in Arthur MacGregor (ed.), *Naturalists in the Field: Collecting, Recording and Preserving the Natural World from the Fifteenth to the Twenty-First Century* (Leiden: Brill, 2018), 441–69.

Geography Militant: Cultures of Exploration and Empire (Oxford: Blackwell, 2001).

'Henry Morton Stanley and His Critics: Geography, Exploration, and Empire', *Past & Present*, 133 (1991), 134–66.

'Hidden Histories Made Visible? Reflections on a Geographical Exhibition', *Transactions of the Institute of British Geographers*, 38 (2013), 420–35.

'Intermediaries and the Archive of Exploration', in Shino Konishi, Maria Nugent and Tiffany Shellam (eds.), *Indigenous Intermediaries: New Perspectives on Exploration Archives* (Canberra: Australian National University Press, 2015), 11–30.

'Missionary Travels: Livingstone, Africa and the Book', *Scottish Geographical Journal*, 129 (2013), 164–78.

Driver, Felix and Sonia Ashmore, 'The Mobile Museum: Collecting and Circulating Indian Textiles in Victorian Britain', *Victorian Studies*, 52 (2010), 353–85.

Driver, Felix and Lowri Jones, *Hidden Histories of Exploration: Researching the RGS-IBG Collections* (London: University of London and the Royal Geographical Society, 2009).

Duchhardt, Heinz (ed.), *Russland, der Ferne Osten und die 'Deutschen'* (Göttingen: Vandenhoeck & Ruprecht, 2009).

Ducker, Sophie C. (ed.), *The Contented Botanist: Letters of W. H. Harvey about Australia and the Pacific* (Carlton, Victoria: Melbourne University Press, 1988).

Eaton, Natasha, 'Tourism, Occupancy, and Visuality in North India, ca. 1750–1858', in Dana Leibsohn and Jeanette Favrot Peterson (eds.), *Seeing Across Cultures in the Early Modern World* (Farnham: Ashgate, 2012), 213–38.

Edney, Matthew H., *Mapping an Empire. The Geographical Construction of British India, 1765–1843* (Chicago and London: Chicago University Press, 1997).

Elliot, Charles M., 'Magnetic Survey of the Eastern Archipelago', *Philosophical Transactions of the Royal Society*, 141 (1851), 287–331.

Endersby, Jim, '"From having no Herbarium." Local Knowledge versus Metropolitan Expertise: Joseph Hooker's Australasian Correspondence with William Colenso and Ronald Gunn', *Pacific Science*, 55 (2001), 343–58.

Imperial Nature: Joseph Hooker and the Practices of Victorian Science (Chicago and London: University of Chicago Press, 2008).

'Joseph Hooker: A Philosophical Botanist', *Journal of Biosciences*, 33 (2008), 163–9.

'Joseph Hooker and India', online article, http://anayglorious.in/science-conservation/collections/joseph-hooker/india/travels.html.

Fabian, Johannes, *Out of Our Minds: Reason and Madness in the Exploration of Central Africa* (Berkeley and London: University of California Press, 2000).

'Remembering the Other: Knowledge and Recognition in the Exploration of Central Africa', *Critical Inquiry*, 26 (1999), 49–69.

Fahrmeir, Andreas, *Die Deutschen und ihre Nation: Geschichte einer Idee* (Ditzingen: Reclam, 2017).

Fara, Patricia, *Sympathetic Attractions: Magnetic Practices, Beliefs, and Symbolism in Eighteenth-Century England* (Princeton, NJ: Princeton University Press, 1996).

Faraday, Michael, *The Correspondence of Michael Faraday*, 6 vols., ed. Frank. J. L. James (London: Institution of Elecrical Engineers, 1991–2012).

Feldmann, Jeffrey D., 'Contact Points: Museums and the Lost Body Problem', in Elizabeth Edwards, Chris Gosden and Ruth Phillips (eds.), *Sensible Objects: Colonialism, Museums and Material Culture* (Oxford: Berg, 2006), 245–68.

Felsch, Philipp, '14.777 Dinge. Verkehr mit der Sammlung Schlagintweit', in Friedrich Balke, Maria Muhle and Antonia von Schöning (eds.), *Die Wiederkehr der Dinge* (Berlin: Kadmos, 2011), 193–207.

'Humboldts Söhne. Das paradigmatische/epigonale Leben der Brüder Schlagintweit', in Michael Neumann (ed.), *Magie der Geschichten. Schreiben, Forschen und Reisen in der zweiten Hälfte des 19. Jahrhunderts* (Konstanz: Konstanz University Press, 2011), 113–29.

Fenske, Hans, 'Ungeduldige Zuschauer. Die Deutschen und die europäische Expansion 1815–1880', in Wolfgang Reinhard (ed.), *Imperialistische Kontinuität und nationale Ungeduld im 19. Jahrhundert* (Frankfurt am Main: Fischer, 1991), 87–124.

Finkelstein, Gabriel, 'Berge von Daten: die Obsession der Gebrüder Schlagintweit', in Philipp Felsch, Beat Gugger and Gabriele Rath (eds.), *Berge, eine unverständliche Leidenschaft* (Vienna: Folio, 2007), 49–73.

'"Conquerors of the Künlün"? The Schlagintweit Mission to High Asia, 1854– 57', *History of Science*, 38 (2000), 179–214.

'Headless in Kashgar', *Endeavour: Review of the Progress of Science*, 23 (1999), 5–9.

Fischer-Tiné, Harald, 'Reclaiming Savages in "Darkest England" and "Darkest India": The Salvation Army as Transnational Agent of the Civilizing Mission', in Carey Watt and Michael Mann (eds.), *Civilizing Missions in Colonial and Postcolonial South Asia* (London: Anthem Press, 2011), 125–65.

'The Ymca and Low-Modernist Rural Development in South Asia, c.1922– 1957', *Past & Present*, 240, 1 (2018), 193–234, https://doi.org/10.1093/ pastj/gty006.

Fisher, Donald, 'Rockefeller Philanthropy and the British Empire: The Creation of the London School of Hygiene and Tropical Medicine', *History of Education: Journal of the History of Education Society*, 7 (1978), 129–43.

Fisher, Michael H., *Counterflows to Colonialism. Indian Travellers and Settlers in Britain 1600–1857* (New Delhi: Permanent Black, 2004).

Fitzpatrick, Matthew P., *Liberal Imperialism in Germany: Expansionism and Nationalism, 1848–1884* (New York: Berghahn Books, 2008).

Fletcher, Joseph, 'Sino-Russian Relations, 1800–1862', in Denis Crispin Twitchett and John King Fairbank (eds.), *The Cambridge History of China*,

378   Bibliography

vol. X: *Late Ch'ing, 1800–1911*, part 1 (Cambridge: Cambridge University Press, 1978), 318–50.

Flüchter, Antje, 'Identität in einer transkulturellen Gemeinschaft? "Deutsche" in der Vereenigde Oost-Indische Compagnie', in Christoph Dartmann and Carla Meyer (eds.), *Identität und Krise? Zur Deutung vormoderner Selbst-, Welt- und Fremderfahrungen* (Münster: Rhema, 2007), 155–86.

Foerster, Frank, *Christian Carl Josias Bunsen: Diplomat, Mäzen und Vordenker in Wissenschaft, Kirche und Politik* (Bad Arolsen: Waldeckischer Geschichtsverein, 2001).

Forbes Watson, John, *A Classified and Descriptive Catalogue of the Indian Department* (London: Spottiswoode and Co., 1862).

Foster, William, *The East India House, Its History and Associations* (London: John Lane, 1924).

Franke, Wolfgang, 'Brandt, Max von', *Neue Deutsche Biographie*, 2 (1955), 531.

Freitag, Ulrich, 'Ferdinand von Richthofens "Atlas von China" (Idee – Durchführung – Ergebnis)', *Die Erde*, 114 (1983), 119–34.

Friedel, Ernst, *Die Gründung preußisch-deutscher Colonien im Indischen und Großen Ocean mit besonderer Rücksicht auf das östliche Asien, eine Studie im Gebiete der Handels- und Wirthschafts-Politik* (Berlin: A. Eichhoff, 1867).

Friedl, Wolfgang, 'Europäische Forscher und Reisende in den Berichten der Herrnhuter Mission. Kontakte und Ergebisse – ein Überblick', in Lydia Icke-Schwalbe and Gudrun Meier (eds.), *Wissenschaftsgeschichte und gegenwärtige Forschungen in Nordwest-Indien* (Dresden: Staatliches Museum für Völkerkunde, 1990), 80–5.

Fritscher, Bernhard, '"Humboldtian Views": Hermann and Adolph Schlagintweit's Panoramas and Views from India and High Asia', in Rudolf Seisig, Menso Folkerts and Ulf Hashagen (eds.), *Form, Zahl, Ordnung* (Munich: Steiner, 2004), 603–13.

'Making Objects Move: On Minerals and their Dealers in 19th Century Germany', *Journal of History of Science and Technology*, 5 (2012), 84–105.

'Zwischen "Humboldt'schem Ideal" und "kolonialem Blick": zur Praxis der Physischen Geografie der Gebrüder Schlagintweit', *Wissenschaft und Kolonialismus. Wiener Zeitschrift zur Geschichte der Neuzeit*, 9 (2009), 72–97.

Furnivall, John S., *Netherlands India: A Study of Plural Economy* (Cambridge: Cambridge University Press, 1939).

Gaenszle, Martin, 'Brian Hodgson as Ethnographer and Ethnologist', in David M. Waterhouse (ed.), *The Origin of Himalayan Studies: Brian Hodgson in Kathmandu and Darjeeling, 1820–1858* (London and New York: Routledge, 2004), 206–26.

Gascoigne, John, *Encountering the Pacific in the Age of Enlightenment* (Cambridge: Cambridge University Press, 2014).

Gelder, Roelof van, *Het Oost-Indisch avontuur: Duitsers in Dienst van de VOC (1600–1800)* (Nijmegen: Uitgeverij SUN, 1997).

Geppert, Dominik, *Pressekriege. Öffentlichkeit und Diplomatie in den deutsch-britischen Beziehungen (1896–1912)* (Munich: Oldenbourg, 2007).

Gesellschaft für Erdkunde zu Berlin, 'Die Thätigkeit des Vorstandes der Gesellschaft für Erdkunde zu Berlin ... Expedition auf die Erforschung Aequatorial-Afrika's hinzuwirken', *ZGE*, 8 (1873), 'Erster Aufruf', 170–2.

Ghatak, Usha Ranjan, 'Introduction', in Usha Ranjan Ghatak (ed.), *Indian Association for the Cultivation of Science: A Century [1876–1976]* (Calcutta: The Association, 1976), 1–27.

Gilman, Daniel C., 'Schlagintweit's Ethnographical Collections', *American Journal of Science and Arts*, 29 (1860), 235–36.

Gilmour, Robin, *The Victorian Period: The Intellectual and Cultural Context of English Literature, 1830–1890* (Abingdon and New York: Routledge, 2013).

Gißibl, Bernhard, 'Imagination and Beyond: Cultures and Geographies of Imperialism in Germany, 1848–1918', in John MacKenzie (ed.), *European Empires and the People: Popular Responses to Imperialism in France, Britain, the Netherlands, Belgium, Germany and Italy* (Manchester: Manchester University Press, 2011), 158–94.

Godsall, Jon R., *The Tangled Web: A Life of Sir Richard Burton* (Leicester: Matador, 2008).

Goedsche, Hermann, *Nena Sahib oder Die Empörung in Indien. Historisch-politischer Roman aus der Gegenwart*, 3 vols. (Berlin: Carl Röhring, 1858–9).

Goldstein, Jürgen, *Georg Forster. Zwischen Freiheit und Naturgewalt* (Berlin: Matthes & Seitz, 2015).

Gollwitzer, Heinz, '"Für welchen Weltgedanken kämpfen wir?" Bemerkungen zur Dialektik zwischen Identitäts- und Expansionsideologien in der deutschen Geschichte', in K. von Hildebrand and R. Pommerin (eds.), *Deutsche Frage und europäisches Gleichgewicht* (Cologne: Böhlau, 1985), 83–109.

Golubief, Captain, 'Observations on the Astronomical Points Determined by the Brothers Schlagintweit in Central Asia', *JASB*, 35 (1867), 46–50.

Good, Gregory A. 'Sabine, Sir Edward, 1788–1883', *Oxford Dictionary of National Biography* (Oxford: Oxford University Press, 2004), online edn, September 2004, https://doi.org/10.1093/ref:odnb/24436.

Gräbel, Carsten, *Die Erforschung der Kolonien. Expeditionen und koloniale Wissenskultur deutscher Geographen, 1884–1919* (Bielefeld: transcript, 2015).

Green, Abigail, *Fatherlands: State-Building and Nationhood in Nineteenth-Century Germany* (Cambridge: Cambridge University Press, 2011).

Grimm, Jacob and Wilhelm Grimm, *Deutsches Wörterbuch von Jacob und Wilhelm Grimm*, 16 vols. (Leipzig: S. Hirzel, 1854–1961).

Grove, Richard H., *Green Imperialism: Colonial Expansion, Tropical Island Edens and the Origins of Environmentalism, 1600–1860* (Cambridge: Cambridge University Press, 1995).

Habermas, Rebekka, 'Intermediaries, Kaufleute, Missionare, Forscher und Diakonissen. Akteure und Akteurinnen im Wissenstransfer', in Rebekka Habermas and Alexandra Pyrembel (eds.), *Von Käfern, Märkten und Menschen. Kolonialismus und Wissen in der Moderne* (Göttingen: Vandenhoeck & Ruprecht, 2013), 27–48.

Halbfass, Wilhelm, *India and Europe. An Essay in Understanding* (Albany: State University of New York Press, 1988).

Hall, Catherine, *Civilising Subjects. Metropole and Colony in the English Imagination, 1830–1867* (Oxford: Polity, 2002).

Handique, Rajib, *British Forest Policy in Assam* (New Delhi: Concept Publishing Co., 2004).

Hannay, S. F., 'Notes on the Iron Ore Statistics and Economic Geology of Upper Assam', *JASB*, 25 (1857), 330–44.

Hansen, Peter H., 'Founders of the Alpine Club (act. 1857–1863)', in *Oxford Dictionary of National Biography* (Oxford: Oxford University Press, 2004), online edn, October 2007, https://doi.org/10.1093/ref:odnb/96327.

Harrison, Mark, 'Networks of Knowledge: Science and Medicine in Early Colonial India, c. 1750–1820', in D. M. Peers and N. Gooptu (eds.), *India and the British Empire* (Oxford: Oxford University Press, 2012), 191–211.

'Tropical Medicine in Nineteenth-Century India', *British Journal for the History of Science*, 25 (1992), 299–318.

Headrick, Daniel, *The Tentacles of Progress: Technology Transfer in the Age of Imperialism, 1850–1940* (New York and Oxford: Oxford University Press, 1988).

The Tools of Empire: Technology and European Imperialism in the Nineteenth Century (New York: Oxford University Press, 1981).

Henderson, Louise C., 'Historical Geographies of Textual Circulation: David Livingstone's Missionary Travels in France and Germany', in Diarmid A. Finnegan and Jonathan Jeffrey Wright (eds.), *Spaces of Global Knowledge: Exhibition, Encounter and Exchange in an Age of Empire* (Abingdon: Routledge, 2015), 227–44.

Herschel, John, 'Memorial of the Committee Appointed … on the Subject of Terrestrial Magnetism', in *Report of the Ninth Meeting of the British Association for the Advancement of Science, Held at Birmingham in August 1839* (London: John Murray, 1840), 33–9.

Hevia, James, *The Imperial Security State: British Colonial Knowledge and Empire-Building* (Cambridge: Cambridge University Press, 2012).

Hobsbawm, Eric, *The Age of Empire, 1875–1914* (London: Weidenfeld and Nicolson, 1987).

Höcker, Wilma, *Der Gesandte Bunsen als Vermittler zwischen Deutschland und England* (Göttingen: Hansen-Schmidt, 1951).

Hodacs, Hanna, 'Circulating Knowledge on Nature: Travelers and Informants, and the Changing Geography of Linnaean Natural History', in Gesa Mackenthun, Andrea Nicolas and Stephanie Wodianka (eds.), *Travel, Agency, and the Circulation of Knowledge* (Münster: Waxmann, 2017), 75–98.

Hodge, Joseph Morgan, 'Science and Empire: An Overview of the Historical Scholarship', in Joseph Morgan Hodge and Brett M. Bennett (eds.), *Science and Empire: Knowledge and Networks of Science across the British Empire, 1800–1970* (Basingstoke: Palgrave Macmillan, 2011), 3–29.

Triumph of the Expert: Agrarian Doctrines of Development and the Legacies of British Colonialism (Athens: Ohio University Press, 2007).

Hodkinson, James and John Walker with Shaswati Mazumdar and Johannes Feichtinger (eds.), *Deploying Orientalism in Culture and History: From Germany to Central and Eastern Europe* (Rochester, NY: Camden House, 2013).

Hoerder, Dirk, 'Segmented Macrosystems, Networking Individuals, Cultural Change: Balancing Processes and Interactive Change in Migration', in Veit Bader (ed.), *Citizenship and Exclusion* (New York: St. Martin's Press, 1997), 81–95.

Holmes, Frederic L., 'The Complementarity of Teaching and Research in Liebig's Laboratory', *Osiris*, 5 (1989), 121–64.

Home, Rod, 'Science as a German Export to Nineteenth-Century Australia', *Working Papers in Australian Studies*, 104 (1995), 1–21.

Hooker, Joseph, *Flora Antarctica: The Botany of the Antarctic Voyage*, part 1 (London: Reeve Brothers, 1847).

The Flora of British India, 7 vols. (London: Lovell Reeve, 1872–97).

Himalayan Journals. Notes of a Naturalist in Bengal, the Sikkim and Nepal Himalayas, The Khasia Mountains, &c., 2 vols. (London: John Murray, 1854).

Hooker, Joseph and Thomas Thomson, *Flora Indica*, vol. I (London: W. Pamplin, 1855).

Hopkins, Anthony G., 'Explorers' Tales: Stanley Presumes – Again', *Journal of Imperial and Commonwealth History*, 36 (2008), 669–84.

Hopkins, Benjamin D., *The Making of Modern Afghanistan* (New York and Basingstoke: Palgrave Macmillan, 2008).

House of Commons, *Fourth Report from the Select Committee on Colonization and Settlement (India), together with the Proceedings of the Committee, Minutes of Evidence, and Appendix* (London, 1858).

Howe, Kathleen, 'Hooper, Colonel Willoughby Wallace (1837–1913)', in John Hannavy (ed.), *Encyclopedia of Nineteenth-Century Photography*, vol. I. (New York and London: Routledge, 2008), 713–14.

Huber, Toni and Tina Niermann, 'Tibetan Studies at the Berlin University: An Institutional History', in Petra Maurer (ed.), *Tibetstudien: Festschrift für Dieter Schuh* (Bonn and Bier: Bier'sche Verlagsanstalt, 2007), 95–122.

Huber, Valeska, *Channelling Mobilities: Migration and Globalisation in the Suez Canal Region and Beyond, 1869–1914* (Cambridge: Cambridge University Press, 2013).

'Multiple Mobilities: über den Umgang mit verschiedenen Mobilitätsformen um 1900', *Geschichte und Gesellschaft*, 36 (2010), 317–41.

Hügel, Carl von, *Das Kabul-Becken und die Gebirge zwischen dem Hindu-Kosch und der Sutlej* (Vienna: K. Hof- und Staatsdruckerei, 1851–2).

Travels in Kashmir and the Panjab (London: John Petheram, 1845).

Humboldt, Alexander von, *Asie Centrale. Recherches sur les chaînes de montagnes et la climatologie comparée*, 3 vols. (Paris: Gide, 1843).

Cosmos: A Sketch of a Physical Description of the Universe, trans. E. C. Otté, 5 vols. (London: H. G. Bohn, 1849–58).

Fragments de géologie et de climatologie asiatique (Paris: Nabu Press, 1831).

Letters of Alexander von Humboldt Written between the years 1827 and 1858 to Varnhagen von Ense. Together with Extracts from Varnhagens Diaries (London: Trübner, 1860).

Personal Narrative of Travels to the Equinoctial Regions of the New Continent, during the Years 1799–1804. By Alexander de Humboldt, and Aimé Bonpland; trans. into English by Helen M. Williams, 7 vols. (London, 1814–29).

Huxley, Leonard (ed.), *Life and Letters of Sir Joseph Dalton Hooker*, 2 vols. (London: John Murray, 1918).

Life and Letters of Thomas Henry Huxley, 2 vols. (London: Macmillan, 1900).

Ireton, Sean Moore and Caroline Schaumann, 'Introduction: The Meaning of Mountains: Geology, History, Culture', in Sean Moore Ireton and Caroline Schaumann (eds.), *Heights of Reflection: Mountains in the German Imagination from the Middle Ages to the Twenty-First Century* (Rochester, NY: Camden House, 2012), 1–19.

Jaehn, Tomas, *Germans in the Southwest, 1850–1920* (Albuquerque: University of New Mexico Press, 2005).

Jarvis, Andrew, 'Indien im Porträt: die Aufnahmen der Brüder Schlagintweit als frühe Dokumente kolonialer Fotografiegeschichte', in Brescius et al. (eds.), *Über den Himalaya*, 161–72.

Jasanoff, Maya, *The Edge of Empire: Conquest and Collecting in the East 1750–1850* (London: Fourth Estate, 2005).

Jeal, Tim, *Stanley: The Impossible Life of Africa's Greatest Explorer* (London: Faber, 2007).

Johnson, Robert, *Spying for Empire: The Great Game in Central and South Asia, 1757–1947* (London: Greenhill Books, 2006).

Johnson, Samuel, *A Dictionary of the English Language* (London: John Williamson & Co., 1839).

Jones, Lowri, 'Local Knowledge and Indigenous Agency in the History of Exploration: Studies from the RGS-IBG Collections' (unpublished PhD thesis, Royal Holloway, University of London, 2010).

Jordanova, Ludmilla, 'Science and Nationhood: Cultures of Imagined Communities', in G. Cubitt (ed.), *Imagining Nations* (Manchester: Manchester University Press, 1998), 192–211.

Kalka, Claudia, '"Ordensjäger". Miscellanea zur Sammlung Schlagintweit im Niedersächsischen Landesmuseum Hannover', in Anna Schmid (ed.), *Mit Begeisterung und langem Atem. Ethnologie am Niedersächsischen Landesmuseum Hannover* (Hanover: Niedersächsisches Landesmuseum Hannover 2006), 89–95.

Keighren, Innes M., Charles W. J. Withers and Bill Bell, *Travels into Print: Exploration, Writing, and Publishing with John Murray, 1773–1859* (Chicago and London: University of Chicago Press, 2015).

Kennedy, Dane, 'Introduction: Reinterpreting Exploration', in Dane Kennedy (ed.), *Reinterpreting Exploration: The West in the World* (Oxford: Oxford University Press, 2014), 1–20.

The Last Blank Spaces. Exploring Africa and Australia (Cambridge, MA and London: Harvard University Press, 2013).

The Magic Mountains: Hill Stations and the British Raj (Berkeley: University of California Press, 1996)

Kiepert, Heinrich, *Neuer Handatlas über alle Teile der Erde* (Berlin: D. Reimer, 1855–60).

Kirchberger, Ulrike, *Aspekte deutsch-britischer Expansion. Die Überseeinteressen der deutschen Migranten in Großbritannien in der Mitte des 19. Jahrhunderts* (Stuttgart: Franz Steiner, 1999).

'Deutsche Naturwissenschaftler im britischen Empire: die Erforschung der außereuropäischen Welt im Spannungsfeld zwischen deutschem und britischem Imperialismus', *Historische Zeitschrift*, 271 (2000), 621–60.

'German Scientists in the Indian Forest Service: A German Contribution to the Raj?', *Journal of Imperial and Commonwealth History*, 29 (2001), 1–26.

'Introduction', in Ulrike Kirchberger and Heather Ellis (eds.), *Anglo-German Scholarly Networks in the Long Nineteenth Century* (Leiden: Brill, 2014), 1–19.

Kirchberger, Ulrike and Heather Ellis (eds.), *Anglo-German Scholarly Networks in the Long Nineteenth Century* (Leiden: Brill, 2014).

Kleidt, Stephanie, 'Lust und Last. Die Sammlungen der Gebrüder Schlagintweit', in Brescius et al. (eds.), *Über den Himalaya*, 113–37.

'Zwischen Dokument und Kunstwerk. Die Zeichnungen und Aquarelle aus Indien und Hochasien', in Brescius et al. (eds.), *Über den Himalaya*, 145–59.

Knight, Charles (ed.), *London*, vol. V (London: Charles Knight & Co., 1843).

Koerner, Lisbet, *Linnaeus: Nature and Nation* (Cambridge, MA: Harvard University Press, 2001).

Königliche Museen zu Berlin (ed.), *Zur Geschichte der Königlichen Museen in Berlin* (Berlin: Reichsdruckerei, 1880).

Kontje, Todd, *German Orientalisms* (Ann Arbor: University of Michigan Press, 2004).

Körner, Hans, 'Die Brüder Schlagintweit', in Claudius C. Müller and Walter Raunig (eds.), *Der Weg zum Dach der Welt* (Innsbruck and Frankfurt am Main: Pinguin, 1982), 62–75.

'Photographien auf Forschungsreisen. Robert Schlagintweit und seine Brüder erforschen die Alpen, Indien und Hochasien (1850–1857)', in Bodo von Dewitz and Reinhard Matz (eds.), *Silber und Salz: zur Frühzeit der Photographie im deutschen Sprachraum (1839–1860)* (Cologne and Heidelberg: Edition Braus, 1989), 310–20.

Kraft, Tobias, *Figuren des Wissens bei Alexander von Humboldt: Essai, Tableau und Atlas im amerikanischen Reisewerk* (Berlin and Boston: De Gruyter, 2014).

Kretschmann, Carsten, *Räume öffnen sich: naturhistorische Museen im Deutschland des 19. Jahrhunderts* (Berlin: Akademie Verlag 2006).

Kreutzmann, Hermann, 'Habitat Conditions and Settlement Processes in the Hindukush-Karakoram', *PGM*, 138 (1994), 337–56.

Wakhan Quadrangle: Exploration and Espionage during and after the Great Game (Wiesbaden: Harrassowitz-Verlag, 2017).

Kuklick, Henrika and Robert E. Kohler (eds.), *Science in the Field*, Osiris 11 (2nd ser.) (Chicago: University of Chicago Press, 1996).

Kumar, Krishan, *Visions of Empire: How Five Imperial Regimes Shaped the World* (Princeton, NJ and Oxford: Princeton University Press, 2017).

Kumar, Prakash, *Indigo Plantations and Science in Colonial India* (Cambridge: Cambridge University Press, 2012).

Kundrus, Birthe (ed.), *Phantasiereiche: zur Kulturgeschichte des deutschen Kolonialismus* (Frankfurt am Main: Campus, 2003).

Laak, Dirk van, *Über alles in der Welt. Deutscher Imperialismus im 19. und 20. Jahrhundert* (Munich: C. H. Beck, 2005).

Lambert, David and Alan Lester (eds.), *Colonial Lives across the British Empire: Imperial Careering in the Long Nineteenth Century* (Cambridge: Cambridge University Press, 2006).

384   Bibliography

Lässig, Simone, 'The History of Knowledge and the Expansion of the Historical Research Agenda', *Bulletin of the German Historical Institute*, 59 (2016), 29–58.

Latour, Bruno, *Science in Action: How to Follow Scientists and Engineers through Society* (Milton Keynes: Open University Press, 1987).

Leask, Nigel, 'Darwin's "Second Sun": Alexander von Humboldt and the Genesis of *The Voyage of the Beagle*', in Trudi Tait and Helen Small (eds.), *Literature, Science, Psychoanalysis, 1830–1970: Essays in Honour of Gillian Beer* (Oxford: Oxford University Press, 2003), 13–36.

'"Travelling the Other Way": The Travels of Mirza Abu Talib Khan (1810) and Romantic Orientalism', in M. J. Franklin (ed.), *Romantic Representations of British India* (London and New York: Routledge, 2006), 220–37.

Leibsohn, Dana, 'Introduction: Geographies of Sight', in Dana Leibsohn and Jeanette Favrot Peterson (eds.), *Seeing Across Cultures in the Early Modern World* (Farnham: Ashgate, 2012), 1–22.

Lenz, Karl, '150 Jahre Gesellschaft für Erdkunde zu Berlin', *Die Erde*, 109 (1978), 67–86.

'The Berlin Geographical Society 1828–1978', *Geographical Journal*, 144 (1978), 218–23.

Lepsius, Karl Richard, *Denkmäler aus Ägypten und Äthiopien*, 12 vols. (Berlin: Nicolaische Buchhandlung, 1849–59).

Lewis, Joanna, *Empire of Sentiment: The Death of Livingstone and the Myth of Victorian Imperialism* (Cambridge: Cambridge University Press, 2018).

Liebersohn, Harry, 'A Half Century of Shifting Narrative Perspectives on Encounters', in Dane Kennedy (ed.), *Reinterpreting Exploration: The West in the World* (Oxford: Oxford University Press, 2014), 38–53.

Music and the New Global Culture: From the Great Exhibitions to the Jazz Age (Chicago: University of Chicago Press, 2019).

The Travelers' World: Europe to the Pacific (Cambridge, MA: Harvard University Press, 2006).

Liebig, Georg von, 'Natural Science in India', *Calcutta Review*, 26 (1856), 211–22.

Lindner, Ulrike, *Koloniale Begegnungen. Deutschland und Grossbritannien als Imperialmächte in Afrika 1880–1914* (Frankfurt am Main: Campus, 2011).

Livingstone, David N., *Putting Science in Its Place: Geographies of Scientific Knowledge* (Chicago: University of Chicago Press, 2003).

Livingstone, David N. and Charles W. J. Withers (eds.), *Geographies of Nineteenth-Century Science* (Chicago and London: University of Chicago Press, 2011).

Livingstone, Justin David, *Livingstone's 'Lives': A Metabiography of a Victorian Icon* (Manchester: Manchester University Press, 2015).

Lucier, Paul, 'The Professional and the Scientist in Nineteenth-Century America', *Isis*, 100, 4 (2009), 699–732.

Ludden, David, 'Orientalist Empiricism: Transformations of Colonial Knowledge', in Carol A. Breckenridge (ed.), *Orientalism and the Postcolonial Predicament* (Philadelphia: University of Pennsylvania Press, 1993), 250–78.

Lüdecke, Cornelia, 'Carl Ritters (1779–1859) Einfluß auf die Geographie bis hin zur Geopolitik Karl Haushofers (1869–1946)', *Sudhoffs Archiv*, 88 (2004), 129–52.

'Für Humboldt ins Hochgebirge: der schulische und universitäre Hintergrund der Brüder Schlagintweit', in Brescius et al. (eds.), *Über den Himalaya*, 173–86.

'"Indian Heat and Storm to the South, and the Deserts of Central Asia to the North". Die meteorologischen Untersuchungen der Brüder Schlagintweit im Himalaya (1854–57)', in Brescius et al. (eds.), *Über den Himalaya*, 209–18.

Lülfing, Hans, 'Amthor, Eduard Gottlieb', *Neue Deutsche Biographie*, 1 (1953), 264.

McCook, Stuart, '"It May Be Truth, but It Is Not Evidence": Paul du Chaillu and the Legitimation of Evidence in the Field Sciences', *Osiris*, 11 (1996), 177–97.

MacDonald, Helen Patricia, *Human Remains: Dissection and Its Histories* (New Haven, CT and London: Yale University Press, 2006).

McGetchin, Douglas T., *Indology, Indomania, and Orientalism: Ancient India's Rebirth in Modern Germany* (Madison, NJ: Fairleigh Dickinson University Press, 2009).

MacKenzie, John, 'Heroic Myths of Empire', in John MacKenzie (ed.), *Popular Imperialism and the Military* (Manchester: Manchester University Press, 1992), 109–38.

'Introduction', in John MacKenzie (ed.), *European Empires and the People: Popular Responses to Imperialism in France, Britain, the Netherlands, Belgium, Germany and Italy* (Manchester: Manchester University Press, 2011), 1–18.

MacKenzie, John (ed.), *European Empires and the People: Popular Responses to Imperialism in France, Britain, the Netherlands, Belgium, Germany and Italy* (Manchester: Manchester University Press, 2011).

MacKenzie, John and T. M. Devine (eds.), *Scotland and the British Empire* (Oxford: Oxford University Press, 2017).

MacKillop, Andrew, 'Europeans, Britons, Scots: Scottish Sojourning Networks and Identities in Asia, c.1700–1815', in Angela McCarthy (ed.), *A Global Clan: Scottish Migrant Networks and Identities since the Eighteenth Century* (London: Tauris Academic Studies, 2006), 19–47.

Mangold, Sabine, *Eine 'weltbürgerliche Wissenschaft'. Die deutsche Orientalistik im 19. Jahrhundert* (Stuttgart: Franz Steiner Verlag, 2004).

Manias, Chris, *Race, Science and the Nation: Reconstructing the Ancient Past in Britain, France and Germany, 1800–1914* (Abingdon: Routledge, 2013).

Manjapra, Kris, *Age of Entanglement: German and Indian Intellectuals across Empire* (Cambridge, MA: Harvard University Press, 2014).

Mann, Gustav, *Progress Report of Forest Administration in the Province of Assam* (Shillong: Government Press, 1874–88).

Mann, Michael, *Flottenbau und Forstbetrieb in Indien, 1794–1823* (Stuttgart: Franz Steiner Verlag, 1996).

Marchand, Suzanne, *Down from Olympus: Archaeology and Philhellenism in Germany, 1750–1970* (Princeton, NJ: Princeton University Press, 1996).

German Orientalism in the Age of Empire. Religion, Race, and Scholarship (Cambridge: Cambridge University Press, 2009).

Markham, Clements Robert, *A Memoir on the Indian Surveys* (London: W. H. Allen, 1871).

Markham, Clements Robert (ed.), 'Report of the Meteorological Committee of the Royal Society, 11.4.1871', in *East India (Meteorological Department)* (London: House of Commons, 1874), 10–12.

Markovits, Claude, Jacques Pouchepadass and Sanjay Subrahmanyam (eds.), *Society and Circulation: Mobile People and Itinerant Cultures in South Asia 1750–1950* (London, New York and Delhi: Anthem Press, 2006).

Martin, Emma, 'Translating Tibet in the Borderlands: Networks, Dictionaries, and Knowledge Production in Himalayan Hill Stations', *Transcultural Studies*, 1 (2016), 86–120.

Mason, Kenneth, 'Kishen Singh and the Indian Explorers', *Geographical Journal*, 62 (1923), 429–40.

Maury, Alfred, 'Rapport sur les travaux de la Société de Géographie, et sur les progrès des sciences géographiques pendant l'année 1858', *BSG*, 17 (1859), 5–110.

Mayer, Christoph, 'Die Gletscherforschungen der Brüder Schlagintweit. Ihre Bedeutung für die Bayerische Akademie der Wissenschaften', in Brescius et al. (eds.), *Über den Himalaya*, 295–304.

Mayr, Helmut, 'Schlagintweit, Emil', *Neue Deutsche Biographie*, 23 (2007), 24–5.

Mazumdar, Shaswati, 'Introduction', in Shaswati Mazumdar (ed.), *Insurgent Sepoys: Europe Views the Revolt of 1857* (New Delhi: Routledge, 2012), 1–15.

'The Jew, the Turk, and the Indian: Figurations of the Oriental in the German-Speaking World', in James Hodkinson and John Walker (eds.), *Deploying Orientalism in Culture and History: From Germany to Central and Eastern Europe* (Rochester, NY: Camden House, 2013), 99–116.

Meissner, Kristin, 'Responsivity within the Context of Informal Imperialism: Oyatoi in Meiji Japan', *Journal of Modern European History*, 14 (2016), 268–89.

Mellilo, Edward, 'Global Entomologies: Insects, Empires, and the "Synthetic Age" in World History', *Past & Present*, 223, 1 (2014), 233–70.

Metcalf, Thomas R., *The New Cambridge History of India*, vol. III.4: *Ideologies of the Raj* (Cambridge: Cambridge University Press, 1995).

Meyer, Hermann (ed.), *Neues Konversations-Lexikon, ein Wörterbuch des allgemeinen Wissens*, vol. VIII (Hildburghausen: Bibliographisches Institut, 1864).

Middleton, Townsend, *The Demands of Recognition: State Anthropology and Ethnopolitics in Darjeeling* (Stanford, CA: Stanford University Press, 2016).

Miller, David Philip, 'Joseph Banks, Empire, and "Centers of Calculation" in Late Hanoverian Britain', in David Philip Miller and Peter Hanns Reill (eds.), *Visions of Empire: Voyages, Botany, and Representations of Nature* (Cambridge: Cambridge University Press, 1996), 21–37.

Mindt, Erich, *Der Erste war ein Deutscher! Kämpfer und Forscher jenseits der Meere* (Berlin: Ebner & Ebner, 1942).

Mitchell, Timothy, *Rule of Experts: Egypt, Techno-Politics, Modernity* (Berkeley: University of California Press, 2002).

Montgomery, Robert, *Maps Accompanying Report on the Trade and Resources of the Countries on the North Western Boundary of British India* (Lahore: Government Press, 1862).

Morrell, Jack and Arnold Thackray, *Early Years of the British Association for the Advancement of Science* (Oxford: Clarendon Press, 1981).

Morris, Deirdre, 'Baron Sir Ferdinand von Mueller', *Australian Dictionary of Biography*, http://adb.anu.edu.au/biography/mueller-sir-ferdinand-jakob-heinrich-von-4266/text6893.

Morrison, Alexander, 'Beyond the "Great Game": The Russian Origins of the Second Anglo–Afghan War', *Modern Asian Studies*, 51 (2017), 686–735.

'Camels and Colonial Armies: The Logistics of Warfare in Central Asia in the Early 19th Century', *Journal of the Economic and Social History of the Orient*, 57 (2014), 443–85.

Russian Rule in Samarkand 1868–1910: A Comparison with British India (Oxford: Oxford University Press, 2008).

Muhs, Rudolf, Johannes Paulmann and Willibald Steinmetz (eds.), *Aneignung und Abwehr: interkultureller Transfer zwischen Deutschland und Grossbritannien im 19. Jahrhundert* (Bodenheim: Philo, 1998).

Müller, Claudius C. and Walter Raunig (eds.), *Der Weg zum Dach der Welt* (Innsbruck and Frankfurt am Main: Pinguin, 1982).

Müller, Frank Lorenz, 'Imperialist Ambitions in Vormärz and Revolutionary Germany: The Agitation for German Settlement Colonies Overseas, 1840–1849', *German History*, 17 (1999), 346–68.

Muir, Ramsey, *The Making of British India, 1756–1858* (Manchester: Manchester University Press, 1915).

Murchison, Roderick, 'Address at the Anniversary Meeting, 24th May, 1852', *JRGS*, 23 (1853), lxii–cxxxviii.

'Address to the Royal Geographical Society of London, 25th May, 1857', *JRGS*, 27 (1857), xciv–cxcviii.

'Address to the Royal Geographical Society of London, 24th May 1858', *PRGS*, 2 (1858), 231–334.

'Address to the Royal Geographical Society of London, 23rd May 1859', *PRGS*, 3 (1858–9), 224–346.

Naranch, Bradley, 'Beyond the Fatherland: Colonial Visions, Overseas Expansion, and German Nationalism, 1848–1885' (unpublished PhD thesis, University of North Carolina, 2006).

National Herbarium Nederland, 'Heyne (or Heine), Benjamin', www.nationaalherbarium.nl/fmcollectors/H/HeyneB.htm#3a.

Nayar, Pramod K., 'Beyond the Colonial Subject: Mobility, Cosmopolitanism and Self-Fashioning in Sarat Chandra Das' *A Journey to Lhasa and Central Tibet*', *New Zealand Journal of Asian Studies*, 14 (2012), 1–16.

Nechtman, Tillman W., *Nabobs: Empire and Identity in Eighteenth-Century Britain* (Cambridge: Cambridge University Press, 2010).

Neuhaus, Tom, 'Emil Schlagintweit und die Tibet-Forschung im 19. Jahrhundert', in Brescius et al. (eds.), *Über den Himalaya*, 219–28.

Tibet in the Western Imagination (Basingstoke: Palgrave Macmillan, 2012).

Nipperdey, Thomas, *Deutsche Geschichte 1800–1866: Bürgerwelt und starker Staat* (Munich: Beck, 1993).

Nüsser, Marcus, 'Natur und Kultur im Himalaya. Die Gletscher- und Siedlungspanoramen der Brüder Schlagintweit', in Brescius et al. (eds.), *Über den Himalaya*, 319–43.

Nyhart, Lynn K., 'Emigrants and Pioneers: Moritz Wagner's "Law of Migration" in Context', in Jeremy Vetter (ed.), *Knowing Global Environments: New*

Historical Perspectives in the Field Sciences (New Brunswick, NJ: Rutgers University Press, 2010), 39–58.

Oesterheld, Joachim, 'Germans in India between Kaiserreich and the end of World War II', in Joanne Miyang Cho, Eric Kurlander and Douglas T. McGetchin (eds.), *Transcultural Encounters between Germany and India: Kindred Spirits in the Nineteenth and Twentieth Centuries* (London: Routledge, 2014), 101–14.

Ogborn, Miles, *Global Lives: Britain and the World, 1550–1800* (Cambridge: Cambridge University Press, 2008).

Olesko, Kathryn, 'Humboldtian Science', in John Heilbron (ed.), *The Oxford Guide to the History of Physics and Astronomy* (Oxford: Oxford University Press, 2005), 159–62.

Oppen, Achim von and Silke Strickrodt, 'Introduction: Biographies between Spheres of Empire', *Journal of Imperial and Commonwealth History*, 44, 5 (2016), 717–29.

Orlich, Leopold von, *Reise in Ostindien in Briefen an A. v. Humboldt und Karl Ritter*, 2 vols. (Leipzig: Mayer/Wigand, 1845).

Osborne, Michael A., 'Applied Natural History and Utilitarian Ideals: "Jacobin Science" at the Muséum d'Histoire naturelle, 1789–1870', in Bryant T. Ragan Jr. and Elizabeth A. Williams (eds.), *Re-creating Authority in Revolutionary France* (New Brunswick, NJ: Rutgers University Press, 1992), 125–43.

Osterhammel, Jürgen, *Die Entzauberung Asiens: Europa und die asiatischen Reiche im 18. Jahrhundert* (Munich: C. H. Beck, 1998).

Europe, the 'West' and the Civilizing Mission (London: German Historical Institute, 2006).

The Transformation of the World: A Global History of the Nineteenth Century (Princeton, NJ: Princeton University Press, 2014).

Osterhammel, Jürgen and Boris Barth (eds.), *Zivilisierungsmissionen. Imperiale Weltverbesserung seit dem 18. Jahrhundert* (Konstanz: UVK, 2005).

Outram, Dorinda, 'New Spaces in Natural History', in N. Jardine, J. Secord and E. C. Spary (eds.), *Cultures of Natural History* (Cambridge: Cambridge University Press, 1996), 249–65.

Paletschek, Sylvia, 'Was heißt "Weltgeltung deutscher Wissenschaft"? Modernisierungsleistungen und -defizite der Universitäten im Kaiserreich', in Michael Grüttner, Rüdiger Hachtmann, Konrad H. Jarausch, Jürgen John and Mattias Middell (eds.), *Gebrochene Wissenschaftskulturen: Universität und Politik im 20. Jahrhundert* (Göttingen: Vandenhoeck & Ruprecht, 2010), 29–54.

Panayi, Panikos, *The Germans in India: Elite European Migrants in the British Empire* (Manchester: Manchester University Press, 2017).

Päßler, Ulrich, *Alexander von Humboldt–Carl Ritter. Briefwechsel* (Berlin: Akademie Verlag, 2010).

Ein 'Diplomat aus den Wäldern des Orinoko'. Alexander von Humboldt als Mittler zwischen Preußen und Frankreich (Stuttgart: Steiner, 2009).

Pathak, Shekhar and Uma Bhatt (eds.), *Asia Ki Peeth Par: Life, Explorations and Writings of Pundit Nain Singh Rawat* [in Hindi], 2 vols. (Nainital: Pahar Pothi, 2006).

Peabody, Norbert, 'Knowledge Formation in Colonial India', in D. M. Peers and N. Gooptu (eds.), *India and the British Empire* (Oxford: Oxford University Press, 2012), 75–99.

Penny, H. Glenn, *Objects of Culture: Ethnology and Ethnographic Museums in Imperial Germany* (Chapel Hill: University of North Carolina Press, 2001).

Peschel, Oscar, *Völkerkunde* (Leipzig: Duncker & Humboldt, 1874).

Petermann, August, 'Dr. D. Livingstone's Reisen in Süd-Afrika, 1841 bis heute', *PGM*, 3 (1857), 91–108.

'Kartenskizze von Africa zur vergleichenden Übersicht der Reisen Dr. Barth's und Dr. Livingstone's', *PGM*, 3 (1857), 'Tafel 3'.

'Die Reisen der Gebrüder Schlagintweit in Indien bis zum 26. Febr. 1856', *PGM*, 2 (1856), 104–8.

'Th. v. Heuglin's Expedition nach Wadai', *PGM*, 6 (1860), 318.

Petrovskii, N. F., *Turkestanskie Pis'ma [Turkestan Letters]*, ed. V. S. Miasniko (Moscow: Pamjatniki Istoričeskoj Mysli, 2010).

Pinchot, Gifford, 'Sir Dietrich Brandis', *Indian Forester*, 35 (1909), 68–80.

Pinney, Christopher, 'Colonial Anthropology in the "Laboratory of Mankind"', in C. A. Bayly (ed.), *The Raj: India and the British 1600–1947* (London: National Portrait Gallery, 1990), 252–63.

Polter, Stefan B., 'Nadelschau in Hochasien: englische Magnetforschung und die Brüder Schlagintweit', in Claudius C. Müller and Walter Raunig (eds.), *Der Weg zum Dach der Welt* (Innsbruck and Frankfurt am Main: Pinguin, 1982), 78–80 and 97–8.

Porter, Bernard, *The Absent-Minded Imperialists: Empire, Society, and Culture in Britain* (Oxford: Oxford University Press, 2004).

Porter, Roy, 'Gentlemen and Geology: The Emergence of a Scientific Career, 1660–1920', *Historical Journal*, 21 (1978), 809–36.

Porter, Theodore M., 'The Fate of Scientific Naturalism: From Public Sphere to Professional Exclusivity', in Bernard Lightman and Gowan Dawson (eds.), *Victorian Scientific Naturalism: Community, Identity, Continuity* (Chicago and London: Chicago Universiy Press, 2014), 265–87.

Pouchepadass, Jacques, 'The Agrarian Economy and Rural Society (1790–1860)', in Claude Markovits (ed.), *A History of Modern India, 1480–1950* (London: Anthem Press, 2004), 294–315.

Prakash, Gyan, *Another Reason: Science and the Imagination of Modern India* (Princeton, NJ: Princeton University Press, 1999).

Pratt, Mary Louise, *Imperial Eyes: Travel Writing and Transculturation*, 2nd edn. (New York: Routledge, 2007).

Quatrefages, Armand de and Ernest T. Hamy, *Les crânes des races humaines* (Paris: J. B. Baillière et fils, 1882).

Raj, Kapil, 'Beyond Postcolonialism … and Postpositivism: Circulation and the Global History of Science', *Isis*, 104, 2 (2013), 337–47.

Relocating Modern Science: Circulation and the Construction of Knowledge in South Asia and Europe, 1650–1900 (New York: Palgrave Macmillian, 2007).

'When Human Travellers Become Instruments: The Indo-British Exploration of Central Asia in the Nineteenth Century', in Marie-Noëlle Bourget, Christian Licoppe and H. Otto Sibum (eds.), *Instruments, Travel and Science: Itineraries of Precision from the Seventeenth to the Twentieth Century* (London: Routledge, 2002), 156–88.

Rajan, Ravi, *Modernizing Nature: Forestry and Imperial Eco-Development, 1800–1950* (Oxford: Oxford University Press, 2006).

Ramasheshan, Sita, 'The History of Paper in India up to 1948', *Indian Journal of History of Science*, 24 (1989), 103–21.

Ranking, James L., *Report upon the Military and Civil Station of Trichinopoly* (Madras: Gantz Bros., 1867).

Ratcliff, Jessica, 'The East India Company, the Company's Museum, and the Political Economy of Natural History in the Early Nineteenth Century', *Isis* 107, 3 (2016), 495–517.

'Travancore's Magnetic Crusade: Geomagnetism and the Geography of Scientific Production in a Princely State', *British Journal for the History of Science*, 49 (2016), 325–52.

Reeve, Lovell, *Portraits of Men of Eminence in Literature, Science, and Art*, vol. I (London: Lovell Reeve & Co., 1863).

Reich, Karin, Eberhard Knobloch and Elena Roussanova, *Alexander von Humboldts Geniestreich: Hintergründe und Folgen seines Briefes an den Herzog von Sussex für die Erforschung des Erdmagnetismus* (Heidelberg and Berlin: Springer, 2016).

Reich, Karin and Elena Roussanova, 'Mit dem Magnetometer in den Himalaya', in Brescius et al. (eds.), *Über den Himalaya*, 193–208.

Reichel, Claudia, 'German Responses: Theodor Fontane, Edgar Bauer, Wilhelm Liebknecht', in Shaswati Mazumdar (ed.), *Insurgent Sepoys: Europe Views the Revolt of 1857* (New Delhi: Routledge, 2011), 19–42.

Reidy, Michael S., 'From the Oceans to the Mountains: Spatial Science in an Age of Empire', in Jeremy Vetter (ed.), *Knowing Global Environments: New Historical Perspectives on the Field Sciences* (New Brunswick, NJ: Rutgers University Press, 2011), 17–39.

Ribbentrop, Berthold, *Forestry in British India* (Calcutta: Superintendent of Government Printing, 1900).

Richter, Hermann M., *Die leitenden Ideen und der Fortschritt in Deutschland von 1860 bis 1870* (Nördlingen: C. H. Beck, 1873).

Richthofen, Ferdinand von, 'Bericht über den internationalen geographischen Congress in Paris', *Verhandlungen der Gesellschaft für Erdkunde zu Berlin*, 2 (1875), 182–94.

Riffenburgh, Beau, *The Myth of the Explorer: The Press, Sensationalism, and Geographical Discovery* (London: Belhaven Press, 1993).

Ritter, Carl, *Die Erdkunde im Verhältniß zur Natur und zur Geschichte des Menschen, oder allgemeine, vergleichende Geographie*, 2. stark vermehrte und verbesserte Ausgabe in 19 Teilen (Berlin: G. Reimer, 1822–59).

'Sitzung der geographischen Gesellschaft zu Berlin, vom 9.1.1858', *Zeitschrift für allgemeine Erdkunde*, 4 (1858), 87–8.

'Ueber die wissenschaftliche Reise der drei Gebrüder Schlagintweit in Indien', *ZAE*, 5 (1855), 148–71.

Robb, Peter, 'Colin MacKenzie's Survey of Mysore, 1799–1810', *Journal of the Royal Asiatic Society*, 8 (1998), 181–206.

Roberts, Lissa, 'Situating Science in Global History: Local Exchanges and Networks of Circulation', *Itinerario*, 33 (2009), 9–30.

Roberts, Peder, 'Traditions, Networks and Deep-Sea Expeditions after 1945', in Marianne Klemun and Ulrike Spring (eds.), *Expeditions as*

Experiments: Practising Observation and Documentation (London: Palgrave Macmillan, 2016), 213–34.

Robinson, Michael F., 'Science and Exploration', in Dane Kennedy (ed.), *Reinterpreting Exploration: The West in the World* (Oxford: Oxford University Press, 2014), 21–37.

Rolf, Malte, 'Einführung: Imperiale Biographien. Lebenswege imperialer Akteure in Groß- und Kolonialreichen (1850–1918)', in Malte Rolf (ed.), *Imperiale Biographien*, special issue of *Geschichte und Gesellschaft*, 40 (2014), 5–21.

Roquette, Jean B. M. A. Dezos de la, 'Note de M. de la Roquette sur des ouvrages offerts par MM. Schlagintweit et sur leur prochain voyage dans l'Inde', *BSG*, 7 (1854), 229–32.

'Rapport sur le prix annuel, pour la découverte la plus importante en géographie pendant le cours de l'année 1856, fait au nom d'une Commission spéciale, par M. de la Roquette', *BSG*, 17 (1859), 226–45.

Rose, Gustav, *Mineralogisch-geognostische Reise nach dem Ural, dem Altai und dem Kaspischen Meere*, 2 vols. (Berlin: Sandersche Buchhandlung, 1837–42).

Ross, James Clark, *A Voyage of Discovery and Research in the Southern and Antarctic Regions during the Years 1839–43* (London: John Murray, 1847).

Roy, Rohan Deb, *Malarial Subjects: Empire, Medicine and Nonhumans in British India, 1820–1909* (Cambridge: Cambridge University Press, 2017).

Royal Commission on Scientific Instruction and the Advancement of Science, *Reports*, vol. II: *Minutes of Evidence, Appendices, and Analyses of Evidence* (London: HMSO, 1874).

Rudwick, Martin J. S., *The Great Devonian Controversy: The Shaping of Scientific Knowledge among Gentlemanly Specialists* (Chicago: University of Chicago Press, 1985).

Rüger, Jan, 'Writing Europe into the History of the British Empire', in Jan Rüger, Matthew Hilton and John H. Arnold (eds.), *History after Hobsbawm: Writing the Past for the Twenty-First Century* (Oxford: Oxford University Press, 2017), 35–49.

Ruprecht, Adrian, 'De-Centering Humanitarianism: The Red Cross and India, *c.* 1877–1939' (unpublished PhD thesis, University of Cambridge, 2018).

Russell, Sir William Howard, *My Diary in India, in the Year 1858–9*, 2 vols. (London: Routledge, 1860).

Ryan, James R., *Picturing Empire: Photography and the Visualization of the British Empire* (London: Reaktion Books, 1997).

Sachs, Aaron, *The Humboldt Current: A European Explorer and His American Disciples* (Oxford: Oxford University Press, 2007).

Safier, Neil, *Measuring the New World: Enlightenment Science and South America* (Chicago and London: University of Chicago Press, 2008).

Said, Edward, *Orientalism: Western Conceptions of the Orient* (New York: Pantheon, 1978).

Saikia, Arupjyoti, *Forests and Ecological History of Assam, 1826–2000* (Oxford: Oxford University Press, 2011).

Sarkar, Oyndrila, 'Science, Surveying and Scientific Authority: The Brothers Schlagintweit in "India and High Asia", 1854–57', *South Asia: Journal of South Asian Studies*, 40 (2017), 544–65.

Schaffer, Simon, 'How Disciplines Look', in Andrew Barry and Georgina Born (eds.), *Interdisciplinarity: Reconfigurations of the Social and Natural Sciences* (London: Routledge, 2013), 57–81.

Schaffer, Simon, Lissa Roberts, Kapil Raj and James Delbourgo (eds.), *The Brokered World: Go-Betweens and Global Intelligence, 1770–1820* (Sagamore Beach, MA: Watson Publishing International, 2009).

Schär, Bernhard C., *Tropenliebe: Schweizer Naturforscher und niederländischer Imperialismus in Südostasien um 1900* (Frankfurt am Main and New York: Campus, 2015).

Schenck, Carl Alwin, *The Birth of Forestry in America: Biltmore Forest School, 1898–1913* (Santa Cruz, CA: Forest History Society, 1974).

Scherzer, Karl von, *The Circumnavigation of the Globe by the Austrian Frigate 'Novara'*, 2 vols. (London: Saunders & Otley, 1861).

'Die deutsche Arbeit in fremden Erdtheilen', *Mittheilungen des Vereins für Erdkunde zu Halle an der Saale* (1880), 1–23.

Scheuchzer, Johann Jacob, *Natur-Historie des Schweizerlandes*, 3 vols. (Zurich: Heidegger, 1716–18).

Schiera, Pierangelo, *Laboratorium der bürgerlichen Welt. Deutsche Wissenschaft im 19. Jahrhundert* (Frankfurt am Main: Suhrkamp, 1992).

Schlager, Edda, 'Vergessene Helden. Pioniere der Hochgebirgsforschung', in Dieter Lohmann and Nadja Podbregar (eds), *Im Fokus: Entdecker. Die Erkundung der Welt* (Berlin: Springer, 2012), 199–210.

Schlagintweit, Adolph, 'Schreiben des Herrn A. Schlagintweit an Herrn A. v. Humboldt. Bombay, 10.11.1854', *ZGE*, 3 (1854), 338–40.

'Summary of the Principal Results of the Investigations of Himself and his Brother into the Vegetation of the Alps', *Proceedings of the Linnean Society of London*, 2 (1855), 102–5.

Ueber die Ernährung der Pflanzen mit besonderer Rücksicht auf die Bedingungen ihres Gedeihens in verschiedenen Höhen der Alpen (Munich: J. A. Barth, 1850).

Untersuchungen über die Thalbildung und die Formen der Gebirgszüge in den Alpen (Leipzig: J. A. Barth, 1850).

Schlagintweit, Adolph and Hermann Schlagintweit, *Neue Untersuchungen über die physikalische Geographie und die Geologie der Alpen* (Leipzig: Weigel, 1854).

Untersuchungen über die physicalische Geographie und die Geologie der Alpen in Beziehungen zu den Phänomen der Gletscher, zur Geologie, Meteorologie und Pflanzengeographie (Leipzig: J. A. Barth, 1850).

Schlagintweit, Adolph, Hermann Schlagintweit and Robert Schlagintweit, *Report [Nos. I–X] on the Proceedings of the Officers Engaged in the Magnetic Survey of India* (Madras and Calcutta, etc.: Chronicle Press, etc., 1855–7).

Schlagintweit, Adolph and Robert Schlagintweit, 'Notices of Journeys in the Himalayas of Kemaon (communicated by Col. Sykes, F.R.S)', *Report of the 25th Meeting of the BAAS* (London, 1856), 152–5.

Schlagintweit, Eduard, 'Militärische Skizzen über England und Frankreich', *Allgemeine Militärische Zeitung*, 34/35 (1861), 268–9; 275–67.

Der spanisch-marokkanische Krieg in den Jahren 1859 und 1860 (Leipzig: F. A. Brockhaus, 1863).

Schlagintweit, Emil, 'Bericht über das Denkmal für Adolf Schlagintweit in Kaschgar', *Sitzungsberichte der philosophisch-philologischen und historischen Classe der k. b. Akademie der Wissenschaften* (1890), 457–72.

'Bericht über eine Adresse an den Dalai Lama, 1902, zur Erlangung von Bücherverzeichnissen aus den dortigen buddhistischen Klöstern', *Abhandlungen der I. Klasse der k. Akademie der Wissenschaften*, 22 (1905), 657–78.

'Ein Besuch der Jägersburg und der Schlagintweit'schen Sammlungen', *Morgenblatt zur Bayerischen Zeitung*, 326–7, 23–26 November 1864, 1110–11; 1114–15 [published anonymously].

Le Bouddhisme au Tibet: précédé d'un résumé des précédents systèmes bouddhiques dans l'Inde, trans. Louis de Milloué (Lyons: Pitrat, 1881).

Buddhism in Tibet Illustrated by Literary Documents and Objects of Religious Worship (Leipzig: Brockhaus; London: Trübner, 1863).

'Englische Forschungsreisen in Centralasien', *Globus*, 25 (1874), 365–6; 376–8.

Die Gottesurtheile der Inder (Munich: Verlag der königlichen Akademie, 1866).

Indien in Wort und Bild. Eine Schilderung des indischen Kaiserreiches, 2 vols. (Leipzig: H. Schmidt and C. Günther, 1880–1).

'Indiens Grenznachbarn gegen Afghanistan', *Globus*, 30 (1876), 105–7; 123–5.

Die Lebensbeschreibung von Padma Sambhava, dem Begründer des Lamaismus, 2 vols. (Munich: Verlag der königlichen Akademie, 1899–1903).

'Schlagintweit', *Allgemeine Deutsche Biographie [ADB]*, 31 (Leipzig: Duncker & Humblot, 1890), 336–48.

'Die Uferstaaten des Persischen Golfs', *Globus*, 30 (1876), 362–5; 379–81.

'Die Völker Ost-Turkistans', *Globus*, 31 (1877) 236–8; 251–4; 263–5.

'Die Zustände in Bhutan', *Globus*, 6 (1864), 330–3.

Schlagintweit, Hermann, 'Bericht über Anlage des Herbariums während der Reisen', *Abhandlungen der mathematisch-physikalischen Classe der k. Bayerischen Akademie der Wissenschaften*, 12 (1876), 133–96.

'Bericht über die ethnographischen Gegenstände unserer Sammlungen', *Sitzungsberichte der mathematisch-physicalischen Classe der k. b. Akademie der Wissenschaften*, 7 (1877), 336–80.

'Numerical Elements of Indian Meteorology [Series 1]', *Philosophical Transactions of the Royal Society*, 153 (1863), 525–42.

Observations sur la hauteur du Mont-Rose et des points principaux de ses environs (Turin: Imprimerie Royal, 1853).

Reisen in Indien und Hochasien: eine Darstellung der Landschaft, der Cultur und Sitten der Bewohner, in Verbindung mit klimatischen und geologischen Verhältnissen, Basirt auf die Resultate der wissenschaftlichen Mission von Hermann, Adolph und Robert von Schlagintweit, 4 vols. (Jena: Costenoble, 1869–80).

vol. I: *Indien* (1869);

vol. II: *Hochasien: I. Der Himalaya von Bhutan bis Kashmir* (1871);

vol. III: *Hochasien: II. Tibet; zwischen der Himalaya- und der Karakorum-Kette* (1872);

vol. IV: *Hochasien: III. Ost-Turkistan und Umgebungen* (1880).

'Remarks on Some Physical Observations Made in India by the Brothers Schlagintweit', *Literary Gazette*, 19 September 1857, 909.

Das Scalenrädchen (Würzburg: F. E. Thein, 1866).

'Ueber das Auftreten von Bor-Verbindungen in Tibet', *Sitzungsberichte der mathematisch-physicalischen Classe der k. b. Akademie der Wissenschaften*, 8 (1878), 505–38.

Über die Vertheilung der mittleren Jahrestemperatur in den Alpen. Habilitationsschrift (Munich: Hofbuchdruckerei, 1850).

'Über Messinstrumente mit constanten Winkeln (Linsen- und Prismenporrhometer)', *Dingler's polytechnisches Journal*, 112 (1849), 334–56.

'Zur Charakteristik der Kru-Neger', *Sitzungsberichte der mathematischphysicalischen Classe der k. b. Akademie der Wissenschaften*, 5 (1875), 183–201.

Schlagintweit, Hermann, Adolph Schlagintweit and Robert Schlagintweit, 'Latitudes, Longitudes and Magnetic Elements, determined in India and High Asia', *Astronomische Nachrichten*, 55 (1861), 161–6.

Results of a Scientific Mission to India and High Asia, Undertaken between the Years 1854 and 1858, by Order of the Court of Directors of the Honourable East India Company, 4 vols. (Leipzig: F. A. Brockhaus; London: Trübner & Co., 1861–6).

vol. I: *Astronomical Determinations of Latitudes and Longitudes and Magnetic Observations: During a Scientific Mission to India and High Asia*, by H., A. and R. de Schlagintweit (1861);

vol. II: *General Hypsometry of India, The Himálaya, and Western Tibet. With Sections across the Chains of the Karakorum and Kuenluen, Comprising, in Addition to Messrs. de Schlagintweits' Determinations, the Data Collected from Books, Maps, and Private Communications*, ed. Robert de Schlagintweit (1862);

vol. III: *Route-Book of the Western Parts of the Himalaya, Tibet, and Central Asia: And Geographical Glossary from the Languages of India and Tibet, Including the Phonetic Transcription and Interpretation*, by H., A. and R. de Schlagintweit (1863);

vol. IV: *Meteorology of India, an Analysis of the Physical Conditions of India, the Himalaya, Western Tibet, and Turkistan …: Based upon Observations Made by Messrs. de Schlagintweit … and Increased by Numerous Additions Chiefly Obtained from the Officers of the Medical Departments* (1866);

accompanying atlas: *Atlas of Panoramas and Views, with Geographical, Physical, and Geological Maps: Dedicated to Her Majesty the Queen of England* (1861–6).

Schlagintweit, Hermann, Adolph Schlagintweit and Robert Schlagintweit, with F. A. Brockhaus, *Prospectus: Results of a Scientific Mission to India and High Asia* (Leipzig: F. A. Brockhaus, 1860).

Schlagintweit, Hermann and Robert Schlagintweit, *Allgemeiner naturgeschichtlicher Catalog der v. Schlagintweit'schen Sammlungen Schloss Jägersburg* (n.p., 1868).

'Aperçu sommaire des résultats de la Mission scientifique dans l'Inde et la Asie', *Extrait des Comptes rendus des séances de l'Académie des sciences*, 45 (1857), 1–7.

'Notices and Abstracts of Miscellaneous Communications to the Sections. Geology: On the Erosion of Rivers in India', in *Report of the Twenty-Seventh Meeting of the British Association for the Advancement of Science, Held at Dublin in August and September, 1857* (London: John Murray, 1858), 90–1.

Official Reports on the Last Journeys and the Death of Adolphe Schlagintweit in Turkistan (Berlin, 1859).

'On Contributions to Statistics by Measurements of Human Tribes', *Report of the Proceedings of the Fourth Session of the International Statistical Congress* (London: E. Eyre and W. Spottiswoode, 1861), 500.

Prospectus of Messrs. de Schlagintweits' Collection of Ethnographical Heads from India and High Asia, 2nd edn. (Leipzig, 1859).

Schlagintweit, Joseph, *Ueber den gegenwärtigen Zustand der künstlichen Pupillenbildung in Deutschland* (Munich: Lentner, 1818).

Schlagintweit, Max, *Afrikanische Kolonialbahnen: Verkehrswege und Verkehrsprojekte* (Munich: Piloty & Loehle, 1907).

Deutsche Kolonisationsbestrebungen in Kleinasien (Munich: K. Hof- und Universitätsbuchdruckerei Dr. C. Wolf & Sohn, 1899).

Militärische und topographische Mitteilungen aus Konstantinopel und Kleinasien (Berlin: Reimer, 1899).

Routen-Aufnahme: praktische Erfahrungen und Anleitung (Munich: Riedel et. al., 1910).

Die Verteidigungsfähigkeit Konstantinopels (Berlin: Süsserott, 1912).

Die Verwaltung des Kongostaates und die deutsche Interessen (Munich: R. Oldenbourg, 1906).

Schlagintweit, Robert, 'Angaben über die Entfernung zwischen den wichtigsten Städten in den westlichen Theilen des Himálaya, Tibets und Central-Asiens', *PGM*, 10 (1862), 50.

Bemerkungen über die physikalische Geographie des Kaisergebirges (Munich, 1854).

'Ein Besteigungs-Versuch des Ibi Gamin Gipfels in Hochasien', *Gaea, Natur und Leben*, 4 (1868), 313–22; 373–8.

'Geographische Schilderung des Himalaya', *Globus*, 12 (1867), 1–9.

Schlagintweit, Robert von, 'On Thermo-Barometers, Compared with Barometers at Great Heights', *Report of the Thirtieth Meeting of the British Association for the Advancement of Science, Held at Oxford in June and July 1860* (London: John Murray, 1861), 50–1.

Robert von Schlagintweit's als Manuscript gedruckter und nur zur Privatvertheilung bestimmter Bericht über die 1000 von ihm zwischen … 1864 und … 1878 in Europa und Nordamerika gehaltenen öffentlichen populärwissenschaftlichen Vorträge (Leipzig: W. Drugulin, 1878).

Schlagintweit, Stefan, 'Die Brüder Schlagintweit: ein Abriß ihres Lebens', in Claudius C. Müller and Walter Raunig (eds.), *Der Weg zum Dach der Welt* (Innsbruck and Frankfurt am Main: Pinguin, 1982), 11–13.

Schomburgk, Richard, *Versuch einer Fauna und Flora von British-Guiana* (Leipzig: J. J. Weber, 1848).

Schröder, Iris, *Das Wissen von der ganzen Welt. Globale Geographien und räumliche Ordnungen Afrikas und Europas 1790–1870* (Paderborn: Schöningh, 2011).

Schröder, Wilfried and Karl-Heinrich Wiederkehr, 'Geomagnetic Research in the 19th Century: A Case Study of the German Contribution', *Journal of Atmospheric and Solar-Terrestrial Physics*, 63 (2001), 1649–60.

Schulte Beerbühl, Margrit, *Deutsche Kaufleute in London: Welthandel und Einbürgerung (1660–1818)* (Munich: Walter de Gruyter, 2007).

Schwarz, Ingo (ed.), *Briefe von Alexander von Humboldt an Christian Carl Josias Bunsen* (Berlin: Rohrwall, 2006).

Schwartzberg, Joseph E., 'Maps of Greater Tibet', in J. B. Harley and David Woodward (eds.), *The History of Cartography*, vol. II, book 2: *Cartography in the Traditional East and Southeast Asian Society* (Chicago: Chicago University Press, 1994), 607–81.

Sèbe, Berny, *Heroic Imperialists in Africa: The Promotion of British and French Colonial Heroes, 1870–1939* (Manchester: Manchester University Press, 2013).

Secord, James A., *Controversy in Victorian Geology: The Cambrian-Silurian Dispute* (Princeton, NJ: Princeton University Press, 1986).

'Knowledge in Transit', *Isis*, 95, 4 (2004), 654–72.

Seemann, Berthold Carl, *The Botany of the Voyage of H.M.S. Herald, under the Command of Captain Henry Kellett ... during the Years 1845–51* (London: Lovell Reeve, 1852–7).

'Review of *Results of a Scientific Mission to India and High Asia*', *The Athenaeum*, 1764, 1861, 215–16.

Sen, Srabani, 'The Asiatic Society and the Sciences in India, 1784–1947', in Uma Das Gupta (ed.), *Science and Modern India: An Institutional History, c. 1784–1947* (New Delhi: Pearson Longman, 2011), 27–68.

Sengupta, Indra, *From Salon to Discipline: State, University and Indology in Germany, 1821–1914* (Heidelberg: Ergon, 2005).

Sera-Shriar, Efram, *The Making of British Anthropology, 1813–1871* (London and New York: Routledge, 2016).

Shapin, Steven, 'Here and Everywhere: Sociology of Scientific Knowledge', *Annual Review of Sociology*, 21 (1995), 289–321.

Sharma, Jayeeta, *Empire's Garden: Assam and the Making of India* (Durham, NC: Duke University Press, 2011).

Sheehan, James J., *Museums in the German Art World: From the End of the Old Regime to the Rise of Modernism* (New York: Oxford University Press, 2000).

Shreider, I. A. A., 'Science and Circus', *Configurations*, 1 (1993), 457–63.

Sievers, Wilhelm, *Asien. Eine allgemeine Landeskunde* (Leipzig: Bibliographisches Institut, 1892).

Sikka, D. R., 'The Role of the India Meteorological Department, 1875–1947', in Uma Das Gupta (ed.), *Science and Modern India: An Institutional History, c. 1784–1947* (Delhi: Pearson Longman, 2011), 381–428.

Sivasundaram, Sujit, 'Introduction: Global Histories of Science', *Isis*, 101 (2010), 95–7.

'Sciences and the Global: On Methods, Questions, and Theory', *Isis*, 101 (2010), 146–58.

Smithsonian Institution, *Annual Report of the Board of Regents* (Washington, DC: Government Printing Office, 1863).

Stafford, Robert A., 'Annexing the Landscapes of the Past: British Imperial Geology in the Nineteenth Century', in John M. MacKenzie (ed.), *Imperialism and the Natural World* (Manchester and New York: Manchester University Press, 1990), 67–89.

Scientist of Empire: Sir Roderick Murchison, Scientific Exploration and Victorian Imperialism (Cambridge: Cambridge University Press, 1990).

Standfield, Rachel, 'Violence and the Intimacy of Imperial Ethnography. The Endeavour in the Pacific', in Tony Ballantyne and Antoinette Burton (eds.), *Moving Subjects: Gender, Mobility, and Intimacy in an Age of Global Empire* (Urbana: University of Illinois Press, 2009), 31–48.

Stern, Philip J., *The Company-State: Corporate Sovereignty and the Early Modern Foundations of the British Empire in India* (Oxford: Oxford University Press, 2011).

'Exploration and Enlightenment', in Dane Kennedy (ed.), *Reinterpreting Exploration: The West in the World* (Oxford: Oxford University Press, 2014), 54–79.

'Politics and Ideology in the Early East India Company-State: The Case of St Helena, 1673–1709', *Journal of Imperial and Commonwealth History*, 35 (2007), 1–23.

Stewart, Gordon, 'The Exploration of Central Asia', in Dane Kennedy (ed.), *Reinterpreting Exploration: The West in the World* (Oxford: Oxford University Press, 2014), 195–213.

Journeys to Empire: Enlightenment, Imperialism, and the British Encounter with Tibet, 1774–1904 (Cambridge: Cambridge University Press, 2009).

Stoddard, D. R., 'The RGS and the "New Geography": Changing Aims and Changing Roles in Nineteenth Century Science', *Geographical Journal*, 146 (1980), 190–202.

Strachey, Henry, 'Note on the Construction of the Map of the British Himalayan Frontier in Kumaon and Garhwal', *JASB*, 17 (1848), 532–8.

Strachey, Richard, 'The Physical Geography of the Provinces of Kumáon and Garhwál in the Himalayan Mountains', *JRGS*, 21 (1851), 57–85.

Stronge, Susan, *Tipu's Tigers* (London: V&A Publishing, 2009).

Subrahmanyam, Sanjay, *Europe's India: Words, People, Empires, 1500–1800* (Cambridge, MA: Harvard University Press, 2017).

Three Ways to be Alien: Travails and Encounters in the Early Modern World (Waltham, MA: Brandeis University Press, 2011).

Suckow, Christian, 'Alexander von Humboldt und Rußland', in Ottmar Ette, Ute Hermanns, Bern M. Scherer and Christian Suckow (eds.), *Alexander von Humboldt: Aufbruch in die Moderne* (Berlin: Akademie, 2001), 247–64.

Sysling, Fenneke, 'Heritage of Racial Science: Facial Plaster Casts from the Netherland Indies', in Susan Legêne, Bambang Purwanto and Henk S. Nordholt (eds.), *Sites, Bodies and Stories: Imagining Indonesian History* (Singapore: NUS Press, 2015), 113–32.

Racial Science and Human Diversity in Colonial Indonesia: Physical Anthropology and the Netherlands Indies, ca. 1890–1960 (Singapore: NUS Press, 2016).

Tammiksaar, E., N. G. Sukhova and I. R. Stone, 'Hypothesis versus Fact: August Petermann and Polar Research', *Arctic*, 52 (1999), 237–43.

Teisch, Jessica B., *Engineering Nature: Water, Development, and the Global Spread of American Environmental Expertise* (Chapel Hill: University of North Carolina Press, 2011).

Terra, Helmut de, *Humboldt: The Life and Times of Alexander von Humboldt, 1769–1859* (New York: Alfred Knopf, 1955).

Théoderides, Jean, 'Humboldt and England', *British Journal for the History of Science*, 3 (1966), 39–55.

Theye, Thomas, 'Der geraubte Schatten. Einführung', in Thomas Theye (ed.), *Der geraubte Schatten. Eine Weltreise im Spiegel der ethnographischen Photographie* (Munich and Lucerne: Bucher, 1989), 8–59.

Thomas, Frederick William, 'Obituary Notices', *Journal of the Royal Asiatic Society of Great Britain and Ireland* (1905), 215–18.

Thomas, Martin, 'The Expedition as a Cultural Form: On the Structure of Exploratory Journeys as Revealed by the Australian Explorations of Ludwig Leichhardt', in Martin Thomas (ed.), *Expedition into Empire: Exploratory Journeys and the Making of the Modern World* (London: Routledge, 2015), 65–87.

'What is an Expedition? An Introduction', in Martin Thomas (ed.), *Expedition into Empire: Exploratory Journeys and the Making of the Modern World* (London: Routledge, 2015), 1–24.

Thomas, Nicholas, *Entangled Objects: Exchange, Material Culture, and Colonialism in the Pacific* (Cambridge, MA: Harvard University Press, 1991).

Thuillier, Henry and R. Smith (eds.), *A Manual of Surveying for India, Detailing the Mode of Operations on the Revenue Surveys in Bengal and the North-Western Provinces* (Calcutta: W. Thacker & Co, 1851).

Tilley, Helen, *Africa as a Living Laboratory: Empire, Development, and the Problem of Scientific Knowledge, 1870–1950* (Chicago: University of Chicago Press, 2011).

Tolen, Rachel J., 'Colonizing and Transforming the Criminal Tribesman: The Salvation Army in British India', *American Ethnologist*, 18, 1 (1991), 106–25.

Topinard, Paul, 'Visite à la collection anthropologique du Muséum d'Histoire naturelle', in Exposition Universelle, Paris 1878, *Congrès international des sciences anthropologiques, tenu à Paris du 16 au 21 août 1878* (Paris: Imprimerie Nationale, 1880), 293–7.

Torma, Franziska, *Turkestan-Expeditionen: zur Kulturgeschichte deutscher Forschungsreisen nach Mittelasien (1890–1930)* (Bielefeld: transcript, 2011).

Torrens, Henry D'Oyley, *Travels in Ladak, Tartary and Kashmir* (London: Saunders & Otley, 1862).

Trentin-Meyer, Maike, 'Die Indien- und Hochasienreise der Brüder Schlagintweit', in Christoph Köck (ed.), *Reisebilder. Produktion und Reproduktion touristischer Wahrnehmung* (Münster: Waxman, 2001), 41–51.

Tucker Jones, Ryan, *Empire of Extinction: Russians and the North Pacific's Strange Beasts of the Sea, 1741–1867* (Oxford and New York: Oxford University Press, 2014).

Tzoref-Ashkenazi, Chen, *German Soldiers in Colonial India* (London and New York: Routledge, 2016).

'German Voices from India: Officers of the Hanoverian Regiments in East India Company Service', *South Asia: Journal of South Asian Studies*, 32 (2009), 189–211.

Uhlig, Harald, 'Das Neue Schloß als Geographisches Institut: frühe geographische Vorlesungen. Die Gießener Geographen Robert von Schlagintweit und Wilhelm Sievers', *Nachrichten der Giessener Hochschulgesellschaft*, 34 (1965), 87–103.

Urry, James, 'Notes and Queries on Anthropology and the Development of Field Methods in British Anthropology, 1870–1920', *Proceedings of the Royal Anthropological Institute* (1972), 45–57.

Valikhanof, Chokan and M. Veniukof, *The Russians in Central Asia: Their Occupation of the Kirghiz Steppe and the Line of the Syr-Daria: Their Political Relations with Khiva, Bokhara, and Kokan: Also Descriptions of Chinese Turkestan and Dzungaria* (London: Edward Stanford, 1865).

Vambéry, Ármin, *Westlicher Kultureinfluss im Osten* (Berlin: Reimer, 1906).

Veracini, Lorenzo, 'Settler Colonial Expeditions', in Martin Thomas (ed.), *Expedition into Empire: Exploratory Journeys and the Making of the Modern World* (London: Routledge, 2015), 51–64.

Vermeulen, Han F., *Before Boas: The Genesis of Ethnography and Ethnology in the German Enlightenment* (Lincoln: University of Nebraska Press, 2015).

Verne, Jules, *Claudius Bombarnac: The Adventures of a Special Correspondent* (New York: Lovell, Coryell & Company, 1894).

Five Weeks in a Balloon (London and New York: George Routledge and Sons, 1876).

Robur the Conqueror (London: Sampson Low, Marston, Searle & Rivington, 1887).

The Steam House (New York: Scribner, 1881).

A Voyage round the World. In Search of the Castaways (Philadelphia: J. B. Lippincott & Co., 1874).

Vetter, Jeremy, 'Introduction', in Jeremy Vetter (ed.), *Knowing Global Environments: New Historical Perspectives on the Field Sciences* (New Brunswick, NJ: Rutgers University Press, 2011), 1–16.

Vicziany, Marika, 'Imperialism, Botany and Statistics in Early Nineteenth-Century India: The Surveys of Francis Buchanan (1762–1829)', *Modern Asian Studies*, 20 (1986), 625–61.

Voelcker, John Augusts, *Report on the Improvement of Indian Agriculture* (London: Eyre & Spottiswoode, 1893).

Wagener, Hermann, 'Schlagintweit', in Hermann Wagener (ed.), *Staats- und Gesellschafts-Lexikon: neues Conversations-Lexikon*, vol. XVIII (Berlin: Heinicke, 1865), 260–4.

Wagner, Moritz, *Reisen in der Regentschaft Algier in den Jahren 1836, 1837 und 1838*, 3 vols. (Leipzig: Voss, 1841).

Waldron, Peter, 'Przheval'skii, Asia and Empire', *Slavonic and East European Review*, 88 (2010), 309–27.

Walker, John, 'The Great Trigonometrical Survey of India', *Calcutta Review*, 16 (1851), 514–40.

Waller, Derek J., *The Pundits: British Exploration of Tibet and Central Asia* (Lexington: University Press of Kentucky, 1990).

Waterhouse, David M., 'Brian Hodgson: A Biographical Sketch', in David M. Waterhouse (ed.), *The Origin of Himalayan Studies: Brian Hodgson in Kathmandu and Darjeeling, 1820–1858* (London and New York: Routledge, 2004), 1–24.

Waugh, Andrew Scott, 'The Great Trigonometrical Survey of India', in *Professional Papers on Indian Engineering*, vol. II (Calcutta, 1865), 285–300.

Webb, William S., 'Memoir Relative to a Survey in Kemaon', *Asiatic Researches*, 13 (1820), 293–310.

Weber, Andreas, *Hybrid Ambitions. Science, Governance, and Empire in the Career of Caspar G. C. Reinwardt (1773–1854)* (Leiden: Leiden University Press, 2012).

Wehler, Hans-Ulrich, *Deutsche Gesellschaftsgeschichte*, vol. III: *Von der 'Deutschen Doppelrevolution' bis zum Beginn des Ersten Weltkrieges 1849–1914* (Munich: C. H. Beck, 1995).

Weigl, Engelhard, 'Acclimatization: The Schomburgk Brothers in South Australia', *Humboldt im Netz*, 4 (2003), 2–13.

Welcker, Hermann, 'Kraniologische Mittheilungen', *Archiv für Anthropologie*, 1 (1867), 89–160.

Werner, Petra, *Himmel und Erde. Alexander von Humboldt und sein Kosmos*, Beiträge zur Alexander-von-Humboldt-Forschung 24 (Berlin: Akademie Verlag, 2004).

Naturwahrheit und ästhetische Umsetzung. Alexander von Humboldt im Briefwechsel mit bildenden Künstlern (Berlin: Akademie, 2013).

'Zum Verhältnis Darwins zu Humboldt und Ehrenberg', *Humboldt im Netz*, 10 (2009), 68–95.

Werner, Wilhelm (ed.), *Das Kaiserreich Ostindien und die angrenzenden Gebirgsländer. Nach den Reisen der Brüder Schlagintweit und anderer neuerer Forscher dargestellt* (Jena: Hermann Costenoble, 1884).

Wessels, Cornélius, *Early Jesuit Travellers in Central Asia: 1603–1721* (The Hague: Martinus Nijhoff, 1924).

White, Daniel, *From Little London to Little Bengal: Religion, Print and Modernity in Early British India, 1793–1835* (Baltimore, MD: Johns Hopkins University Press, 2013).

Whitfield, Peter, *New Found Lands: Maps in the History of Exploration* (London: Routledge, 1998).

Wippich, Rolf H., *Japan als Kolonie? Max von Brandts Hokkaido-Projekt 1865/67*, 2nd edn. (Hamburg: Abera, 1997).

Withers, Charles W., 'Mapping the Niger, 1798–1832. Trust, Testimony and "Ocular Demonstration" in the Late Enlightenment', *Imago Mundi: The International Journal for the History of Cartography*, 56 (2004), 170–93.

'On Enlightenment's Margins: Geography, Imperialism and Mapping in Central Asia, c.1798–c.1838', *Journal of Historical Geography*, 39 (2013), 3–18.

'Voyages et crédibilité. Vers une géographie de la confiance', *Géographies et Cultures*, 33 (2000), 3–17.

Wokoeck, Ursula, *German Orientalism: The Study of the Middle East and Islam from 1800 to 1945* (London: Routledge, 2009).

Wood, Frances, *The Silk Road: Two Thousand Years in the Heart of Asia* (London: The British Library, 2003).

Woodward, B. B. 'Sykes, William Henry (1790–1872)', rev. M. G. M. Jones. *Oxford Dictionary of National Biography* (Oxford: Oxford University Press, 2004), online edn, September 2004, https://doi.org/10.1093/ref:odnb/26871.

Wright, Thomas, *The Royal Dictionary-Cyclopædia, for Universal Reference*, vol. III (London and New York: London Printing and Publishing Company, 1862).

Zantop, Susanne, *Colonial Fantasies: Conquest, Family, and Nation in Precolonial Germany, 1770–1870* (Durham, NC: Duke University Press, 1997).

Zimmerman, Andrew, *Anthropology and Antihumanism in Imperial Germany* (Chicago: University of Chicago Press, 2001).

'Die Gipsmasken der Brüder Schlagintweit. Verkörperlichung kolonialer Macht', in Brescius et al. (eds.), *Über den Himalaya*, 241–50.

Zimmermann, W. F. A., *Malerische Länder- und Völkerkunde ... unter besonderer Berücksichtigung der neuesten Entdeckungsreisen von ... Humboldt, Schlagintweit, Barth, Livingstone, Vogel ...*, 7th edn. (Berlin: Hempel, 1867).

Index